BRADY

Fire Suppression Practices and Procedures

Second Edition

Eugene F. Mahoney, MS, BS

Retired Battalion Chief, LAFD
Retired Fire Chief, Arcadia Fire Department
Retired Professor of Fire Science

Tracy E. Rickman, MPA

California State Fire Marshal Certified Instructor
Fire Inspector I, International Code Council (#1067094-66)
Hazmat Response Instructor, California Specialized Training Institute (#797)
Coordinator, Rio Hondo Community College
Assistant Professor, California State University, Long Beach, CA

Gerald Wallace

Professor (Tenure) Rio Hondo College (Current)
Fire Chief City of South Pasadena (Current)
Firefighter/Chief from 1964-1996 City of Lynwood (Fire Chief)
Master Instructor (State of California)
Certified Chief Officer
Certified Fire Officer
State Certified Arson Investigator
State Certified Fire Prevention Officer

PEARSON
Prentice Hall

Upper Saddle River, New Jersey 07458

Library of Congress Cataloging-in-Publication Data
Mahoney, Gene, (date)
    Fire suppression practices and procedures / Eugene Mahoney, Tracy Rickman, Gerald
        Wallace. —2nd ed.
        p. cm.
    ISBN 0-13-151773-2
    1. Fire extinction.   I. Rickman, Tracy.   II. Wallace, Gerald.   III. Title.
TH9151, M25 2007
628.9'25—dc22

                                                                                    2006051890

Publisher: Julie Levin Alexander
Publisher's Assistant: Regina Bruno
Senior Acquisitions Editor: Stephen Smith
Associate Editor: Monica Moosang
Editorial Assistant: Patricia Linard
Director of Marketing: Karen Allman
Executive Marketing Manager: Katrin Beacom
Marketing Coordinator: Michael Sirinides
Marketing Assistant: Wayne Celia, Jr.
Managing Production Editor: Patrick Walsh
Production Liaison: Julie Li
Production Editor: Lisa S. Garboski, *bookworks*
Media Product Manager: John Jordan
Media Project Manager:
    Tina Rudowski
Manufacturing Manager: Ilene Sanford

Manufacturing Buyer: Pat Brown
Senior Design Coordinator:
    Christopher Weigand
Cover Designer: Christopher Weigand
Director, Image Resource Center:
    Melinda Patelli
Manager, Rights and Permissions:
    Zina Arabia
Manager, Visual Research: Beth Brenzel
**Manager, Cover Visual Research &
    Permissions:** Karen Sanatar
Image Permission Coordinator:
    Ang'john Ferreri
Composition: Techbooks
Printing and Binding: Courier Westford
Cover Printer: Phoenix Color Corporation

Pearson Education Ltd.
Pearson Education Singapore, Pte. Ltd.
Pearson Education Canada, Ltd.
Pearson Education—Japan
Pearson Education Australia Pty. Limited

Pearson Education North Asia Ltd.
Pearson Educación de Mexico,
    S.A. de C.V.
Pearson Education Malaysia, Pte. Ltd.
Pearson Education, Inc., Upper Saddle River,
    New Jersey

ISBN-13: 978-0-13-151773-8
ISBN-10: 0-13-151773-2

This book is dedicated to the memories and spirits of those individuals who lost their lives as a result of the disastrous attack on America that occurred on 9/11/2001—and to the thousands of husbands, wives, girlfriends, boyfriends, fathers, mothers, brothers, sisters, children, grandparents, grandchildren, and great-grandchildren who grieve so often and miss them so dearly.

Contents

Preface

This second edition of *Fire Suppression Practices and Procedures* is the revision of the text written by author and retired Fire Chief Eugene Mahoney. In addition to Chief Mahoney, this edition has been co-authored by Chief Gerald Wallace and Chief Tracy Rickman and has been updated to provide instructors with an up-to-date resource to teach current trends, terminology, and fire-related techniques to students in Fire Science classrooms throughout the country.

This book has been created to be the primary classroom text for Fire Behavior and Combustion courses or as a primary text in basic Fire Principles and Practices courses. This text and the corresponding instructor resources can be used as a comprehensive "stand alone" package that fits perfectly within a 16- to 18-week semester course within Fire Technology programs. It has been written to a level that allows students to grasp new concepts and procedures easily. The authors' practical experience also provides the veteran fire service professional with a valuable resource for any fire department library.

DEVELOPMENT AND ORGANIZATION OF THIS TEXTBOOK

Fire Practices and Procedures was originally written in 1992 by Eugene Mahoney. The success and popularity of Chief Mahoney's writing style and decades of experience in the fire service have made this text a valuable and respected resource for more than a decade. Gerald Wallace, current fire chief of South Pasadena Fire Department, California, and Tracy Rickman, current chief of the Rio Hondo Fire Academy and former assistant chief at Northrop Grumman Fire Department, B-2 Division, were brought in to aid in the collaborative development of this new edition.

The second edition is organized into eight chapters that encompass many aspects of today's evolving fire service. New coverage, including NIMS (National Incident Management System) in Chapter Three, brings the latest information available on this critical new topic area. Fire departments nationally have added responsibilities in responding to all categories of terrorist incidents and natural disasters. Recent events dominate the headlines and have been the worst in our country's history. Chapter Five covers GIS, or Geographical Information System, information as well GPS (Global Positioning System). Terms such as RIT (rapid intervention teams), information on different types of fires and how they burn, and what a firefighter can expect to observe based on given factors are also discussed in Chapters Seven and Eight. Pertinent information related to wildland firefighting has been added to Chapter Eight and includes the current

National Wildland Coordinating Group (NWCG) terms and protocols for fighting these types of fires.

The student will be able to use this book as a reference in other Fire Technology–related courses. It provides a broad background on several topics that relate to fire, the understanding of how fire occurs, the suppression of different types of fires, and the procedures (tactics and strategy in most cases) to deal best with confinement and extinguishments of fire.

This text briefly describes many newer fire extinguishing concepts and procedures, particularly dealing with response. The book is "broad based" in scope and provides a very well rounded introductory program.

Readers are invited to contact the authors with any comments or feedback. Please send email to either Tracy Rickman at trickman@riohondo.edu or Gerald Wallace at Gwallace@riohondo.edu.

Gerald Wallace, Fire Chief	Tracy E. Rickman, MPA
South Pasadena Fire Department	EMT, Wildland and Fire Coordinator
Assistant Professor, RHC	Rio Hondo College
Gwallace@riohondo.edu	trickman@riohondo.edu

ACKNOWLEDGMENTS

It is impossible to acknowledge all those individuals who have contributed to this book. There is no doubt that many of the officers and firefighters that I have worked with over the years have unknowingly contributed to various portions of these chapters. The instructors of courses I have taken and the students of classes I have taught have definitely had an influence on the book's contents. The authors of books I have researched, such as Emanuel Fried, James Casey, William Clark, Charles Walsh, Leonard Marks, and Lloyd Layman, have all provided bits of information, as have the contributing authors of training manuals and magazines published by such organizations as the National Fire Protection Association, the International Fire Service Training Association, Fire Engineering, the U.S. Fire Administration, and the Los Angeles Fire Department. To each of the individuals and organizations who have unknowingly contributed, I offer my sincere appreciation.

There are some individuals, however, who have made a substantial contribution in the final steps of preparation. First credit goes to my wife, Ethel, who read each chapter as it was completed and made constructive recommendations for improvement. The efforts and concrete improvements made by the updating, revising, and adding of material by Tracy Rickman, associate professor, Rio Hondo College, and Chief Gerald Wallace, South Pasadena Fire Department, are sincerely appreciated.

No book such as this can be considered complete without the addition of a substantial amount of art. My appreciation is extended to each organization and individual who contributed photographs for inclusion in the manual. Special thanks are extended to:

Rick McClure, Los Angeles Fire Department
District Chief Chris E. Mickal, New Orleans Fire Department
The Las Vegas Fire Department
The Clark County, Nevada, Fire Department
Chris Anders, Spencer Fire Department, Oklahoma
Federal Emergency Management Agency (FEMA}

Special thanks are extended to Kierra Kashickey who devoted so many hours and days to coordinating the efforts of so many individuals and organizations, and to Production Editor Lisa Garboski who was responsible for all portions of the final product.

Last but not least, I thank the staff members of Brady/Prentice Hall who played an important part in producing the book. Special thanks are extended to Stephen Smith, senior acquisitions editor, and Kierra Kashickey, who spent many hours coordinating the final product.

Gene Mahoney

As the coordinator of the Fire Academy at Rio Hondo College Fire Academy, I have learned that it takes teamwork, cooperation, patience, and an understanding of how the fire service operates to make firefighters out of people who have desire. Those individuals that enter our doors, like so many other fire training organizations throughout the country, come to us looking for leadership, guidance, and direction. They look to us as fire service instructors to help them along their own journey to become part of a larger group known today as firefighters. Firefighters come from all walks of life, from small organizations to large departments. They are sometimes volunteers, some are paid, some firefighters work in municipalities, and some work for private organizations. Some firefighters work at a crash rescue fire department and some do wildland firefighting while others work on fire boats. However, they all have one thing in common. Their mission is *to save lives and property* from the destruction of fire. All firefighters strive to do their best to help others in their time of need. This book can be used as an aid as it presents a perspective into the fire service based on over 100 years of combined experience between the now three authors. We hope you will build on what you read, as you continue to encircle your life with those people with whom you associate, those people called firefighters.

Chief Wallace and I would also like to thank the following people who have supported us in this textbook revision. First and most important to both of us, we would like to thank our wives and families for their support and encouragement along this journey. We would like to thank Chief Neal Welland of the Santa Fe Springs Fire Department for his opinion, advice, and consultation. Chief James Brakebill for filling in for us when Gerald and I needed time together to work on specific parts of this book. Division Chief Cliff Hadsell for his insight in today's fire service management and Battalion Chief Richard Beckman for his review of certain chapters contained within this book. Last, Gerald and I would like to thank the dean of Public Safety at Rio Hondo Community College, Chief Joseph Santoro.

We would be remiss if we didn't say "thanks" to Chief Mahoney. He has been gracious, supportive, and helpful to both Jerry and me during this book revision. We thank him for the opportunity and hope that we have accomplished what the chief was looking for in this revision of *Fire Suppression Practices and Procedures*.

Tracy E. Rickman, MPA
Associate Professor of Fire Technology
EMT, Wildland and Fire Coordinator
Rio Hondo College
trickman@riohondo.edu

REVIEWER LIST

Bruce Evans, MPA
Captain/Paramedic Henderson Fire
Fire Programs Coordinator Community College of Southern Nevada
Henderson, NV

Chief William Goswick, BA, EFOP
Assistant Professor, Fire and Emergency Services
Tulsa Community College
Tulsa Fire Department (ret'd)
Tulsa, OK

Jeffrey T. Lindsey, PhD, CFO, EMT-P
Executive Officer
Estero Fire Rescue
Estero, FL

Richard A. Marinucci
Fire Chief, City of Farmington Hills
Farmington Hills, MI

Mark Martin, Division Chief
City of Stow Fire Department
Stow, OH

Jerry A. Nulliner
Fishers Fire Department
Fishers, IN

Dr. Mark A. Rivero
Las Vegas Fire & Rescue
Las Vegas, NV

Lieutenant Michael G. Stanley
Aurora Fire Department
Aurora, CO

Mario H. Trevino, Chief of Department
San Francisco Fire Department
San Francisco, CA

Captain Allen Walls
Colerain Township Department of Fire & EMS
Cincinnati, OH

About the Authors

Gene Mahoney was released to inactive duty as a dive-bomber pilot from the U.S. Navy in 1946. He served an additional 18 years in the Reserve, retiring as a lieutenant commander. He flew both reciprocating engine and jet aircraft during his time with the navy.

Gene joined the Los Angeles Fire Department as a firefighter in 1947 and retired as a battalion chief in 1969. During his time with the department, he was assigned to various areas of the city including five years in the downtown area, five years in the harbor area, and five years in the south-central area of the city. As a battalion chief, he served five years in the most active firefighting battalion in the city, served additional time in the high-rise area of the city, and acted as commander in charge of the firefighting forces at the Los Angeles International Airport. His special-duty assignments included several years in the training section. At the time of his retirement, he was responsible for the public relations section of the department.

Gene retired from the Los Angles Fire Department to accept the position of fire chief for the city of Garden Grove, California. He was later advanced to the position of public safety director. This position was upgraded to that of assistant city manager for public safety. In these positions, he was responsible for the operation of both the fire and police departments. He left the city of Garden Grove to accept the position of fire chief for the Arcadia, California, Fire Department. He retired from this position in 1975.

Gene, together with another Los Angeles Fire Department captain, was responsible for the development of the fire science curriculum at Los Angeles Harbor College, Wilmington, California. He served there as a part-time instructor for 12 years. He also taught Fire Administration courses at Long Beach State College, Long Beach, California, for two years. On retiring as fire chief from the city of Arcadia, he accepted the position of fire science coordinator at Rio Hondo College, Whittier, California. While there, he developed the fire science curriculum into one of the most complete programs in the United States. The program included a Fire Academy that provided all the training required for certification as a Fire Fighter I in California. He retired from Rio Hondo College as a professor of Fire Science in 1988.

While with the Los Angeles Fire Department, Gene attended the University of Southern California, where he received his B.S. degree in Public Administration with a minor in Fire Administration in 1956 and three years later his M.S. degree in Education.

In addition to authoring several articles in professional magazines, Gene has authored a number of textbooks and study guides in the field of fire science. The textbooks include *Fire Department Hydraulics, Introduction to Fire Apparatus and Equipment, Fire Department Oral Interviews: Practices and Procedures,* and *Fire Suppression Practices and Procedures.* The study guides include one for his text *Introduction to Fire Apparatus and Equipment;* one entitled *Firefighters Promotion Examinations;* and one on *Effective Supervisory Practices.* He also had a novel published, entitled *Anatomy of an Arsonist.*

During his career, Gene has been very active in professional and service organizations. He has served as:

District Chairman, Boy Scouts of America
President, United Way
District Chairman, Salvation Army
President, International Association of Toastmasters
President, Rio Hondo College Faculty Association

Tracy E. Rickman began his firefighting career in the U.S. Air Force in 1982. While assigned to the Grand Forks Air Force Base in North Dakota, he earned the Commendation Medal, Good Conduct Medal, and Achievement Medal. While on active duty, Tracy earned his associate degree in Fire Science from the Community College of the Air Force and a bachelor's degree from Park College from Parkville, Missouri. On being honorably discharged from the air force he was hired as the training captain for the Northrop Grumman Fire Department (B-2 Division) and was promoted to assistant chief of Operations and Training. He left Northrop Grumman in 1997 and was hired as the fire prevention specialist for the City of South Pasadena Fire Department and was offered a full-time tenure-track teaching position at Rio Hondo Community College. Tracy has taught courses at California State University, Los Angeles, Cerritos College, Citrus Community College, and currently teaches at California State University at Long Beach within the Professional Studies program. He has earned a master's in Public Administration from the University of LaVerne and has five children and two grandsons. He enjoys the outdoors, is an avid fisherman, and enjoys snowboarding. Tracy is a California State Certified Instructor, a Hazardous Material Response Instructor with the California Specialized Training Institute, and an Internationally Certified Fire Inspector I.

Gerald "Jerry" Wallace, a California native, started his fire career with the City of Lynwood in 1965. He went through the ranks to the position of fire chief and retired late in 1996. He has also been a fire service instructor with Rio Hondo College from 1978 to the present. He is currently the fire chief for the City of South Pasadena, California.

The following grid outlines the course requirements of the Fire Behavior and Combustion course developed as part of the FESHE Model Curriculum. For your convenience, we have indicated specific chapters where these requirements are located in this text.

Course Requirements	*1*	*2*	*3*	*4*	*5*	*6*	*7*	*8*
Identify physical properties of the three states of matter.	X							
Categorize the components of fire.	X	X		X				X
Recall the physical and chemical properties of fire.	X							
Describe and apply the process of burning.	X	X	X	X				X
Define and use basic terms and concepts associated with the chemistry and dynamics of fire.	X	X						X
Describe the dynamics of fire.	X	X				X		X
Discuss various materials and their relationship to fires as fuel.	X	X				X	X	X
Demonstrate knowledge of the characteristics of water as a fire suppression agent.	X	X		X		X		X
Articulate other suppression agents and strategies.	X				X	X	X	
Compare other methods and techniques of fire extinguishments.	X			X	X	X	X	X

Fire Behavior and Combustion

Key Terms

Objectives

The objective of this chapter is to acquaint the reader with the basic principles of fire chemistry, how fire behaves, and fire extinguishment. Upon completing this chapter, the reader should be able to:

- Define what fire is and explain how it behaves.
- Explain the difference between an endothermic reaction and an exothermic reaction.
- Explain the difference between complete combustion and incomplete combustion.
- List the three sides of the fire triangle and explain the relationship of each side to fire.
- Explain the methods of heat movement.
- List the stages of fire behavior.
- Explain the causes of backdrafts, flashovers, and rollovers together with the steps to take in order to help prevent each one from occurring.
- Explain the characteristics of carbon monoxide and its effect on the human body.
- List the four sides of the fire tetrahedron and explain how each side contributes to the extinguishment of fire.
- Describe the types of extinguishing agents and how they are utilized in extinguishing different types of fires.
- Explain the four classifications of fire.

◆ INTRODUCTION

Fire, electricity, and the wheel have probably had more influence on the progress of humankind than any other combination of factors. Eliminate these three items from our environment and humans would probably still be identified as nomads, moving seasonally from cave to cave in their constant fight for survival. Add to this list the elimination of the primary extinguishing agent for fire (water) and this planet would possibly be no more than a pockmarked sphere, moving through space with no signs of life aboard.

Fire has contributed its part to people's progress by providing the warmth and comfort essentially needed for our very existence. Additionally, it has produced the power required to move our vehicles and turn our machinery. Yet, when it has been on the rampage, it has turned on us with a vengeance that has been unmatched by any other enemy. This can be illustrated by the Cocoanut Grove incident in Boston in 1942 where fire snuffed out the lives of 492 people; the General Slocum disaster on the East River in New York City in 1904 where it took the lives of 1,030 people; and, even more appalling, the destruction of the Tokyo and Yokohama areas in 1923 where it ended the lives of 91,344 people and destroyed 500,000 buildings. At times it has chosen to destroy some of the major cities of the world, such as Rome in A.D. 64, London in 1666, and San Francisco in 1906; and it has demonstrated its power during the early part of the twenty-first century as it swept through thousands of acres of beautiful wildland in Colorado, Arizona, and California. (See Figure 1.1.)

In order to cope with this beneficiary that has a personality like Dr. Jekyll and Mr. Hyde, people have had to organize for the purpose of controlling and killing this potential monster whenever it chooses to vent its anger. Those who have been given the responsibility for carrying out these tasks recognize that this tyrant has the capability of seriously injuring or destroying them if they make so much as a simple

FIGURE 1.1 ◆ A wildfire is a good example of a fire on the rampage. *(Courtesy of FEMA)*

mistake. Consequently, it has been necessary for these people to understand thoroughly the characteristics of fire and to be trained and capable of putting into effect the tactics required to destroy it when necessary.

Fire chemistry and fire behavior are complex subjects; however, they are not so complex that the basic principles cannot be mastered by the average firefighter. Both are influenced by such facts as pressure, temperature, the potential fuel's surface exposure, a chain reaction, and many others. Fire chemistry and fire behavior will be approached in this text in a simple manner that will provide firefighters and others dedicated to fire protection with the knowledge required to do an effective job both on the fire ground and in fire prevention.

◆ FIRE CHEMISTRY

There are many ways to define what fire is. Some of those definitions include the following:

- ◆ Fire is a chemical process that produces heat and light and combines with oxygen and a substance known as fuel to undergo a chemical process change.
- ◆ Fire, also known as combustion, is a simple chemical reaction. This persistent, rapid chemical change releases heat and light and is accompanied by flame.
- ◆ Fire is a rapid, self-sustaining oxidation process that emits heat and light.

In the simplest terms, fire might be defined as rapid oxidation accompanied by heat and flame (or heat and light). It is a chemical process that gives off energy and variable products of combustion at the same time. A student would normally have little trouble memorizing this and when asked the definition of fire he or she should be able to give a satisfactory answer; however, in reality, the student might still have little understanding of what fire really is because the definition includes the word "oxidation," which may be like a foreign word to him or her.

Oxidation is a chemical process whereby an atom from one material combines with an atom of oxygen from another material to form a new product. An example of a product of oxidation is rust; however, in this process the oxidation takes place extremely slowly, whereas in fire the process is relatively rapid.

Most chemical processes involve the absorption or release of heat. Heat is merely a form of energy in motion. Heat is the most common form of energy encountered by humankind. The amount of heat present or given off is referred to as **temperature.** Temperature is measured in degrees. The degrees may be stated in Celsius, Fahrenheit, Kelvin, or Rankine. The Fahrenheit scale is generally used in the United States when referring to fires.

A material will absorb heat as it changes from a solid to a liquid or from a liquid to a gas. This is referred to as **latent heat.** When it is absorbed as it changes from a solid to a liquid, it is called the **latent heat of fusion.** When it is absorbed as it changes from a liquid to a gas, it is called the **latent heat of vaporization.** If heat is absorbed during a reaction, it is referred to as **endothermic reaction.** If heat is given off during a reaction, it is referred to as an **exothermic reaction.** The oxidation involved with fire is exothermic in nature.

Most providers of fuel contain atoms of carbon. It is a carbon atom that attaches to oxygen atoms from a different source to complete the process of oxidation in the chemical process of fire. It is necessary for a carbon atom to unite with two atoms of oxygen in order to have complete combustion. The resultant product is carbon

oxidation   A chemical process whereby an atom from one material combines with an atom of oxygen from another material to form a new material.

temperature   The amount of heat present or given off.

latent heat   The heat absorbed as a material changes from a solid to a liquid or from a liquid to a gas.

latent heat of fusion   The amount of heat absorbed as a material changes from a solid to a liquid.

latent heat of vaporization   The amount of heat absorbed as a substance passes between the liquid and gaseous phases.

endothermic reaction   When heat is absorbed during a reaction.

exothermic reaction   When heat is given off during a reaction.

FIGURE 1.2 ◆ The products of combustion and incomplete combustion.

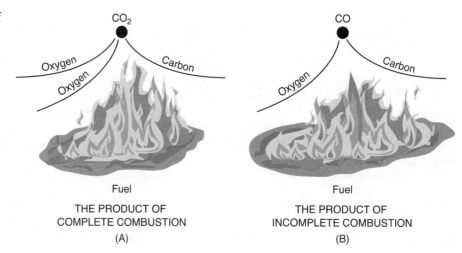

THE PRODUCT OF
COMPLETE COMBUSTION
(A)

THE PRODUCT OF
INCOMPLETE COMBUSTION
(B)

carbon dioxide The product of complete combustion.

dioxide (CO_2). Hence, it can be said that the product of complete combustion is **carbon dioxide.** (See Figure 1.2.)

The carbon atom will reach out and combine with a single atom of oxygen whenever there is an insufficient supply of oxygen atoms available for complete combustion to take place. In this case, the product produced is carbon monoxide (CO). (Note that the prefix *mono-* indicates one.) Hence, it can be said that the product of incomplete combustion is **carbon monoxide.**

carbon monoxide The product of incomplete combustion.

There are several other products of combustion, depending primarily on the composition of the material that is said to be burning. However, of the common products of combustion, the most dangerous to a firefighter is carbon monoxide. Carbon monoxide is a colorless, odorless, tasteless, and invisible gas. There are two characteristics of this gas that make it a killer.

First, carbon monoxide is a poisonous gas. Its toxicity is based on the fact that it combines readily with the hemoglobin in the red blood cells. Unfortunately, the hemoglobin that carries the oxygen through the body likes carbon monoxide better than it likes oxygen. As a result, it will combine with carbon monoxide 250 to 300 times more readily than it will combine with oxygen. In a very short time, a firefighter exposed to carbon monoxide may find himself or herself in a situation in which the oxygen-carrying hemoglobin in his or her bloodstream is so filled with carbon monoxide that he or she is no longer able to function adequately. However, with the safety rules presently used by the fire service, a structural firefighter should never be found in an atmosphere containing carbon monoxide without being fully equipped with a self-contained breathing apparatus (SCBA) and related personal protective equipment, or PPE.

Although a structural firefighter should never be exposed to an atmosphere containing carbon monoxide without the protection of an SCBA, he or she should be aware that the process of producing carbon monoxide takes place in an insidious manner. Sometimes it gives an individual who is exposed to it little or no warning that he or she has absorbed a dangerous amount of the substance. Remember that wildland firefighters and those structural firefighters working on wildfires do not normally have the protection from carbon monoxide that structural firefighters normally enjoy. All firefighters should therefore be aware that a feeling of fatigue, a headache, or disorientation might indicate that carbon monoxide is being absorbed into the body. Absorption might also be indicated in the fingernails. The skin under the nails, which is normally pinkish in color, will turn blue as the body is deprived of oxygen. A firefighter noticing this change should immediately remove himself or herself to fresh air because he or she may be on the point of collapse and not be aware of any other signs of disablement.

The second characteristic of carbon monoxide that makes it a killer is that it is a flammable gas. In fact, it has a flammable range of approximately 12.5 to 74 percent. The meaning of flammable range will be described later in this chapter in the discussion of the fuel side of the fire triangle. At this point it is important to know that carbon monoxide will ignite in a ball of flame when the proper conditions exist. Those who have seen pictures of oil tank fires, or actually had the experience of watching one burn, may have noticed the process. Frequently, large balls of fire can be seen rolling through the heavy black smoke above the tank. These balls of fire are generally carbon monoxide burning. A similar display takes place in building fires in a condition known as a backdraft. The cause and effects of backdrafts will be discussed later in this chapter.

◆ **THE FIRE TRIANGLE**

The "fire triangle" is an older term used to describe the relationships of the components of fire. Although no longer technically correct, it will be used to explain the birth of fire.

The initiation and extinguishment of fire are two opposing chemical reactions that basically involve the same elements. For years the fire triangle has been used to introduce both children and adults to the theory of fire. People have been taught that fire occurs whenever three elements combine in the proper proportions. Basic as it may seem, these three elements (fuel, oxygen, and heat) are best illustrated by the use of the fire triangle.

The three sides of the triangle are shown in Figure 1.3. However, to understand the initiation of a fire fully, it is imperative for an individual to have a thorough grasp of what is meant by the proper proportions of each side of the triangle. Variations in the balance amounts of the elements represented by the three sides will determine whether the fire will smolder and spread slowly or whether it will spread rapidly.

THE OXYGEN SIDE OF THE TRIANGLE

Although it is possible for a fire to occur in some atmospheres other than oxygen (e.g., chlorine gas and its compounds), for practical purposes, it should be considered that oxygen is necessary in order to have a fire. Oxygen is an element that has an atomic weight of 16, which means that it is 16 times heavier than hydrogen. Although it is essential to the burning process, oxygen itself does not burn.

The primary source of oxygen for a fire is the air. The air normally contains approximately 21 percent oxygen, 78 percent nitrogen, and 1 percent other gases. If there were such a thing as a "normal" fire, it would burn best in an atmosphere of 21 percent oxygen. A fire will no longer burn normally whenever the percentage of oxygen available is above or below the 21 percent figure. When the oxygen concentration is above 21 percent, it is known as an oxygen-enriched atmosphere. Materials exhibit different burning characteristics when exposed to an oxygen-enriched atmosphere.

The flames from a fire will begin to diminish in size and intensity whenever the concentration begins to be reduced to less than 21 percent. The fire will appear to die when

FIGURE 1.3 ◆ The fire triangle.

FIGURE 1.4 ◆ A firefighter preparing to enter a dwelling with an indication of a possible smoldering fire inside. *(Courtesy of Rick McClure, LAFD)*

the concentration reaches approximately 16 percent, and it will normally be completely extinguished by the time the percentage is reduced to 13 percent. Therefore, it can be concluded that a fire needs an atmosphere containing at least 16 percent oxygen to live. Ironically, the human body also requires an atmosphere containing at least 16 percent oxygen in order to survive. From this it is easy to develop an important correlation that is valuable to a firefighter. *If the atmosphere is such that a fire cannot burn, it cannot support life.* A firefighter should be aware that if the atmosphere inside a structure is such that a fire cannot freely burn, the situation is ripe for a backdraft. (See Figure 1.4.)

Opposite to the effect of a fire diminishing in size and intensity whenever the concentration of oxygen is less than 21 percent is the concept that a fire will intensify whenever the supply is above 21 percent. Oxygen-enriched atmospheres not only affect the burning characteristics of materials but also present a safety factor to firefighters operating in them.

A product called Nomex® is a material used in the construction of much of the protective gear worn by firefighters. This material will not burn when the oxygen concentration is normal at 21 percent. However, Nomex will ignite and burn vigorously in an oxygen-enriched atmosphere of 31 percent or greater. The conclusion is that the intensity of a fire depends on the rate at which oxygen is provided. Materials that are normally not flammable may become flammable in an oxygen-enriched atmosphere. A demonstration may help the reader understand these concepts. What happens to a burning cigarette when hit with 100 percent oxygen? The cigarette is burning normally in a 21 percent oxygen atmosphere before the pure oxygen is supplied. The pure oxygen enriches the atmosphere and the entire cigarette will be immediately consumed, from one end to the other.

A disastrous example of the hazard of an oxygen-enriched atmosphere occurred at Cape Kennedy, Florida, on January 27, 1967. Astronauts Lieutenant Colonel Grissom, Lieutenant Colonel White, and Lieutenant Commander Chafee were strapped in position in the cockpit of *Apollo 1* while rehearsing for a scheduled launching when a small fire (presumably electrical) broke out in the cockpit. The atmosphere in the cockpit at the time was 100 percent oxygen. The result was the instant cremation of all three. It took 13 seconds from the time one of the astronauts said, "Fire. I smell fire," until they were all dead. It is estimated that the fire may have reached a temperature of 2500°F.

Oxidizers.   Air is the primary source of oxygen for a fire, but there are other sources. The most important from a firefighting standpoint is a group of chemicals called **oxidizers.** These chemicals contain oxygen in their makeup and release it under the process of decomposition. Some of the more common oxidizers are chlorates, nitrates, nitrites, nitric acid, and peroxides.

> **oxidizers** Chemicals that contain oxygen in their makeup and release it under the process of decomposition.

The most common of these chemicals are nitrates and chlorates. At times, these chemicals are involved in a fire situation by providing additional oxygen to a fire that is already in progress. The result is generally the development of an extremely intense, hot fire. An example might be a fire on a wooden loading dock where chlorates had previously been spilled and absorbed into the wood. The fire would intensify as the chlorates decompose and release oxygen.

At other times the chemicals provide the oxygen needed to get the fire started. A good example would be in a closed container of nitrocellulose film. While nitrocellulose film is no longer used, the older films were made of nitrocellulose. The film itself contains an oxidizer. Oxygen will be provided in a sufficient quantity to possibly complete the oxygen side of the fire triangle as the material starts its process of decomposition. If successful, a fire will be initiated without the film being exposed to any other oxygen source. (See Figure 1.5.)

FIGURE 1.5 ◆ Firefighters preparing to make a direct attack on a fire searching for oxygen.
(Courtesy of District Chief Chris E. Mickal, New Orleans Fire Department—Photo Unit)

THE HEAT SIDE OF THE TRIANGLE

When a person checks his or her present surroundings, he or she will most likely find that all three sides of the fire triangle are present. Fuel is readily available, there is sufficient oxygen in the room, and the room is probably comfortably warm, indicating the presence of heat. However, there is not sufficient heat available to cause all the combustibles in the room to burn. If all the combustibles in a room are heated to a certain point, all of the combustibles will break into flames and a condition called a **flashover** will occur.

Flashover is defined as that fire phenomenon in which all combustible materials in a given area reach their ignition temperature simultaneously.

The point of reference is called the **ignition temperature.** The ignition temperature can be defined as the minimum temperature that will cause self-sustained combustion of a material, or the temperature to which a material must be heated in order to burn. It should be noted that a material does not have to be in contact with the heating sources in order for combustion to take place. All that is needed is for it to be heated to its ignition temperature.

Those who have been firefighters for any length of time have probably watched this process develop at a fire. Occasionally, while firefighters are laying lines to protect an exposure, the entire side of the exposure will break into flames before their eyes. The ignition temperature of the exposed material has been reached and ignition has taken place, even though the exposure was not in direct contact with the flames from the original fire. It might be added that an **exposure** is anything in close proximity to the fire that is not burning but that might start burning if some type of corrective action is not taken quickly. (See Figure 1.6.)

On the other hand, most firefighters have at one time or another witnessed an opposite result. They have seen an easily ignitable material not ignite because the temperature of the heat source it came in contact with was too low. A good example is the spilling of a small amount of oil on an engine. The general result is a lot of smoke but no fire. The reason for this is because the temperature of the outside surface of the engine is below the ignition temperature of the oil.

Ignition temperatures vary from one material to another and have little relationship to the potential hazard of the material. For example, gasoline (which is considered

flashover When all of the combustibles in a room break into flames.

ignition temperature The minimum temperature that will cause self-sustained combustion of a material.

exposure Anything in close proximity to the fire that is not burning but that might start burning if some type of corrective action is not taken quickly.

FIGURE 1.6 ◆ Protective action being taken to prevent the exposed structure to the right from breaking into flames. *(Courtesy of District Chief Chris E. Mickal, New Orleans Fire Department—Photo Unit)*

one of the most hazardous materials associated with our daily lives) has an ignition temperature of approximately 800°F, whereas most woods have ignition temperatures in the 400° to 500° range. On the other hand, some extremely hazardous materials have low ignition temperatures. A good example is carbon disulfide, one of the most hazardous materials known, which has an ignition temperature in the low 200s. Agriculture dusts, which are considered extremely hazardous, have ignition temperatures in the same range as woods. Even the ignition temperature of a material appears to change over a period of time when exposed to a certain environment. In reality, however, the ignition temperature of the material does not change but the material itself changes. For instance, wood subjected to a low heat source over a long period of time will have the moisture driven out and the wood will change to charcoal. Charcoal has an extremely low ignition temperature.

Ignition temperatures should always be considered an approximation. Different results will be obtained from the same material depending on when it is tested, the test conditions, and the form of the tested material. A different ignition temperature will be found when testing wood in small chips than when it is tested in larger pieces. The ignition temperature of gasoline will vary according to the octane rating. Regardless, the resultant principle remains the same. When all other conditions are right, a material will start to burn when its ignition temperature is reached.

Heat Sources.   Although the open flame is generally considered the primary heat source for a fire, there are many others.

The sun itself may be a source. It is not normally thought of as providing sufficient heat to initiate a fire, but when its rays are magnified, the resultant heat can be sufficient to raise the temperature of a material above its ignition point. A number of grass fires have occurred as a result of the sun's rays penetrating through the glass from a broken bottle. The broken bottle is shaped in such a manner as to act like a magnifying glass, concentrating the resultant heat on a batch of dry grass. Fires of this type are very rare, but their potential should not be ignored.

Friction is one of the oldest known sources of heat. Although early humans did not really understand the process involved, they did comprehend that sufficient heat could be developed to start a fire if they rubbed two sticks together vigorously for a long enough period of time. This knowledge proved to be a key factor in the daily survival of the tribes of nomads roaming the earth. Today we have to be careful of this phenomenon in our industrial processes as we guard against slipping belts that are used to run machinery.

All the marvels of electricity carry with them the constant potential for the development of heat. The mere passage of amperage through wires causes a heat buildup. Heat can be developed to the point of raising insulation or surrounding material to above its ignition temperature whenever an attempt is made to push more amperage through a wire than the wire is capable of carrying. Additional resistance placed in the line can have the same result. The arcing that occurs from the breaking of a circuit, by opening a switch either directly or accidentally, can produce a temperature of over 2000°F. Similar temperatures can be found in the sparks caused by an electrical short circuit. Several serious fires have occurred as a result of a person carrying a cell phone receiving a call while filling the tank of his or her vehicle.

Of course, nature's sources of electrical current can also result in temperatures that can ignite combustible mixtures. The two most common are static electricity and lightning. Safeguards are built into hazardous areas such as operating rooms to protect the environment against the potential danger of static electricity, and additional

FIGURE 1.7 ◆ Lightning like this is responsible for many wildland and structure fires. *(© Copyright Joseph Matthews)*

devices are used to protect entire buildings from lightning when a thunderstorm occurs. (See Figure 1.7.) Lightning has been the cause of a number of serious wildland fires when a bolt from the sky struck a tree.

Mixing two chemicals can also produce dangerous levels of heat. Some reactions are slow; others are instantaneous. Some years ago a number of fires were reported in drugstores in a large metropolitan area. These fires were started by school children who had learned in a chemistry class that mixing a certain type of hair cream with another common product found in drugstores could develop heat. An individual would enter a store, mix the two materials together on a shelf, and leave. A small fire would break out a short time later.

A much faster reaction takes place when other types of chemicals are mixed. A good example is the propellant system used in some rockets. Two different chemicals are carried in separate storage tanks. Controlled mixing of the two occurs in the combustion chamber. The mixing of the two causes instant ignition, completely eliminating the need for another ignition source in the rocket. This type of response is referred to as a **pyrophoric reaction.**

pyrophoric reaction When the mixing of two materials causes instant ignition.

A process referred to as spontaneous ignition causes another source of heat. Generally, spontaneous ignition is caused by oxidation. Those combustible materials that are capable of combining with oxygen will, in most cases, become involved in the oxidation process when they are exposed to the air. The result of the oxidation is heat generation; however, the heat is usually dissipated as fast as it is developed. Materials that are

FIGURE 1.8 ◆ This roof fire will continue to spread by direct contact if no action is taken to stop it.
(Courtesy of Rick McClure, LAFD)

capable of spontaneous ignition under proper conditions build up heat faster than it is lost to the surroundings. The time required for the heat to build up to the point where ignition takes place varies with the condition under which the heating takes place.

Heat Movement.   Most fires in combustible materials are initiated when the burnable material comes in direct contact with a heating source. In such cases the heating source is above the ignition temperature of the burnable material. The most prevalent type of direct contact is the open flame, with the match playing a big part. A match burns at a temperature of approximately 2000°F, which is well above the ignition temperature of most combustible materials. When held in contact with a burnable material, it will raise the material to its ignition temperature unless the heat is dissipated within the material faster than it is being formed at the contact point. The same principle holds true regardless of the heating source. The heat buildup at a contact point must be faster than it is being dissipated to the surroundings if ignition is to occur. (See Figure 1.8.)

There are three heat sources other than direct flame contact that can raise the heat level of a material to above its ignition temperature: radiation, conduction, and convection. These are methods by which heat moves from one location to another. They are samples of nature's attempt to maintain a heat balance by movement of heat from a warmer to a colder body. (See Figure 1.9.)

Radiation: Heat is radiated from a burning fire in all directions. Some of the radiated heat is transferred back into the fire and helps accelerate the chain reaction process. The rays travel in a straight line and would travel at the speed of light if they were moving in a vacuum. The heat travels as electromagnetic waves, similar to light, radio waves, and X-rays. Radiated heat is a problem at all fires.

The rays travel in any direction and continue to move in that direction until their energy is dissipated or they are stopped by an intervening body. It is generally radiated heat that penetrates the exposures in single-story structures and ground-level open fires. When streams from lines laid at fires are used to wet down adjacent structures, they are so used to prevent the radiated heat from the fire from raising the temperatures of the structures to above their ignition temperatures. (See Figure 1.10.)

FIGURE 1.9 ◆ The three methods of heat movement.

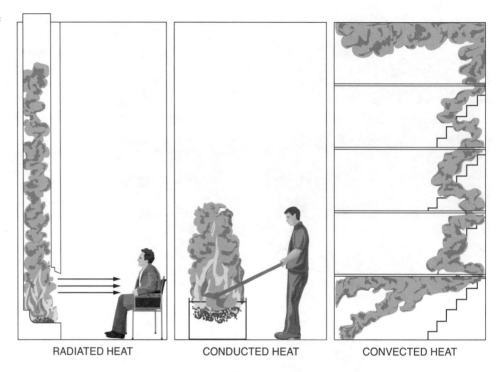

RADIATED HEAT   CONDUCTED HEAT   CONVECTED HEAT

FIGURE 1.10 ◆ Radiant heat is being projected in all directions from this fire.
(Courtesy of District Chief Chris E. Mickal, New Orleans Fire Department—Photo Unit)

In some types of small fires, such as a fire in methyl alcohol, only about 10 percent of the heat produced is dispersed as radiated heat. However, in large fires involving ordinary fuels, it has been estimated that as much as 30 to 50 percent of the total amount of heat produced is dispersed as radiation.

Conduction: The second method of heat moving from a source to an exposure is called conduction. The principle of conduction is simple. When two materials having different temperatures are placed in contact with one another, the material having the greater heat will transmit some of it to the material having the lower heat. The rate of the heat transfer depends on the temperature difference of the two bodies and the conductivity of the path of travel. The temperatures of the two materials will become identical over a period of time. The transfer of heat cannot be completely stopped by any heat-insulating material.

The principle involved can be demonstrated by a person holding a metal rod with the opposite end of the rod placed in an open fire. The rod will become hot enough after a period of time that the person will have to drop it. What happens is that all the molecules within the rod are in contact with other molecules. Those in the fire start transmitting heat to adjacent molecules, and movement continues up the rod until the person holding it senses the heat. (See Figure 1.9.)

The principle of conduction can causes fires to start inside walls of structures, but it plays a more important role when the structures are made of metal. A good example is a fire aboard a ship. A fire in one hold can be transmitted to the cargo in another hold by heat conducted from one hold to the other through the metal bulkheads. A fire in a stateroom can be transmitted to an adjacent stateroom. In fact, it can move in one of six directions or perhaps in all six at the same time—to the areas adjacent to the four bulkheads, the area above, and the area below.

In addition to contributing to fire spread, the principle of conduction can be a primary factor in the weakening of a structure. Unprotected steel loses its strength as it is heated. It is estimated that unprotected steel will have lost approximately 90 percent of its strength at 1400°F. The critical temperature for structural steel is 1000°F. This is the temperature at which steel can no longer be considered as a supporting element. In the opinion of the authors, this factor may have played a major part in the collapse of the twin towers of the World Trade Center on 9/11/2001. The intensity of the fire at those levels where the planes struck was extreme, and the impact may have removed the covering of the protective material over the vertical steel supports, leaving the steel exposed to the severity of the fire. It is possible that the transfer of heat by conduction down the steel risers could have reached a point where the steel at the lower floors could no longer support the weight of the buildings. Although this theory cannot be proven, for the safety of firefighters, it should be considered whenever it is necessary to send them into a modern high-rise building when a severe fire exists on an upper floor.

Convection: The third method for transferring heat from one location to another is called convection. Convected heat moves in air currents. The currents travel in an upward direction from the fire. They continue in this direction until stopped by a ceiling or other obstruction. Then they move along the ceiling, searching for another vertical opening. If they find one, they will again start in an upward direction. If they do not find a vertical opening, they will move until they hit a wall and then start a downward movement. The vertical rise of the heat and products of combustion, their banking along the ceiling, and their travel down the walls is referred to as the **mushroom effect.**

mushroom effect   The vertical rise of the heat and products of combustion, their banking along the ceiling, and their travel down the walls.

Convected heat is the culprit in multiple-story fires. The heat will generally continue in an upward direction as long as it can find a vertical opening. A vivid example of its movement was demonstrated at a five-story hotel fire in the Los Angeles area. The fire started in a room adjacent to an unprotected stairwell on the third floor. The heat from the fire left the room through an open door and traveled up the stairwell to the top (fifth) floor. It spread across the entire length of the hall on the top floor and through open transoms into the rooms. Seven people died on this floor. It spread approximately halfway down the length of the hall on the fourth floor. Damage from the natural spread on the third floor was limited to approximately a fourth of the length of the hall area.

Although convected heat is normally thought of as moving in an upward direction, the wind or drafts can change its path of travel. Many times the drafts are induced by firefighters in their ventilation procedures. A poor choice for cutting a hole in a roof can cause the fire to be transmitted to uninvolved areas of the building through horizontal openings.

THE FUEL SIDE OF THE TRIANGLE

Most people who first cast their eyes on the fire triangle form the opinion that the fuel side would be the most easily understood; however, they are generally mistaken. They have little problem conceptualizing that fuel comes in three forms—solids, liquids, and gases—but this erroneous concept can cause a certain amount of confusion. With the exception of certain metals, solids and liquids do not burn. Both of these give off vapors when subjected to heat. It is the vapors that burn and not the material itself. Therefore, they should not be considered as fuel but rather as fuel providers. This principle can be observed by watching a candle. Note that the flame does not actually touch the wax. What happens is that the heat from the flame reduces the wax to a vapor. The vapor rises and burns. This same phenomenon can be observed by taking a close look at a fire in a fireplace. From a distance the flames appear to be in contact with the wood, but a closer look will show that they are not. What is happening is that the vapors given off by the wood are burning a short distance above the wood. For practical purposes, then, it can be said that only vapors are involved in the burning process. Another term the reader should become familiar with is pyrolysis. When solid and liquid types of fuels are converted to their gaseous state (decompose or vaporize), we set the stage for combustion. Thus, pyrolysis is chemical change brought about due to heat. One thing to remember here is that "only *gases* burn." Through pyrolysis, we bring about needed change by adding heat, allowing combustion or fire to occur.

But all vapors do not burn. This becomes apparent on those occasions when a person tries to start his or her car and it will not cooperate, even though there are vapors in the cylinder and an adequate spark for ignition is being provided. At times there are insufficient vapors in the cylinder to burn and at other times the concentration is too great. Whenever there are insufficient vapors available, the mixture is said to be too "lean" to burn. When there is an excessive amount of vapors, the mixture is too "rich" to burn.

flammable limits   The limits within which a vapor will burn.

explosive limits   The limits within which a vapor will burn.

These examples illustrate the principle that vapors will burn only when conditions are right. They will burn within certain limits, which are referred to as the **flammable limits** or **explosive limits.** The flammable limits are expressed as a percentage of a vapor/air mixture. For example, if the flammable limits of a certain flammable liquid are 2.0 to 8.0, it means that the vapors will burn only when the vapor/air

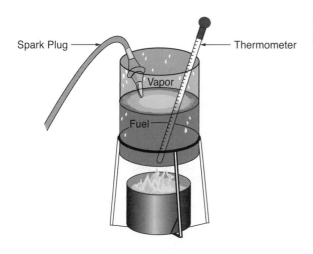

FIGURE 1.11 ◆ An open cup flash point tester.

Spark Plug — Thermometer

Vapor

Fuel

mixture is between 2 percent vapor and 98 percent air, and 8 percent vapor and 92 percent air. To understand this concept thoroughly, it is necessary to be familiar with the evolution process that takes place as a flammable liquid changes from a liquid to a vapor.

Figure 1.11 illustrates an open cup flash point tester. Open cup testers are used for some liquids, and closed cup testers are used for others. One of the most commonly used open cup testers is the Cleveland Open Cup apparatus. Different flash points will be obtained for the same liquid if tested by both the closed cup and open cup methods; however, the concept of the testing and the principles involved are the same for both instruments.

A thermometer has been placed in the liquid to be tested (Figure 1.11). It is used to determine the temperature of the liquid at two points in the testing process. The spark plug above the liquid is constantly providing a spark that has a temperature well above the ignition temperature of the liquid. At some point, a flash will occur across the face of the liquid and go out. The temperature at which this occurs is called the **flash point.** Heating of the liquid is continued and a few degrees above the flash point (usually 2 to 4) a flash will occur above the liquid and the resultant fire will continue to burn. The temperature at which this occurs is called the **fire point.**

Examine what happened. Vapors were being given off from the liquid before the first flash occurred; however, the mixture formed by these vapors with the air would not burn. The mixture was too lean. (See Figure 1.12.) When the first flash occurred it indicated that the vapor/air concentration was burnable. Therefore, the flash point was the temperature at which the vapors first entered the flammable range. The flash across the top of the liquid burned off all the available vapors but more vapors were developed as the temperature of the liquid continued to rise. At the second flash

flash point The temperature at which a flash will occur across the face of a liquid and go out.

fire point The temperature at which a flash will occur above a liquid and the resultant fire will continue to burn.

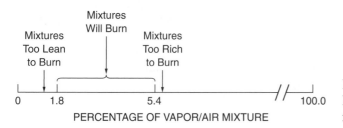

Mixtures
Will Burn

Mixtures
Too Lean
to Burn

Mixtures
Too Rich
to Burn

0   1.8   5.4   100.0
PERCENTAGE OF VAPOR/AIR MIXTURE

FIGURE 1.12 ◆ Percentage of vapor/air mixture.

(fire point) the vapors were within the flammable range and were being given off at a sufficient rate to continue the burning.

Figure 1.12 should put all this in perspective. Remember that it was stated earlier that all sides of the fire triangle had to be in the proper proportions in order to have a fire. The proper proportion for the fuel side of the triangle is for the fuel/air vapor from the material that will be said to burn to be within the flammable range.

To illustrate, the flammable liquid shown in Figure 1.12 has a flammable range from 1.8 to 5.4. Burning will not take place (the mixture will be considered to be too lean to burn) when the concentration of vapors is below 1.8. Nor will burning take place when the concentration is above 5.4 (in this case the mixture will be too rich to burn). Burning can occur only when the vapor/air concentration is between 1.8 and 5.4.

The flammable limits for both liquids and gases are measurable and vary from one product to another. They also vary with the ambient temperature. The limits for some of the common liquids and gases at 70°F are:

Acetylene	2.5–81.0
Butane	1.9–8.5
Carbon monoxide	12.5–74.0
Gasoline	1.4–7.6
Kerosene	0.7–5.0

The flammable range of liquids and gases can be measured, but the burnable limits for gases given off from solids cannot. It is assumed, however, that these gases also burn only within predetermined limits and that the principle involved is the same as that of measurable fuels.

The term "flammable limit" may also be expressed by the term *"explosive limits."* Many books will also refer to these terms as "LEL" (lower explosive limit), "UEL (upper explosive limit), "LFL" (lower flammable limit), and "UFL" (upper flammable limit). Different terminology but the same meaning is expressed.

Last, the fire tetrahedron is the term used to explain the fourth element to the fire triangle. That last element is the chemical chain reaction that occurs when fuel is broken down by heat. As long as there are fuel, oxidizers, and heat energy within the appropriate amounts and as long as the chemical chain reaction is not interrupted, combustion will continue to occur. The fire triangle is thus expanded to the new terminology called the fire tetrahedron when the chemical chain reaction of fire is included.

◆ FIRE BEHAVIOR

Fires generally start small and continue to grow until one of the sides of the fire tetrahedron begins to disappear. The fire will go through at least three definite stages and perhaps five stages during this process. The five stages are defined as the ignition stage, the growth stage, the flashover stage, the fully developed stage, and the decay stage. Following are some brief descriptions of the five stages.

It should be kept in mind that the five stages are defined only for the purpose of understanding the general progress of a fire. There is no clear line of demarcation from one stage to another under actual fire conditions. For example, when the fire first begins to burn in a closed room, the heat from the fire will collect at ceiling level and be transmitted down to other combustible material in the room.

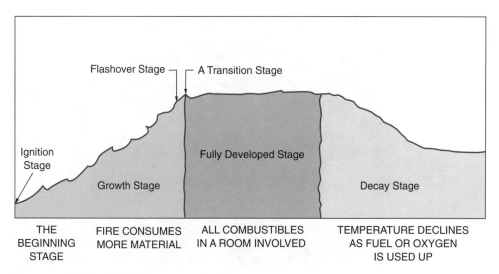

FIGURE 1.13 ◆ The five stages of a fire.

When the fire is outside or in a large building, the heat will rise unrestricted and not have the same heating effect on other combustibles in the immediate area. In addition, a fire outside a building can be increased in size by winds and sloping terrain. (See Figure 1.13.)

THE IGNITION STAGE

This is the beginning stage. In this stage the four elements later described for a tetrahedron will come together to ignite the combustible material and combustion will begin. The combustible material is heated, normally from an outside source. It will start growing when the heat produced is greater than that being lost. Remembering that only gases burn, the ignition stage of fire can be effectively controlled if not extinguished in many situations if the fire is caught within this first stage.

THE GROWTH STAGE

During this stage the fire will quickly start consuming more material as the chain-reaction process takes place. (See Figure 1.14.) The speed of the growth and ultimate size of the fire will primarily depend on how much burnable material or fuel is available. The amount of oxygen available will have an impact on the speed and size at which the fire grows. The size of the container or space will also affect the growth of a fire. The gaseous products of combustion such as carbon dioxide and carbon monoxide will move in an upward direction with smoke being produced. The temperature of the fire will rise to above 1000°F, and if the fire is in a room, the room will begin to heat up. The oxygen content during this phase will remain at or near 21 percent. The fire will move into the flashover stage if the heat continues to increase and the oxygen content in the room is sufficient. Heat that is radiated back into unburned areas will affect the growth of a fire as well. Insulation of a structure can help trap heat and serve the fire tetrahedron by maintaining the chemical chain process.

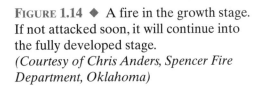

FIGURE 1.14 ◆ A fire in the growth stage. If not attacked soon, it will continue into the fully developed stage. *(Courtesy of Chris Anders, Spencer Fire Department, Oklahoma)*

THE FLASHOVER STAGE

The flashover stage is a transition stage that exists between the growth of a fire and a fully developed fire. A flashover does not take place with every fire. It is an environment that has changed from one dominated by the first material ignited to one that involves all the exposed surface area of all combustible material in a room. The gases that were developed during the growth stage and the combustibles in the room are heated to their ignition temperature by the heat from the fire and the radiant heat projected from the gas layer at the ceiling level. No exact temperature can be assigned to when the flashover takes place, but estimates have placed it in the range between 900°F and 1200°F.

Temperatures begin to increase rapidly during this stage if sufficient oxygen is available. The flames increase in size and intensity. If the fire is in an open area with sufficient fuel and sufficient oxygen, it will continue in this stage and maintain its growth until positive action is taken to stop it. The oxygen concentration remains at or near 21 percent and the products of combustion are carried upward and away from the fire. However, if the fire is in a closed area the temperature at the ceiling will increase to above 1300°F, while the products of combustion will begin to build at the ceiling level and start banking down the walls and the oxygen content in the room will start diminishing as conditions approach the fully developed stage.

THE FULLY DEVELOPED STAGE

This stage occurs when all combustible materials within the perimeter of the fire's boundaries are involved in the fire. (See Figure 1.15.) During this stage the burning materials are releasing the maximum amount of heat possible. The heat produced will depend on whether the fire is in a relative tight compartment such as a room or whether it is outside or in a compartment where ventilation openings are available. If in an airtight compartment, the available oxygen will soon be used up and the fire will pass into the decay stage. If outside, it will pass into the decay stage when all the combustible material available to the fire is burning and the fire begins to release less heat.

FIGURE 1.15 ◆ A good example of a fire in the fully developed stage.
(Courtesy of Chris Anders, Spencer Fire Department, Oklahoma)

THE DECAY STAGE

The temperature of the fire will begin to decline as the available oxygen or available fuel is used up. The fire will use up more and more of the oxygen as it continues into the decay stage in a closed room. It will use up more and more of the fuel as it continues into the decay stage when it is located outside.

In a closed room, carbon monoxide will form a greater part of the gases of combustion, and flame propagation will start diminishing in size. The flames will have been reduced to a flickering stage when most of the oxygen in the room has been consumed. At this point the oxygen content is at or below 16 percent. The flames may disappear completely and the fire will appear as nothing but a glow if allowed to continue in this condition for any length of time. In this situation, there is generally still sufficient fuel in the room to burn and the entire room will be in a superheated condition. At this point the conditions in the room are at the potential backdraft level.

◆ THE PHENOMENA OF THE FIRE TRIANGLE

There are three phenomena of the fire triangle that should be thoroughly understood by every firefighter. The cause of each is different, but the overall result is almost identical—*all are potential killers.*

BACKDRAFTS

The first phenomenon of the fire triangle is referred to as a backdraft. It is also referred to as a smoke explosion. A backdraft occurs when the fuel side and the heat side of the triangle are in proper proportions in an area and the oxygen side is suddenly supplied. A typical example is a tight room in which the fire has been burning for some time. It has reached the decay stage and there is insufficient oxygen in the room for burning to continue. The gases in the room are superheated. In fact, the temperature in the room is above the ignition temperature of the materials in the room. If a door is opened into

FIGURE 1.16 ◆ Constantly be on the alert for a potential backdraft.
(Courtesy of District Chief Chris E. Mickal, New Orleans Fire Department—Photo Unit)

the room, air will rush in and complete the fire triangle. The air mixes with the carbon monoxide in the heavy smoke, resulting in a large ball of flame. The overall effect resembles an explosion, with the explosive force normally leaving the room through the same entry that was used by the air to enter the room. The person who opened the door and started to enter, or has entered the room, would probably be killed or seriously injured. Anyone who had been in the room when the fire started and was asleep or unconscious is most likely already dead. (See Figure 1.16.)

The time it takes for a backdraft to occur once oxygen is provided into an area varies according to the size of the area where the fire is located. The reaction is almost instantaneous when the fire area is small. In large areas, it may take as much as two minutes for a sufficient amount of oxygen to enter the area to complete the fire triangle. A couple of examples will illustrate this point.

A fire occurred in a small closet at the end of a long hall in a dwelling. It had burned for a sufficient period of time to reach the smoldering stage. The hall stretched from the front door of the dwelling to the closet. The firefighters took a booster line in through the front door and advanced to the closet. The firefighter handling the nozzle opened the closet door without first checking for signs of a potential backdraft. The backdraft was instantaneous. The resultant explosion blew all three firefighters down the hall, out the front door, across the lawn, and it rolled them onto the front yard of the house across the street. Fortunately, no one was seriously injured.

Another fire occurred in a large manufacturing warehouse. On arrival at the scene, the company officer went inside to size up the situation. The officer found a small fire at the back of the warehouse. He returned to the apparatus and reported to the dispatch center that the firefighters had a small fire that they could handle. The officer instructed the crew to take a booster line inside to extinguish the fire. When the crew entered the building, they found the entire warehouse involved with fire. A backdraft had occurred in the officer's absence.

At another backdraft situation the firefighters were not as fortunate. Two firefighters carrying salvage covers had entered a large warehouse for the purpose of covering the stock. They had been throwing covers for approximately two minutes when they saw the backdraft coming. The first signs they had were flashes of blue light resembling a neon light at ceiling level. One of the firefighters dropped and covered his head with his turnout coat. The other one tried to run. The explosion went over the head of the man who had dropped but the firefighter who ran was killed.

Although signs of a potential backdraft are not always present in large areas, in small areas they generally are. Before entering any closed area where a fire is burning or suspected, a firefighter should check the door with a bare hand to see if it is hot. If it is, he or she should not enter the room. This is especially true if air is being sucked into the room or small puffs of smoke can be observed coming out from under the door or from other small openings in the compartment. The small puffs of smoke are caused by air entering the room under the door and backdrafting out.

There are a couple of other indicators of a potential backdraft condition. One is that the black smoke becomes a dense gray yellow. There is little or no visible flame, and the windows of the compartment may be smoke/stained.

If possible, vertical ventilation should be started before the room is entered whenever a potential backdraft condition is encountered. Vertical ventilation is the creation of an opening above the fire to allow the heat and smoke to escape. (More information on vertical ventilation will be provided later in the text.) If vertical ventilation is not possible, it may be practical to cut a small opening into the room and stick a fog nozzle through the hole. Swirl the nozzle or direct the stream to the ceiling level and continue flowing water until the temperature of the interior of the room has been reduced below the ignition temperature of the material inside. One indication that this has taken place is a reduction in the generation of steam.

In summary, backdrafts are potential killers. Always be on the alert for signs indicating their potential. Some firefighters get careless because so many of their routine fires closely resemble a backdraft situation and they have not encountered any problem. Heavy black smoke is usually present inside a building and heat is always present. The heavy smoke and heat are also present in a potential backdraft; however, the elements are in the proper proportion to complete the triangle. Always check. Never take what appears to be a safe condition for granted. It is much smarter to assume that conditions are ripe for a backdraft than to assume the opposite. (See Figure 1.17.)

FLASHOVERS

The second phenomenon of the fire triangle is referred to as a flashover. The two sides of the fire triangle that are present in the proper proportions immediately before a flashover occurs are the oxygen side and the fuel side. Note that this is not any different from what occurs when a match is used to start a fire in a small pile of papers.

FIGURE 1.17 ◆ This attic fire shows all signs of a potential backdraft.
(Courtesy of District Chief Chris E. Mickal, New Orleans Fire Department—Photo Unit)

The dissimilarity is that with the papers the fire starts small and relatively slowly increases in size, whereas with a flashover a large area becomes involved in fire instantaneously. The result is that everything in the flashover area starts to burn, including firefighters' protective gear if they are in the area.

Perhaps the best way to visualize the process is to imagine a room approximately 12 feet by 12 feet in size in which there is sufficient oxygen for a fire and sufficient fuel. When a fire first starts, it can easily be extinguished by the occupant with a fire extinguisher. If the fire is not discovered and extinguished, it will continue to grow. A hot smoke layer will accumulate beneath the ceiling and the temperature at the ceiling level will gradually increase. Somewhere during this growth and the point at which the entire room is superheated, is a point at which a flashover is possible. This point cannot be exactly defined or illustrated, but it is important that firefighters are aware of the process. If the entire room is slowly heated, the temperature in the room eventually will reach the ignition temperature of everything in the room, and the total contents will start to burn at the same time. It would then be said that the room "flashed over." Almost immediately the temperatures within the flashover area will escalate from a few hundred degrees to as much as 2000 degrees. The overall result would be similar to a backdraft but be lacking the explosive nature of a pressure wave. Anyone who happened to be in the room at the time would probably be killed or at the minimum severely burned.

Once a fire reaches the flashover stage, it becomes a serious threat to every part of the building. Smoke and flames can travel rapidly to other rooms through interior or exterior pathways for long distances and pose a threat to all occupants of the building. Many firefighters have been killed or severely injured by flashovers. The most prominent indicator of an impending flashover is a rapid and substantial heat buildup in a room or building. If a firefighter is in a fire building and is forced to the floor by extreme heat, he or she should immediately open the nozzle and direct the stream into the upper part of the room. If not equipped with a hose line, the firefighter should exit the room immediately by the nearest exit, as this is an extremely dangerous situation. Once a flashover has occurred, a firefighter has no more than two seconds to escape the room without injury. If the firefighter is not within five feet of an exit, it could be fatal. Remember, a building with a high ceiling, such as a large warehouse, can mask the signs of an impending flashover because the dangerous heat levels can be well above the firefighters.

In addition to different sides of the triangle being supplied to cause a backdraft or a flashover, there is another difference that contributes to the hazard. Generally, some positive warning signs are present whenever an area is ripe for a backdraft. Although there are warning signs of a potential flashover, the signs are not as positive. All of the signs are in the same category—that the heat in the area is very intense and increasing at a dangerous rate. Some unburned articles in the area might start to smoke and fog streams relatively close to the nozzle start turning to steam. The fire is burning freely, which indicates that there is oxygen, fuel, and lots of heat. However, in day-to-day firefighting, the heat at the floor level in an area is normally well below the ignition temperature of the contents of the room.

In flashover situations, heat and smoke from a fire in a building will travel to other rooms and areas of the building. The heat will start building up at ceiling level, sometimes in a small room and sometimes in large areas. The heat buildup is generally above the heads of firefighters, who are down low, advancing lines. A flashover will occur when the temperature of the material and smoke at ceiling level reaches the ignition temperature. The first instantaneous flash will generally be above the heads of the firefighters; however, it will quickly spread to lower levels due to the tremendous volume of heat involved.

It is difficult to establish any positive tactics to protect firefighters from flashovers because one of the early principles they are taught is not to open up a hose line until they see the fire. The reason for this principle is to keep water damage to a minimum. However, the safety of the firefighters is much more important than causing a little water damage.

There are some positive steps that firefighters can take to help reduce the potential of a flashover or provide them with information that a retreat should be made from the area immediately. One is to make it a practice to open the nozzle and direct a few short bursts at the ceiling level when the advancement of a line is first made into the fire area. This action can be repeated periodically if the temperature in the area seems to be increasing. Unfortunately, the protection provided to firefighters today has eliminated heat from reaching parts of the body that could provide an indication of increased temperature. The ears, the hands, and the face are now completely covered. As a precautionary measure, however, it is good practice for the company officer to remove a glove and raise a hand over his or her head periodically to test the temperature of the upper level. If the officer finds it necessary to return his or her hand to a lower level because of the heat, a wide angle fog stream should immediately be directed to the ceiling area and an evaluation should be made by the company officer as to whether it is necessary to remain in the area. If there is any doubt, the officer should follow the policy of protecting the crew members at all times and play it safe by taking them to the outside.

On February 23, 2003, a training session was conducted in an empty dwelling for the Spencer and Forest Park Fire Departments located near Oklahoma City. During the training session, a flashover occurred. The following three photos taken by Chris Anders of the Spencer Fire Department show the progress of the flashover.

The first photo (Figure 1.18) shows conditions in the dwelling prior to the flashover.

FIGURE 1.18 ◆ Fire prior to flashover.
(Courtesy of Chris Anders, Spencer Fire Department, Oklahoma)

The second photo (Figure 1.19) was taken about 10 seconds after the first and shows the flashover. Fortunately, no firefighters had entered the room.

The third photo (Figure 1.20) was taken about 10 seconds after the flashover occurred.

These three photos clearly demonstrate the need for firefighters to be constantly on the alert for a flashover. This was a training exercise. However, if it had been an actual fire, the flashover probably could have been prevented if a strong fog stream had been directed at the ceiling level prior to the flashover occurring.

FIGURE 1.19 ◆ Fire nearing flashover.
(Courtesy of Chris Anders, Spencer Fire Department, Oklahoma)

FIGURE 1.20 ◆ Flashover condition.
(Courtesy of Chris Anders, Spencer Fire Department, Oklahoma)

ROLLOVERS

The third phenomenon of the fire triangle is referred to as a rollover. The phenomenon is called this because the resultant fire manifests itself in a rolling motion, normally at ceiling level and ahead of the main fire, sometimes by as much as 10 to 20 feet. It is caused by the smoke at the ceiling level suddenly igniting as a result of being heated to its ignition temperature and mixing with sufficient oxygen to complete the fire triangle. A typical case follows.

FIGURE 1.21 ◆ A rollover condition. The fire rolls out of the door, over the heads of the firefighters, and down the hall.

Firefighters have advanced a line to a room on fire. The door to the room may have been left partially open, allowing smoke to enter the hallway or allowing smoke to escape from the room and start moving down the hallway as the firefighters crack open the door. They are down low with their line charged but the nozzle closed. When they crack open the door they see that the room is filled with heavy black smoke. They have tested the door prior to opening it to see if the condition inside is ripe for a backdraft. When they determine that it is not, they push open the door—prepared to enter. Suddenly the smoke over their heads and in the hall breaks into flames and appears to roll down the hallway. They open their line and attack the fire.

Rollovers occur more often than do backdrafts or flashovers. They can be killers if firefighters enter the room standing up or if their line is not charged when the rollover occurs. To protect themselves against this phenomenon, firefighters should make it a practice to enter fire areas crouched down low with lines charged. If they feel that there is a threat of a possible rollover, there should be no hesitation in opening their line and direct it at the ceiling level in a spray form. (See Figure 1.21.)

◆ FIRE EXTINGUISHMENT

The fire triangle was used to discuss the theory of fire. It was explained that it is necessary to have all three sides of the fire triangle present in the proper portions and the chain reaction in order to have a fire. It follows that once a fire is burning, the removal of one side of the triangle would cause the fire to go out. For many years this simple explanation was accepted; however, time has proven that there is more to it than that.

Shortly after the end of World War II, experiments were conducted using dry chemicals as an extinguishing agent. Much to the surprise of those conducting the experiments, some of the experimental fires were extinguished quicker than expected. In fact, some of the fires were extinguished when it was clear to the experimentalists that none of the sides of the fire triangle had been removed. It was therefore concluded that another element was involved in the extinguishment that was previously unknown. This other element became known as the **chain reaction.** It became apparent that the process of the fire once it was initiated involved a chain reaction, and if the chain reaction could be broken, the fire would go out. This conclusion led to the design of the fire tetrahedron. As described earlier in this chapter, the fire tetrahedron added a fourth element to the fire triangle that was previously used to explain both the initiation and the elimination of a fire.

chain reaction A process that occurs during the growth of a fire that can be compared with the process that occurs with the transmission of heat during the process of conduction.

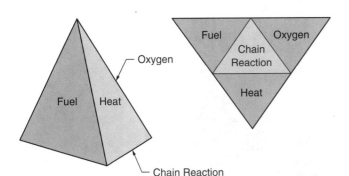

FIGURE 1.22 ◆ The fire tetrahedron.

◆ **THE FIRE TETRAHEDRON**

The fire tetrahedron is a four-sided figure that incorporates the fire triangle (see Figure 1.3) and the added feature of the chain reaction. The tetrahedron is shaped like a pyramid. The familiar fire triangle forms the three sides of the pyramid and the base represents the chain reaction. (See Figure 1.22.) A fire may be extinguished by removing any side of the fire triangle or by breaking the chain reaction. The side to be eliminated during extinguishment and the material used for extinguishment, depend on the type of fire involved.

In general,

The heat side of the fire tetrahedron is removed by the reduction of the temperature of the fire.
The fuel side is removed by the removal of the fuel.
The oxygen side is removed by the expulsion of the oxygen.
The chain-reaction side is removed by the breaking of the chain reaction. (See Figure 1.23.)

CLASSIFICATION OF FIRES

Fires have been classified by type primarily for the purpose of identifying the type of material required for extinguishment and what extinguisher to use on each type of fire. Unfortunately, there is no universal method of identifying fire classes. The classifications of fires in countries such as Germany, England, and Australia are different from those used in the United States. Some classifications are established by companies making extinguishers. As an example, one manufacturer identifies fires as Class A, Class B, Class C, Class D, and Class K. Another uses Classes A, B, C, D, E, and F. For the purpose of this book, the following classifications will be used:

- *Class A.* These are fires involving ordinary combustible material such as wood, paper, cloth, and so on.
- *Class B.* These are fires involving flammable liquids, gases, and greases.
- *Class C.* These are fires involving energized electrical equipment.
- *Class D.* These are fires involving combustible metals such as magnesium, titanium, sodium, and potassium.

FIGURE 1.23 ◆
Removing the heat/fuel/oxygen sides and breaking the chain reaction.

FIREFIGHTER APPLYING WATER TO FIRE

FIREFIGHTER SHUTTING OFF VALVE TO LINE CARRYING THE FLAMMABLE LIQUID

Removing the Heat Side

Removing the Fuel Side

Removing the Oxygen Side

Breaking the Chain Reaction

FIREFIGHTER USING FOAM ON FIRE

FIREFIGHTER USING DRY CHEMICAL EXTINGUISHER ON FIRE

EXTINGUISHMENT OF CLASS A FIRES

Class A fires involve ordinary combustible material such as wood, paper, cloth, and the like. (See Figure 1.24.) During the burning process of Class A materials, a certain amount of heat is released, depending on the type of material involved. For example, in the process of complete combustion one pound of wood will give off approximately 8,000 to 9,000 Btus, and one pound of paper will give off approximately 7,500 Btus. A Btu (British thermal unit) is a heat quantity of measurement and is normally defined as the quantity of heat required to raise the temperature of one pound of water one degree Fahrenheit. The rate at which the heat is released is directly related to the intensity of the fire.

Some of the heat released during the burning process is absorbed by the air or surrounding materials; however, for purpose of extinguishment it should be considered that it is necessary for the extinguishing agent to absorb heat faster than it is being generated. The final objective is to reduce the temperature of the burning material and the surrounding atmosphere to below the ignition temperature of the burning material.

Water. The majority of all fires encountered by fire departments are of the Class A type. Class A fires are generally extinguished by removing the heat side of the fire tetrahedron. The material most commonly used for removing of heat is water. Water is most often used, not only because of its availability but also because of the characteristics of its composition. (See Figure 1.25.)

FIGURE 1.24 ◆ Class A fires are the most common type encountered by firefighters.
(Courtesy of District Chief Chris E. Mickal, New Orleans Fire Department—Photo Unit)

FIGURE 1.25 ◆ Firefighters using water from a wagon battery to attack a Class A fire.
(Courtesy of Rick McClure, LAFD)

There are those who claim that water is also used because it is relatively cheap. This theory has been challenged by a number of individuals. When the cost of laying water mains and properly distributing hydrants is brought into the picture, the cost of supplying one gallon of water for use on fires that have occurred within some communities is no doubt astronomical.

Another factor that has compounded the problem of using water on fires is the changing face of America. America is rapidly moving from the cities to suburban areas. Large single-family homes are being built in areas far remote from a well-designed water supply system. In many parts of the country, cities have not been able to extend their main systems into these areas due to financial problems. The result is the adverse effect it has had on some departments' ability to supply the needed water in these areas. To help correct the problem, a number of departments have ordered new apparatus and equipment designed to help alleviate the problem by transporting water from an available source to the fire. The overall result is that the cost of providing water for fires has shifted from a city's budget to a fire department's budget. Regardless, in the final analysis the cost of a gallon of water for firefighting purposes has continued to remain high but is still paid for primarily by home owners.

specific heat The ability of an extinguishing agent to absorb heat.

thermal heat The ability of an extinguishing agent to absorb heat.

heat capacity The ability of an extinguishing agent to absorb heat.

The ability of an extinguishing agent to absorb heat is referred to as its **specific heat, thermal heat,** or **heat capacity,** the terms being synonymous. The specific heat of an agent is expressed by the number of Btus absorbed by one pound of the agent as its temperature is raised one degree Fahrenheit. Water has a specific heat of 1.0. This is higher than that of most other materials. For instance, ice has a specific heat of approximately 0.5, only half that of water. There is no significant difference between the specific heat of freshwater and that of salt water.

To place the numbers in perspective, 100 Btus will be absorbed by 100 pounds of water as the temperature of the water is increased 1°F. Because water weighs approximately 8.33 pounds per gallon, 8.33 Btus will be absorbed by one gallon of water as the temperature of the water is increased 1°F. Looking at it another way, if the temperature of one gallon of water is raised from 60°F to 212°F during the extinguishment process, it will absorb approximately 1,266 Btus (8.33 × 152). Unfortunately, in most cases the temperature of the water is not raised to 212°F and a good portion of the water applied will not display any significant rise in temperature. In fact, it is estimated that approximately 90 percent of the water applied by straight streams is involved in runoff, which means that the water has utilized a maximum of only 10 percent of its absorption capabilities. (See Figure 1.26.)

In order for water to absorb heat directly from the burning material, there must be surface contact between the burning portion of the material and the water. Consequently, the greater the surface contact of a given amount of water with the burning material, the greater the absorption capability. It follows that spray streams and fog streams have the capability of absorbing more heat than straight streams; however, it should not be concluded that these streams are superior for firefighting under all conditions. It must be remembered that it is necessary for the water to reach the burning material before heat absorption commences, and at times the reach of the stream and the penetration ability of the stream will take priority over the size of the droplets.

Although the heat absorption capability of water is extremely effective when compared with other materials, it is even more effective as the water changes from a liquid to a gas. The absorption that takes place during this process is referred to as the *latent heat of vaporization*. The latent heat of vaporization is defined as the amount of

FIGURE 1.26 ◆ The water runoff from the use of fire streams is substantial at some fires.
(Courtesy of Rick McClure, LAFD)

heat absorbed as a substance passes between the liquid and gaseous phases. The latent heat of vaporization of water at its boiling point (212°F) is 970.3 Btus per pound of water. This means that 970.3 Btus will be absorbed by one pound of water as it changes from a liquid to steam. Compare this with the 152 Btus that are absorbed as one pound of water is raised from 60°F to 212°F. The result is that approximately 6.4 times more heat is absorbed as water is changed from a liquid to steam than when the temperature of water is raised from 60°F to 212°F.

It should be noted that heat is absorbed as water is changed from a liquid to steam; however, the temperature of the liquid does not increase. The absorption of the additional heat reduces the volume of the liquid, with volume reduction continuing until the last drop of water has been converted to steam.

Combining the effectiveness of water due to its specific heat and its latent heat of vaporization, it can be seen that if one pound of water is raised from 60°F to 212°F and then completely vaporized into steam, it would absorb approximately 1122.3 Btus (152 + 970.3). Because water weighs approximately 8.33 pounds per gallon, it is theoretically possible to absorb approximately 9,349 Btus (8.33 × 1,122.3) per gallon during this complete process. Ironically, this is about the same amount of heat that is given off by one pound of wood during the process of complete combustion. Absorption of 9,349 Btus from a gallon of water is theoretically possible, but it should be kept in mind that it is seldom achieved during firefighting operations.

In additon to the heat absorption capability of water as it is changed to steam, there is an extra value to be gained that aids in extinguishment once the change has been completed. This is due to the tremendous expansion of steam that takes place. The expansion ratio is approximately 1,700 to 1 at a normal atmospheric pressure of 14.7 psi and a temperature of 212°F. However, the expansion ratio is a function of temperature. The ratio is approximately 2,400 to 1 when the fire area is 500°F and approximately 4,200 to 1 when the temperature reaches 1200°F. This expansion forces smoky and noxious gases from the involved area and reduces the amount of oxygen available to support combustion. It should be noted, however, that although the primary value of the steam expansion is the purging of the air and the resultant reduction of the available oxygen needed to support combustion, fires in Class A materials are normally extinguished by the absorption of heat, not by the smothering effect created by the steam. The smothering effect has the tendency to suppress flaming, but it is the cooling effect that extinguishes the fire.

Water Additives.   Over the years, there has been a number of chemicals added to water in an attempt to increase its effectiveness as an extinguishing agent. Materials have been added to lower the freezing temperature, to thicken the water, to reduce its surface tension, and to reduce its friction loss. The two most common products in use from a firefighting standpoint are wet water and Class A foam.

Wet Water: Wet water is water to which an additive has been introduced in order to reduce the surface tension of the water. Plain water has a relatively high surface tension, which reduces its effectiveness in deep-seated fires such as those found in couches, mattresses, bales of cotton, and stacked hay.

The effect of the high surface tension of water can be observed when water is poured from a glass. Some of the water will stick to the inside of the glass in the form of droplets. When water is applied to deep-seated fires this surface tension reduces the ability of the water to penetrate. The result is that the fire is extinguished on the surface but continues to burn beneath the surface. Wet water, on the other hand, will penetrate beneath the surface and extinguish the deep-seated and hidden fires. Wet water not only is effective for use in these types of fires but also has proven to be effective on ordinary combustibles because it will penetrate the subsurface and assist in the prevention of rekindles.

Wet water may be produced by premixing the additives with the water prior to the fire or by using proportioning equipment when the demand for its use arises on the fire ground. However, more and more departments are using Class A foam in lieu of wet water for this purpose.

Class A Foam: Water additives known as Class A foam have become increasingly popular during the twenty-first century. More and more departments have adopted Class A foam as a regular part of their standard operating procedures.

There has been some confusion circulating as to whether Class A foam is a retardant, a surfactant, or a wetting agent. Class A foam is a foam that has been specially designed for use on Class A combustibles. The foam is made from hydrocarbon-based surfactants that lack the strong filming properties of Class B foam. However, they exhibit excellent wetting properties.

Class A foam can be applied by an eductor or by a Compressed Air Foam System (CAFS). Both of these systems either can be built into an apparatus when it is designed or can be added later as a permanent part of the apparatus. Applying Class A foam by the use of a CAFS has proven to be particularly effective. A full explanation of the benefits of the system is explored in the Extinguishing the Fire section of Chapter Four.

High-Expansion Foam: High-expansion foam consists of masses of air or carbon dioxide bubbles dispersed in water. The air or carbon dioxide bubbles are entrapped in the water by the addition of stabilizing agents that are introduced at the time the foam is being developed. The foam will expand during the development phases, the amount depending on the materials used for forming the foam. Low-expansion foams have an expansion ratio up to about 20 to 1 and are used primarily to extinguish fires in flammable liquids. High-expansion foams have an expansion ratio as high as 1,000 to 1 and are used in the extinguishment of fires in ordinary combustibles. (See Figure 1.27.)

FIGURE 1.27 ◆ High-expansion foam can be used on several types of fires.
(Courtesy of Chicago Fire Department)

High-expansion foams are used to fill an enclosed area completely. Ventilation to the enclosed area must be provided prior to the use of the foam in order to obtain the proper distribution of the foam. Extinguishment takes place by a dual process. The foam carries water to the fire and provides a cooling effect, and it replaces the air, which helps in the extinguishing process by blanketing the fire. It is particularly effective on basement or cellar fires but is also useful on fires in holds of ships or in ships' engine rooms. Whenever foam is used on this type of fire, every effort should be made to maintain the area where the fire is located in as much an airtight compartment as possible after the foam has been distributed.

Dry Chemicals.   Multipurpose dry chemical extinguishers can be used effectively to extinguish small fires in Class A material. Extinguishment takes place by breaking the chain reaction and by the decomposition of the agent, which leaves a sticky residue on the burning material. The sticky residue in effect assists in the extinguishing process by eliminating the oxygen side of the fire tetrahedron through the formation of a seal between the burning material and the air. The seal not only extinguishes the fire but also assists in the prevention of reignition. Dry chemicals, however, have a limited cooling capability; consequently, they do not extinguish deep-seated fires. Water should be used as a follow-up when fires of this type are encountered.

Multipurpose fire extinguishers also have the capability of extinguishing Class B and Class C fires. Additional information on these extinguishers is included in the next sections.

EXTINGUISHMENT OF CLASS B FIRES

Class B fires are those fires involving flammable liquids, gases, and greases. (See Figure 1.28.) Class B fires are generally extinguished by removing the oxygen side of the fire tetrahedron or by breaking the chain reaction. There are several extinguishing agents available that have the capability of achieving one or both of these objectives.

Dry Chemicals.   Dry chemicals are powders that are formed by grinding dry crystals and adding substances that will cause the powder to flow easily and resist moisture and caking. The developed product is nontoxic; however, breathing too much of the material can cause irritation to the throat and lungs. Dry chemicals shouldn't get confused with "dry powders" that are used exclusively on Class D fires. Most dry chemicals contain sodium or potassium bicarbonate or, in the case of multipurpose dry chemicals, monoammonium phosphate. Dry powders include those agents that are effective in fighting the different types of combustible metal fires.

Dry chemicals are extremely effective on Class B fires. When the material is projected into the fire area, the fire goes out almost immediately. Although it is known that smothering and radiation shielding contribute to the extinguishment of the fire, the quickness with which flames are eliminated suggests that the principal factor in extinguishment is the breaking of the chain reaction.

When dry chemical extinguishers are used on a Class B fire, the discharge should be commenced some distance from the fire. The powder is initially released under considerable force. If it is discharged too close to the burning material it may cause the fire to spread before it can be brought under control.

It should be remembered that dry chemicals have a relatively limited cooling capability; consequently, the extinguisher operator should be on the alert for possible reignition if the Class B material is burning in a metal container. Reignition is the result of the

FIGURE 1.28 ◆ Class B fires are those involving flammable liquids, gases, and greases.

container being hotter than the ignition temperature of the material involved. Reignition is much more likely if burning has been taking place for any length of time.

The residue from a dry chemical extinguisher is a white powder that will cover everything it touches. The residue from a multipurpose extinguisher should be thoroughly removed from all metal parts as quickly as possible after the fire has been extinguished. Failure to do so will cause corrosion of some metals. If metals are not involved, the residue can be washed away with water or possibly removed by the use of a vacuum cleaner or shop vacuum.

It is good public relations for the fire department to remove the residue. In some instances this cannot be done because it may require disassembling the equipment in order to remove the residue adequately. In such cases, the owner or manager should be advised of the decontamination needed.

Carbon Dioxide. Carbon dioxide is a nontoxic, noncorrosive, nonconductive, colorless, tasteless, odorless gas. When used as an extinguishing agent it is normally stored at a pressure of 800 to 900 psi, which causes it to be converted to a liquid. The pressure under which it is stored provides the power for its discharge. Carbon dioxide is effective on both Class B and Class C fires. It was the primary extinguishing agent used during World War II for flammable liquid fires. Its use on this type of fire, however, has given way to dry chemicals due to their more effective extinguishing capabilities.

Carbon dioxide is discharged as a large white cloud that contains particles of dry ice. It is effective as an extinguishing agent primarily because of its smothering effect, which displaces the oxygen content of the atmosphere and therefore removes the oxygen side of the fire tetrahedron. It has a vapor density of 1.5, which causes it to settle over the burning material when discharged above the fire.

Carbon dioxide should be discharged above and directly on the fire. The range of the discharge is extremely limited, which means that a close approach to the fire is mandatory if the fire is to be extinguished. When carbon dioxide is discharged, the "snow" it produces provides some degree of cooling action, but the cooling effect is very limited. Once the fire is extinguished, the same caution should be taken as with the use of dry chemicals when Class B fires are encountered in metal containers. The operator must be alert for a possible reflash as the carbon dioxide cloud is dispersed.

Foam. Firefighting foams are produced by mixing a foam concentrate with water in the proper concentrations. The mixture is then aerated and agitated to form the bubble structure. Some foams are thick and viscous whereas others are thinner, which results in a solution that can be spread more rapidly. In either case, the foam is lighter than the flammable liquids it is designed to extinguish, which allows it to be floated over the liquid to form an air-excluding, cooling blanket that is capable of halting or preventing combustion.

Foams are defined by their expansion ratio. The expansion ratio is the ratio of the final foam volume to the original foam solution volume before adding air. The foams used for the extinguishment of Class B fires are defined as low-expansion foams. Low-expansion foams have an expansion ratio up to 20 to 1.

Air foams that are designed to extinguish Class B fires are not effective for use on flammable liquid fires involving fuels that are water soluble, water miscible, or "polar solvents." Some examples of substances that fall into these categories are alcohol, enamel and lacquer thinners, and acetone. Special foaming agents have been developed for use on fires involving these materials. These foams are referred to as alcohol-type concentrates.

Aqueous Film Forming Foam (AFFF):   This foam is more commonly referred to as "AFFF." This extremely effective foam is recommended for extinguishing fires in hydrocarbon fuels. AFFF is a synthetically produced material that consists of a liquid concentrate made from fluorochemical and hydrocarbon surfactants combined with high-boiling-point solvents and water with suitable foam stabilizers. The foams generated from AFFF solutions have a high viscosity, are fast spreading, and like other foams act as a surface barriers to fuel vaporization.

Fluoroprotein Foam (FPF):   This foam is widely used in protecting fuel tanks and petroleum processing facilities. Its unique qualities make it desirable for subsurface injection.

Use of Foam: Foam is effective on Class B fires in containers as well as Class B spill fires. It extinguishes the fire by covering the fire and blanketing it from the atmosphere. Once extinguished, the fire is relatively safe from a reflash because the blanket of foam will exclude the air for some time—depending on the thickness of the foam blanket and the composition of the foam.

In open containers of Class B fires, foam should be applied by projecting the foam stream to the *far side* of the container and allowing it to cascade over the burning liquid. Spill fires in the open should be attacked by directing the foam stream on the surface in *front* of the fire, which will push the foam blanket over the fire. The foam stream should not be directed into the fire, whether the fire is in a closed container or as a result of a spill. To do so would most likely result in the fire spreading. (See Figure 1.29.)

Foam can also be used to prevent Class B fires under certain circumstances. It can be extremely effective in blanketing a large spill to prevent a fire and eliminate a hazardous condition. It also can be used to blanket a runway in anticipation of a fire in an aircraft that is forced to land with its wheels up; however, its usefulness for this purpose is limited because of the time involved.

EXTINGUISHMENT OF CLASS C FIRES

Class C fires are those involving energized electrical equipment. Once the equipment is de-energized, the fire becomes a Class A or Class B fire and can be handled accordingly. The materials normally used for extinguishing these fires are nonconductive in order to protect the user from dangerous electrical shock.

FIGURE 1.29 ◆ Application of foam to a fire.

FIGURE 1.30 ◆ Class C fires are those involving energized electrical equipment.

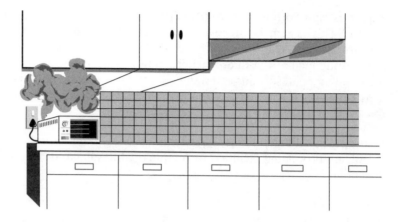

There is no available extinguishing agent for use on Class C fires only. All the agents used on Class C fires are also effective on either Class A fires or Class B fires, or both. The nonconductive agents most commonly used are carbon dioxide and dry chemicals. (See Figure 1.30.)

EXTINGUISHMENT OF CLASS D FIRES

Class D fires are those involving combustible metals. Note that the classification does not refer to metal fires but to fires in combustible metals. Most metals will burn if sufficient heat is applied; however, fires in ordinary metals such as iron, steel, aluminum, and copper can be extinguished by cooling with water. The metals referred to in the classification of combustible metals are those metals that require special extinguishing agents or special application of water in order to achieve extinguishment. These metals are magnesium, sodium, potassium, uranium, titanium, lithium, thorium, hafnium, zirconium, and plutonium.

Extinguishing agents for these fires can be found in extinguishers and in cans located on the premises of some industrial plants where these metals are processed, used, or stored. Unfortunately, no extinguisher or material is effective on all the different types of metals. Equally unfortunate is the fact that the only extinguishing agent found on firefighting apparatus that can be used on any of the metals is water. The common extinguishing agents found in extinguishers on apparatus are not suitable for these fires and can result in dangerous or toxic reactions if used.

Of all the metals mentioned, the one most likely to be encountered by firefighters is magnesium. It may be found in finely divided form, in chips or chunks, or in finished products. Although the general principle is not to use water on magnesium fires, water can be used if proper methods are utilized and the reactions expected are understood.

When magnesium burns, it burns as a molten mass. When water is applied, the water penetrates the surface of the metal and expands. Water has an expansion ratio of 1,700 (or greater) to 1. This means that a single drop of water after penetrating the surface of magnesium will expand to at least 1,700 times its original size. The result is an explosion that throws burning magnesium particles in all directions. These particles often burn through a firefighter's protective clothing. Firefighters should expect violent reactions when straight streams are applied to burning magnesium. Eye goggles or SCBA should always be worn when applying water on magnesium fires.

The success for using water on magnesium fires depends on how quickly a large amount of water can be applied to the fire. It is best when attacking small fires to use fog streams. High-pressure fog laid gently on the surface of the magnesium will restrict penetration, and, although popping will occur, the overall effect will be that the magnesium will be cooled and extinguished. Regardless, as a precautionary measure it is a good idea to operate from as great a distance from the fire as possible.

Large fires in magnesium present a much greater problem. It is impossible to extinguish the fire without a great number of violent reactions occurring. Firefighters should operate from as great a distance from the fire as possible within the range of the hose streams. The initial attack should be made with large quantities of water. It may be necessary and wise to use heavy streams. If an insufficient amount of water is applied, it will only intensify the fire and possibly spread it over a wide area. When adequate quantities of water are used, after the initial explosion and fireworks, the magnesium should be cooled until extinguishment is achieved.

It is good practice not to use water on any of the other so-called combustible metals. It is also good practice always to wear protective clothing, including breathing apparatus, when working on combustible metal fires, regardless of their size.

◆ **SUMMARY**

In this chapter we have explored fire behavior, fire combustion, and some basic principles of fire travel and how fire spreads. In order to anticipate what might happen on any given fire situation, a good firefighter will study those things pertaining to fire ignition and learn "how" and "why" fire behaves the way that it does. We have also looked at the different types of fire extinguishing agents and their individual effects on fire. The classification of fires was described and an overview of the fire triangle and the newer terminology "fire tetrahedron" was reviewed. Use Chapter One as a resource when needed and refer to it often to help you understand the basics of fire behavior and the combustion process.

■■■

Review Questions

1. What are the three sides of the fire triangle?
2. What is the definition of fire?
3. What is the definition of ignition temperature?
4. What is the definition of flammable limits?
5. What is the difference between an endothermic reaction and an exothermic reaction?
6. What are the three methods of heat movement?
7. What are the three stages of fire behavior?
8. What is the difference between flash point and fire point?
9. What is the cause of backdrafts and what can be done to keep one from occurring?
10. What occurs to cause a flashover and what action might be taken to prevent one from occurring?
11. What is a rollover?
12. What are the four sides of a fire tetrahedron?
13. What are the four classifications of fire?
14. What is the primary material used for the extinguishment of a Class A fire?
15. What other materials can be used to extinguish a Class A fire?
16. What materials are used to extinguish Class B fires?
17. What materials are used to extinguish Class C fires?
18. What materials are used to extinguish Class D fires?
19. How should water be used to extinguish fires in magnesium? What is the composition of air?

CHAPTER 2

Fire Ground Tactical Tasks

Key Terms

Objectives

The objective of this chapter is to introduce the reader to the tactics utilized on the fire ground. Upon completing this chapter, the reader should be able to:

- List the seven tactical tasks that are essential for operating at a fire.
- Discuss the various facets of rescue operations.
- Explain the various types of internal and external exposures.
- Outline how overhaul operations should be conducted.
- Explain the various methods used to provide vertical and horizontal ventilation.
- Discuss the importance of salvage operations.
- List some of the various methods available for protecting a building and its contents from water damage.

In its simplest terms, strategy is the overall plan that is developed for controlling an emergency. Many years ago, Lloyd Layman suggested some strategic areas that should be considered when developing the strategy to be used at an emergency. He identified these areas as: rescue, exposures, confinement, extinguishment, overhaul, ventilation, and salvage. From these the term "RECEO VS" has been formulated. RECEO VS is recognized as an excellent basis for the development of strategy. Each of the terms represented by RECEO VS will be examined carefully in this chapter.

Following is a list of the common tactical tasks that must be performed on the fire ground. They are not listed in priority order but rather in a format that will simplify their explanation. Not all of the tasks will have to be performed at every fire, but common sense dictates that thought should be given to each individual task prior to its abandonment. The list can be used by an Incident Commander as a mental checkoff list. The letters in the term refer to:

R Rescue
E Exposures
C Confinement
E Extinguishment
O Overhaul
V Ventilation
S Salvage

◆ **RESCUE**

With the exception of an immediate need to save life that is apparent on the arrival of the first unit, **rescue** operations should not be commenced until a rapid intervention team (RIT), also known as RIC (rapid intervention crew), is in place at the emergency. The RIT concept will be analyzed in Chapter Three.

rescue The process of removing people from burning buildings or buildings likely to become involved to a place of safety.

Rescue attempts always should be completed by a team using the buddy system. The members of the team must maintain voice contact or be in visual contact with each other at all times while conducting the rescue operations.

Part of a rescue effort may be conducted in a large portion of the building that is clear of smoke. However, it is very likely that at least part of the rescue operation will involve working in areas of reduced visibility. When working in these atmospheres, it is better that team members conduct the search on their hands and knees. This reduces the chance of members falling into stairways or other openings in floors. When moving up or down stairs, they should move up head first and feet first when descending. Not only is working on hands and knees safer, but the air near the floor is cooler and normally clearer. (See Figure 2.1.)

It is important that rescue teams maintain radio contact with their supervisor at all times and that they report periodically their progress and where they plan to go next. When they have completed the search of an entire floor of a building, it is vital that this information be relayed to their superior so that the information can be passed on to the Incident Commander.

FIGURE 2.1 ◆ Search teams should work in a group of at least two firefighters. It is better to search on hands and knees because the visibility is better at that level and the temperature is lower.
(Courtesy of Chris Anders, Spencer Fire Department, Oklahoma)

The rescue function would probably be assigned to the rescue unit if the department responds one to the emergency; otherwise it would most likely be given to the truck company.

Rescue is the first priority at any incident, regardless of the size of the building. In a limited sense, rescue is the process of removing people from burning buildings or buildings likely to become involved to a place of safety. In a broader sense, rescue operations include not only the removal of people but also the prevention of further injury and the loss of life at fires. Rescue operations are the first consideration that should be given by the first arriving officer at a fire and should constantly be in the mind of every officer and firefighter on the scene. Although an aggressive attack on the fire may be the quickest and most efficient method of saving life, certain questions should be answered by the first arriving companies regarding rescue operations:

1. Is anyone in the building?
2. If so, are they in immediate danger?
3. If people are in danger, is it possible to rescue them or to prevent them from becoming injured or killed?
4. If so, what is the best method of making the rescue or preventing further injury or loss of life?

Many times the question of whether anyone is in the building is answered immediately by cries for help. At other times there is silence, but information can be obtained from individuals who have escaped from the building or perhaps from the neighbors. Neighbors are generally a good source of information. They will normally be familiar with how many people live in a house and whether people are home. They will also be of great assistance in identifying anyone in the crowd who may have escaped from the burning building. It is good practice to follow up on all information given, but keep in mind that the information is not always reliable. The rescue team should not rely entirely on the information they receive but should still make a complete search of the building even if the information they receive from escaped occupants and neighbors is that everyone is out of the building.

At other times there are no cries for help and no information available from those at the fire, but the conditions indicate that people are probably in the building. For example, if a unit arrives at a single-family dwelling at 3:00 A.M. and finds the house full of smoke, all doors and windows locked, and no one outside of the structure, it is a pretty good indication that people are still in the building and that immediate rescue operations are required. Of course, the people who live in the house may be out of town on vacation, but all operations must proceed as if the building is occupied. Every structure should be assumed to be occupied until a complete search proves otherwise. This includes vacant buildings. It is much better to be wrong than sorry.

Occasionally it is apparent when the first unit arrives that several people are in need of rescue. At this time it is necessary for the Incident Commander mentally to establish a priority list of the order in which people are to be rescued—unless, of course, there is a sufficient number of firefighters at the scene to make all rescues simultaneously. The priority system should be established on the basis of which person is in the most danger. This basically means that an evaluation must be made of who is in the greatest danger and who will probably die first if not rescued. Care must be taken during this evaluation not to let the loudness of the screams or the begging of relatives, friends, or neighbors influence the decision. It should also be kept in mind that who is likely to die first does not necessarily depend on the order in which the fire will reach the people. Many times panic plays an important part. Due to panic, a person might jump from an upper story while a person who is in much greater danger from the fire will remain calm.

The emotional state of individual occupants of the building is difficult, if not almost impossible, to determine; however, where it is obvious that panic has placed one person in more danger than another, then this factor must be considered when establishing a priority list.

Care must be taken when establishing a priority list to consider all occupants of the building. It is quite easy to limit an evaluation to those people or conditions that are observed from the front of the building on arrival. It is not unusual during the stress of an emergency to neglect the fact that a building has six sides. Often people in the back of the building or on the roof or in the basement who are in more danger than those in front are overlooked because of the problems facing the Incident Commander at the front of the building. (See Figure 2.2.) It is wise to keep in mind that the building may have six sides.

PANIC

Panic is one of the primary contributing factors to loss of life when a large number of people are placed in danger at a fire situation. This has been proven time and time again by the large number of people who die while trying to escape from a place of public assembly. A pile of bodies often may be found at an exit where people panicked and piled on top of one another.

panic An emotional reaction to fear.

Panic is an emotional reaction to fear. At times fear robs a person of the ability to reason. A good example is the Cocoanut Grove nightclub fire that occurred in Boston in 1942. Nearly 500 people died in that fire. They died because as a group they panicked and would not stand back and allow room for an inward-swinging door to be opened. Of course, even young schoolchildren logically know that a door cannot be opened if several people are pushing their weight against it. However, in a panic situation almost all logical response ceases to exist.

FRONT OF BUILDING   BACK OF BUILDING

FIGURE 2.2 ◆ Don't forget the back.

Another factor contributing to panic in theaters, hotels, and places of public assembly is the occupants' lack of familiarity with the general floor layout. When people are in a strange building, they tend to leave by the way they entered rather than try to determine the quickest and easiest route to the outside.

From a fire rescue viewpoint, it is important to remember that people should be expected to act in an abnormal manner in fear-producing situations. They will jump when they shouldn't jump, they will hide in places where they shouldn't hide, and they will try to escape from rooms when they should remain there; and more appalling is the fact that they will drag others into danger with them. It does not take a life-threatening situation to trigger panic. Any trifling cause or misapprehension of danger, if allowed to continue, can contribute to panic. It is important early in the emergency to make every effort possible to reassure anyone who appears to be in any degree of danger, and to continue reassuring him or her until a rescue can be made.

SEARCH PROCEDURES

The importance of establishing a priority system for the order in which people should be rescued was previously discussed. Care must be taken not to limit the priority list to those people who can be seen. Many times the people in immediate danger are those who cannot be seen. Most people who die in fires die prior to the arrival of the fire department. Many die of smoke inhalation and not directly from the fire itself. However, some people die after the department arrives on the scene. They may be unconscious at the time of arrival, they may not be able to make their presence known due to physical handicaps, or they may be trapped. It is important that these people be found quickly. It can best be done through a systematic and thorough search.

There are several tools that search teams should take in with them when preparing to conduct a search. One is a forcible entry tool. This tool may not be required but it certainly is beneficial when it is desperately needed and the rescue members would have to return to the outside to get it. Not only may the team need it during the search procedure, but it may also be needed for them to escape from the building in the event they become trapped.

If it is available, a thermal imaging camera might prove to be particularly useful. (See Figure 2.3.) It may reduce the time required to make a search when the team enters an area that is sparely furnished. Other beneficial tools include chalk, crayon marks, or masking tape that can be used for marking areas that have been searched. Masking tape is also valuable for taping over door latches to prevent the team from being locked in.

The search should commence in the place where people are the most likely to be in the most danger. It is the bedrooms in single-family dwellings during sleeping hours. In multiple-story buildings such as hotels and apartment houses, it is generally the fire floor, the floor above the fire floor, and the top floor, in that order (see Figure 2.4).

FIGURE 2.3 ◆ This is a good example of the effectiveness of a thermal camera. *(Courtesy of International Safety Instruments)*

FIGURE 2.4 ◆ Where people are most likely in danger.

One method of ensuring that the search will be conducted in a thorough and systematic manner is to start at the fire perimeter and work away from the fire, making a room-by-room search. Care should be taken to ensure that each room is empty prior to proceeding to the next room. It is good practice for rescue teams to carry chalk or other marking tools and make a large X on a room they have searched. Some people suggest that it is also helpful to make a single line on the area to be searched prior to entering. This will alert other rescue teams that the room is being searched or has already been covered. (See Figure 2.5.) Coordination should be established by those conducting the search.

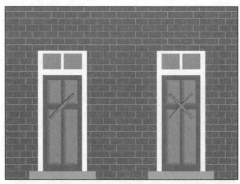

FIGURE 2.5 ◆ Start and finish of search.   START OF SEARCH   SEARCH COMPLETE

A good search is made by using the senses of sight, touch, and sound. Sound is particularly beneficial if the victim is conscious. By listening carefully, the search team may hear weak cries for help, groans, or other sounds originating from the victim. It is good practice to tap a wall or pipes while making the search. This may attract the attention of the victim and thus allow him or her an opportunity to identify his or her location. Even if the individual is unable to make a sound to assist in discovering his or her location, the noise will reassure the victim that help is on the way.

Sound alone will not normally do the job when smoke is down to the floor in a room. It is also necessary to feel and listen. The search should include feeling the tops of all chairs, sofas, beds, and the horizontal flat areas of other pieces of furniture. Perhaps the victim was occupying one of these when he or she became unconscious. The floor around each piece of furniture should be searched since the person may have toppled over. Searchers should always check the area near doors and under windows. Victims are often found in these locations because they were able to get only that far in their search for fresh air before they were overcome. The inside of all closets must be searched. People may have entered the closet thinking it was an exit, or small children may have entered the closet in an attempt to hide from the fire. Searchers should also check under all beds, particularly in children's rooms. Children have a habit of hiding from the fire in places they use when playing the game of hide-and-seek. The room should be searched thoroughly and quickly, but the rescue team should never leave a room until both members are positive no one is in it. (See Figure 2.6.)

Every room of a building should be searched. This includes bathrooms, kitchens, basements, and even rooms that have padlocks on them. On more than one occasion firefighters have found children locked in bedrooms. Parents who had left the house for the evening had placed the children there. In one known case, children were found sleeping on a mattress in an attic with the entrance to the attic locked. Their mother who had gone to a nightclub for the evening had placed them there. Halls should be given the same consideration as rooms. It is important to remember to check hall closets, as people have been known to enter these thinking they were exits.

In hotels, apartment houses, office buildings, and the like it is a good idea to check the roof. This is particularly important if a stairwell leads to the roof from the interior of the building. Occupants will go to the roof to seek help when escape to the lower floors has been cut off. It is also good practice to observe the roofs of adjoining buildings while making the search. People may have jumped to this nearby area.

The only assurance that the fire department has that all people in danger in a building are out of the building is by way of a thorough and systematic search. It should

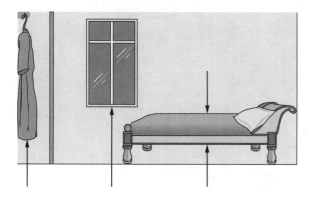

FIGURE 2.6 ◆ Search all areas.

never be assumed on arrival that everyone is out of the building, regardless of what information is received from people at the fire. This also applies to abandoned buildings. Abandoned buildings are havens for the homeless. It is a good idea to keep in mind that someone was most likely inside to start it if a fire starts in an abandoned building.

It is better that the search be conducted using two teams if it is required to search all rooms on a floor with a long hallway. One team can search one side of the hallway while the other searches the opposite side. In the event there is only one team available to search the entire floor, then the procedure should be to search one side of the hallway first and then the other side as the team works its way back to where it started.

The same procedure should be used in each room when searching individual rooms. The team members should turn left or right as they enter the room and follow the wall. When they return to the door, they should cross the room diagonally to make sure that no one is lying in the center of the room. (See Figure 2.7 on search path.) They should close the door as they exit the room, make their search marks on the door, and then turn in the same direction they were moving prior to entering the room.

If the room to be searched is extremely small, it is best that one member of the team remain at the door while the other member searches the room.

For safety purposes, the two members should maintain a dialogue during the search. The two meet at the door on completion of the search and move on to the next room.

REMOVAL OF VICTIMS

When victims are found, they should be carefully removed to a place of safety. How much care is required depends on the condition of a victim when he or she is found. If the victim is seriously injured but not in immediate danger, more time can be taken than if he or she is in jeopardy. The chief danger in moving a victim too quickly is the possibility that the injury may be a spinal injury or other injury that could be further aggravated by movement. However, if the person has been overcome by smoke and asphyxiation is imminent, it is extremely important to get him or her to a location where resuscitation measures can be started immediately.

The technique used to move a person to a place of safety also depends in part on the victim's condition when he or she is found. If the individual is not injured and is able to walk, it may only be necessary to guide him or her to a place of safety. In fact, several people can be taken at one time under these conditions.

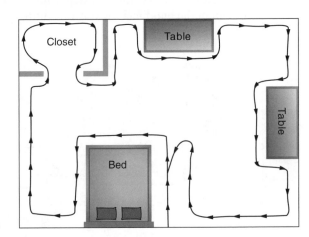

FIGURE 2.7 ◆ Search procedure.

In some cases it will be necessary to carry the victim to safety. Generally, one-person carries should be limited to those situations in which the victim is not too seriously injured, not too heavy, and the distance to safety is not too great. Where possible, two or more firefighters should be used when it becomes necessary to carry a person to safety. Greater weights can be carried longer distances and it is easier to prevent aggravating an injury. A stretcher or an improvised stretcher should be used if possible; otherwise it will be necessary to use one of the two- or three-person methods of carrying.

Occasionally it will be necessary to remove a victim by dragging. This method is most suitable when the victim is not suffering from any injury and only one firefighter is available for removal. As a search team works in pairs, this situation is very unlikely but could manifest itself if the team discovers two victims that can be moved at the same time. Of course, if two victims are in immediate danger, then the two members of the team can be used to drag the victims out regardless of the injuries, the size of the individuals, or the distance to safety. It is generally best to drag a person headfirst. The individual can be dragged by his or her own clothing or by the wrists; however, it is sometimes easier and less hazardous if the person is placed on a turnout coat, blanket, or some other object that can be pulled rather than pulling directly on the victim.

Removal of people from multiple-story buildings presents a much greater challenge than does removal from single-story buildings. Sometimes it is only a matter of guiding people to safety; at other times their removal may be extremely complex. People are more easily moved in situations in which they are familiar. They are used to walking down stairs, but not with walking down fire escapes or climbing down ladders. Consequently, when possible, keep evacuees on familiar territory. For example, if people are trapped several stories above the fire floor in a building that has an interior stairwell and a fire escape, it is best to guide them down the stairway to the floor above the fire, then out onto the fire escape, and as soon as possible after passing the fire floor, take them back inside and proceed down the stairway.

Stairways are the quickest and easiest means of removing a person from an upper floor. Care should be taken to evaluate the situation completely prior to moving a person toward the ground. For example, if a person were waiting in a window of an upper floor to be rescued when the first unit arrives, it might appear that the best means of effecting the rescue would be to raise an aerial ladder to the window and allow the victim to climb down. However, if possible, it would be safer and easier for a firefighter to climb the aerial, go in the window, and guide the person down the stairway, if it is safe to do so.

While some departments use elevators as a means of evacuation, consideration should be given to the fact that heat and flame could have entered the shaft and weakened the cables or damaged the hoisting machinery. Even if the fire is some distance from the elevator shaft, any interruption in the electrical power could render the elevator useless. In such a case, people in the elevator not only would be placed in danger but also would most likely have a tendency to panic. As a result of the danger involved, some fire departments have a standing rule that elevators are not to be used in the evacuation procedure when a serious fire exists in a building. Some firefighters have been killed while using elevators at multistory building fires when the elevator stopped at the fire floor.

Using ladders for rescue work should be one of the last resorts. This is particularly true of aerial ladders. Bringing victims down ladders is dangerous. However, if people are showing at windows on various floors when the department first arrives, it may be necessary to use ladders to cope with the immediate problem. The first ladders should be raised to the highest floor possible. There have been instances where people have jumped toward a ladder raised to a floor below them.

FIGURE 2.8 ◆ Although stairways are the quickest and easiest means of removing a person from an upper floor, it is sometimes necessary to use ladders. *(Courtesy of Chicago Fire Department)*

If it becomes absolutely necessary to use ladders for rescue work, it is generally best that the victim climb down using his or her own power with a firefighter descending in front of the victim for the purpose of providing guidance and stability. (See Figure 2.8.)

One final thought on the removal of victims. When members of a rescue team are removing a victim, it means that the search previously being conducted by the rescue team has been temporarily abandoned. Yet it is still extremely important that all rooms and sections of the building be completely searched. It is possible that the search team removing the victim may be reassigned to pick up the search where they left it; however, it may be necessary for a different search team to be given the assignment. This brings out the importance of search teams carrying chalk and marking the rooms they have searched; otherwise a different team will not know exactly where their search should begin. Even if the original team is reassigned, there might be doubt in their minds as to where they left off.

SUMMARY

Rescue operations take first priority at a fire, but they are challenging and demanding. Many times evacuation and rescue race through the mind of the first officer on the scene when the situation is one in which people are in danger. However, it should be kept in mind that in many situations an immediate and rapid attack on the fire is the best means of saving lives and preventing further injury to the occupants.

◆ EXPOSURES

This function is generally assigned to an engine company but can be assigned at a fire to any available group or company.

An exposure may be defined as anything in close proximity to the fire that is not burning but that might start burning if some type of corrective action is not taken quickly. It should be remembered that if sufficient oxygen is available, a burnable material will start burning once its temperature is raised to its ignition temperature.

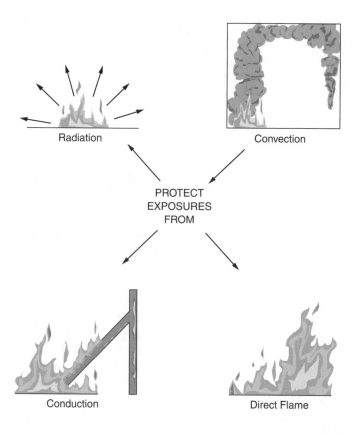

FIGURE 2.9 ◆ How fire can travel to exposures.

Heat can travel from a burning material to a nonburning material by direct contact, radiation, conduction, or convection. (See Figure 2.9.)

Consequently, to protect an exposure it is necessary to protect it against all four sources that are capable of raising it to its ignition temperature. An exposure may be found a few feet or several hundred feet from the main body of the fire, depending on the size and type of fire. Exposures are divided into two general types: interior exposures and external exposures.

INTERNAL EXPOSURES

Internal exposures are those located inside a building. They may be found relatively close to the main body of fire, in the same room as the main body of fire, in adjacent rooms on the fire floor, or on floors above or below the fire floor.

internal exposures Those exposures located inside a building.

Direct Contact.   In most cases the protection of exposures from direct contact is made by an aggressive attack on the fire together with the cooling of the exposures by the use of water. An example is a room partially involved with fire. Somewhere in the room the fire is making direct contact with an exposure. Normal procedure is to enter the room and knock down the fire as rapidly as possible. The exposure problem is therefore eliminated, providing, of course, that the fire has not spread to other areas by conduction, convection, or radiation.

It is possible in some small fires for an exposure that is in direct contact with the fire to be protected by moving it away from the fire. This type of protection of exposures from direct contact, however, is generally limited to very small fires.

Radiation. Other than by an aggressive attack on the fire, interior exposures are generally protected from radiation by cooling, moving the object, or closing openings to other rooms or areas. Protection of interior exposures by cooling is normally accomplished during the general attack on the fire.

An example of protecting interior exposures from radiation is a burning piece of furniture that is threatening other furniture around it. It is generally a simple matter of moving the other furniture away from the fire until a line can be brought in to extinguish the fire.

Exposures in rooms adjacent to a fire area can be protected from the radiated heat by closing a door. This method is used more often by occupants of a building or by built-in protection than it is by fire personnel. An occupant who discovers a fire in a room, closes the door, and then runs to call the fire department has protected the exposures outside of the room from radiated heat. The closing of a fire door by the melting of a fusible link accomplishes the same thing. Although built-in devices or actions of the occupants normally provide this type of protection, fire personnel should not ignore it if it can prove to be effective at a fire.

Conduction. Exposures are generally protected from conducted heat by moving them out of contact with the heated body. This type of protection of exposures occurs more often on ships than it does in buildings. An example is a fire in the hold of a ship. Materials in the adjacent holds are protected by moving them away from the metal bulkheads. The same procedure may occur with a well-involved fire in a stateroom. In this case it may be necessary to move objects away from the four adjacent bulkheads surrounding the room and from the decking above and below the stateroom.

At times it is difficult to protect interior exposures from conducted heat because the resultant fires generally occur in concealed areas. A partition fire is a good example. It should be kept in mind, however, that exposures are ignited by conduction, and a thorough check should be made of all areas where extension of the fire by this source is possible.

Convection. Protection of interior exposures from convection presents the biggest problem for firefighters. Exposures may be ignited in the fire area, in adjacent rooms, and sometimes many floors above the fire floor by convected heat. The truck company generally plays a major part in protecting exposures from this source of heat by controlling the flow of the heat through ventilation. In multistory buildings, however, it is necessary for the engine company to take positive action early to gain control of all vertical openings. This requires extending lines to the floor above the fire floor and using hose streams to protect the upper floors from the convected heat. Tactics will be discussed later in the book, but at this point it is worth mentioning that these tactics are risky and should be carried out only with the primary consideration of safety to personnel.

EXTERNAL EXPOSURES

external exposures Those exposures that are located outside the fire building.

External exposures are those that are outside the fire building, such as another building, a vehicle, a tree, and so on. They may be located relatively close to the main body of fire or in some cases a considerable distance away.

Direct Contact. Protection of exposures from direct contact with the fire generally manifests itself when a building or a large part of a building is well involved with fire and exposures are relatively close to the fire building. When this occurs it is generally

FIGURE 2.10 ◆ This fire will spread rapidly to the second floor by direct contact if quick action is not taken. *(Courtesy of District Chief Chris E. Mickal, New Orleans Fire Department—Photo Unit)*

best to cool down the exposures with a generous amount of water followed by or co-ordinated with an aggressive attack on the fire. (See Figure 2.10.)

Radiation.   Exposures threatened by radiated heat from the fire can be classified as movable or nonmovable. Nonmovable exposures are buildings, trees, and the like, whereas movable exposures are vehicles, boats, or other objects that are not station-ary. Wetting down or setting up a water curtain generally protects nonmovable expo-sures, whereas movable exposures can be wetted down or moved. (See Figure 2.11.)

When lines are laid to a fire having exposures that are in immediate danger of igniting, the first lines should normally be used to protect the exposures and water should not be projected onto the fire until all the exposures are adequately protected. Where possible, those exposures capable of being moved (cars, buses, boats, etc.) should be moved. Although wetting them down can protect movable exposures, it is better to move them. The exposure problem is eliminated once they are moved, and the energy and lines required to keep them wetted can be used elsewhere.

It is possible for radiated heat to enter exposed buildings through openings such as windows. Sending firefighters into the exposed buildings to close the openings can aid in the elimination of this hazard.

On large fires or some multiple-story fires it may be necessary to set up water cur-tains to protect the exposures from radiated heat. It may even be necessary to use heavy stream appliances for this operation. In some cases it is possible to protect the

FIGURE 2.11 ◆ The dwelling on the right is a critical exposure. A line is being advanced to protect it. *(Courtesy of District Chief Chris E. Mickal, New Orleans Fire Department—Photo Unit)*

exposure by setting up a water curtain from inside the exposure. Lines can be taken inside the building and the water curtain set up by using spray streams out the windows of the exposed buildings.

Conduction.   Protection of exterior exposures from conducted heat is normally not a problem; however, it may manifest itself in large fires that are throwing burning embers high into the air. These embers present an exposure hazard when they land on combustible roofs. This type of problem is generally present at wildland fires that are burning in close proximity to built-up areas. The problem will most likely exist downwind. Downwind is the direction toward which the wind is blowing. Protection of exposures from this cause is provided by having engine companies patrol downwind from the fire. However, it should be remembered that placing engine companies downwind from a fast-moving wildland/urban interface fire could put those personnel in substantial danger, especially if the wind speed suddenly increases.

Convection.   Protection of exposures from convected heat is normally not a problem if the exposures are limited to one or two stories in height. It can become a problem, however, if the exposures are multiple-story buildings. The methods used to protect the exposures from this heat source are the same as those for radiated heat.

◆ CONFINEMENT

This functional task will be thoroughly examined in Chapter Four.

◆ EXTINGUISHMENT

Extinguishment is the primary responsibility of an engine company. However, extinguishment may become the responsibility of any company or any group of firefighters at any time. (See Figure 2.12.) The methods for extinguishing fires were thoroughly examined in Chapter One.

Figure 2.12 ◆ A handheld hose being used to project water onto a fire. *(Courtesy of Rick McClure, LAFD)*

◆ **OVERHAUL**

The function of **overhaul** is generally a responsibility of a truck company. However, it may be assigned to any company, particularly in those departments that do not have the luxury of a truck company. (See Figure 2.13.)

Overhaul is the final task performed by firefighters at the scene of a fire. Although the primary objective of overhaul is to ensure that the fire is out, it generally includes doing whatever is necessary to leave the premises in as safe and secure a condition as possible.

The amount and degree of overhaul work done at a fire will vary considerably from one department to another, and many times will vary from one district to

overhaul Overhaul is the final task performed by firefighters at the scene of a fire The primary objective of overhaul is to ensure that the fire is out; however, it generally includes doing whatever is necessary to leave the premises in as safe and secure a condition as possible.

Figure 2.13 ◆ Now the work begins. The next step is the overhaul. *(Courtesy of Las Vegas Fire & Rescue, P10)*

another within the same department. Some departments and some chief officers consider that their job is to extinguish the fire and leave. Others look at overhaul operations as an excellent opportunity to build good public relations. Concepts introduced in this portion of the chapter are based on the second principle.

OBJECTIVES

There are four primary reasons for considering good overhaul practices as an essential part of firefighting operations.

1. The primary purpose is to ensure that the fire is out. If the fire is not completely extinguished it might result in a rekindle. A rekindle is a situation in which the fire department leaves the scene thinking the firefighters have extinguished the fire, and they are called a second time to the location because they did not completely extinguish the fire the first time. Sometimes the fire is larger when they return than it was on their initial arrival. A rekindle is an unpardonable sin. It leaves a stigma on the record of any officer and is an embarrassment to all firefighters who worked at the fire. Consequently, it is important that any remaining fire be located and completely extinguished before the department leaves the scene. (See Figure 2.14.)

2. Eliminate any additional water damage that could be caused by extinguishing operations or the weather.

3. Make sure that all portions of the building are accessible to anyone entering from the outside. This means that passageways must be provided so that the owner, insurance adjuster, or any other person having a legal right to enter can inspect all portions of the premises.

FIGURE 2.14 ◆ A firefighter getting ready to use a pike pole to open the ceiling to ensure there is no hidden fire. *(Courtesy of Rick McClure, LAFD)*

4. Make sure that all portions of the building are left in as safe and habitable condition as possible. Operations to accomplish this objective include removing hanging timbers, piping, and wiring; pulling down unsafe walls or chimneys; and clearing all windows and doors of broken glass.

PLANNING OVERHAUL OPERATIONS

Overhaul operations should normally be started as soon as the fire has been brought under control and adequate personnel are available. However, it may be necessary to delay these operations until the arson squad has made an investigation if the fire is of a suspicious nature. Once operations have commenced, they should not be carried out in a haphazard manner. Planned operations should be so well thought out that they could be accomplished in as systematic and professional manner as possible.

The first step in the planning process is for the Incident Commander to size up the situation to determine what has to be done and the most effective way for getting it done. The size-up generally begins by the Incident Commander making a "walk-through" of the premises to make sure that it is structurally safe for work parties to enter and commence working. During the walk-through the Incident Commander should determine where it would be best to store salvageable material and which spot would be best for dumping debris. Some of the other factors that should be determined are:

1. Are the firefighters at the scene physically capable of doing the work? The stress and physical exertion required during the extinguishment phase may have fatigued personnel to the point where it would be wise to call for fresh crews to do the overhaul work.
2. What is the condition of the exterior walls and chimneys? Do they present a hazard to people who may approach or walk next to the building or adjoining property? If so, the condition must be corrected.
3. Are there any unsafe conditions that should be corrected? Consideration should be given to broken glass in windows or doors, loose-hanging timbers or ceilings, wet plaster on the ceiling, holes in the floor, or unprotected vertical openings such as burned-away staircases.
4. Is there a water removal problem? Is so, what will be the best way to eliminate it?
5. Is there equipment on the premises such as forklifts or skip loaders? Are there any employees on the premises who are trained in the use of the equipment that might be able to assist?
6. What tools and equipment will be needed to complete the work? Are they available on the scene or will they have to be ordered?
7. Does the department have special on-call equipment that might be useful during this particular overhaul operation? Thought should be given to the possible need for an air wagon to refill air bottles, if one is not already on the scene. If the department does not have the needed special equipment, it may be available from the established mutual aid plan or from other departments in the jurisdiction.

Once the survey has been completed, or perhaps prior to making the survey if time permits, the Incident Commander should attempt to contact the owner of the premises if he or she is not at the fire. This is particularly important if the fire is in a commercial occupancy. It is always best that the owner be on the premises before operations commence in the office area.

PUTTING THE PLAN INTO OPERATION

When a plan of operation has been formulated and the Incident Commander is assured that there are adequate personnel and the necessary tools and equipment to do the job, the plan can be put into operation. It is normally better to divide the available personnel into work groups and place each work group under the command of a responsible officer. Sometimes the necessary work groups are already divided by company, under the command of an experienced company officer. The work groups should be assigned to the various tasks that must be done or to various portions of the building in such a manner that the work could be done freely, without everyone getting in each other's way.

It is good practice to continue the use of blowers during the overhaul operations to keep the carbon monoxide level below the acceptable level. The working atmosphere should be checked with carbon monoxide monitoring instruments to confirm that the carbon monoxide level is below the acceptable limit prior to the removal of breathing apparatus. If the Incident Commander is unable to confirm that the atmosphere is below the acceptable level, then all personnel should wear breathing apparatus during the overhaul phase.

Care must be taken during the overhaul operations to make sure that unnecessary additional damage is not done. Salvageable material must be separated from the unsalvageable and placed in a predetermined location.

Particular care must be taken when conducting overhaul operations in an office area. If the owner has been contacted but has not arrived on the scene, overhaul operations should be limited to ensuring that the fire is out. Care must be taken to save any partially burned records as they may contain the information needed to determine the inventory, the amount of insurance carried, and the financial condition of the business. The information contained in the records may be vital to the owner in proving his or her loss and may be invaluable to arson investigations in the event the fire is of a suspicious nature.

During cleanup operations, the firefighters should be on the alert for any valuables that might be mixed in with the debris. This is particularly important in residential fires. The importance of setting aside partially burned items should not be overlooked. Many insurance companies make a habit of not paying off on insured personal property unless there is proof that it was destroyed during the fire. The meager remnants of a burned object, such as a fur coat or a painting, may be all the evidence an owner needs to collect for his or her loss. (See Figure 2.15.)

FIGURE 2.15 ◆ Firefighters removing furniture from the structure in preparation to overhaul.
(Courtesy of Rick McClure, LAFD)

The amount of work necessary to ensure that the fire is out will depend a great deal on the burning characteristics of the material involved. Debris should be separated on the basis of burning characteristics. Material such as wood can be collected and thoroughly wet down as it is removed from the building and thrown into a pile. Smoldering mattresses and overstuffed furniture should be removed from the building and overhauled outside. Fires in these objects are usually deep seated. Once they have been taken outside, any burned material should be removed and immersed in a bucket or tub of water. Such material should not be piled with other debris until it is certain that no fire remains. Caution should be taken in removing material capable of deep-seated fires from upper floors of buildings. If an elevator is used for this purpose, it is best to wrap the object in such a manner as to eliminate air getting to the fire area. It is also best to have a water-type extinguisher in the elevator. (See Figure 2.16.)

If space is available to conduct the final extinguishment portion of the overhaul process inside the building in large industrial or mercantile occupancies, it may be advantageous to do so. This does not, however, eliminate the need for the burned material to be turned over and thoroughly wetted down. In residential occupancies it is generally best to move the debris to the outside of the building. Regardless of the choice, the spot where the debris will be piled should be chosen prior to any movement. It is unwise to have to handle the burned material twice. If it is decided to pile the debris outside the building, a spot should be chosen that is a safe distance from the building and on the fire building property unless it is impossible to do so. The

FIGURE 2.16 ◆ Firefighters conducting overhaul operations. *(Courtesy of Rick McClure, LAFD)*

proper authorities should be notified if it becomes necessary to pile the debris on the sidewalk or street so that adequate barricades might be set up. Small lines should be laid out and at least one firefighter left to staff the location where the debris will be piled. The debris should be turned over and thoroughly wetted down as it is removed from the building and thrown on the pile.

Many times removing the debris becomes a problem. With some fires, it may take longer to remove the debris and wet it down than it did to extinguish the fire. Many truck companies carry rubbish carriers or carryalls that can be used for removing the debris. This operation is cumbersome and normally ties up four firefighters. If the overhaul operation is extensive, it is best to utilize wheelbarrows and even skip loaders if they are available. Regardless of what is used, it is better that the debris be removed from the premises in a safe manner rather than being thrown out.

It should be remembered that removal of debris from a fire is for the purpose of making sure the fire is out. If there is an extensive amount of debris inside a building, and the Incident Commander is sure the fire is out and the premises are safe, then the removal of the debris is the responsibility of the owner. Under these circumstances, removal of the debris should be under the instruction and supervision of the building owner. The owner may use his or her employees, or contract with a private company that is properly equipped and trained in this type of operation.

Overhaul operations include the removal of water from a building, but the removal of all water in the building is not the responsibility of the fire department. The objective of water removal from the building is to prevent further damage. For example, water puddles on a concrete floor will not normally cause any additional damage, whereas the same size puddle on a hardwood floor could cause extensive damage if not removed immediately. Discretion should be used by those responsible for overhaul operations, because arguing who is responsible for the removal of a small amount of water does not necessarily improve public relations.

There are a number of additional thoughts that should be considered during overhaul operations.

1. All avenues through which heat may have extended during the fire should be thoroughly checked for hidden or smoldering fires. Attic scuttle holes should be located and a firefighter sent into the attic to check for hot spots. Do not overlook the tight space next to the eaves. Birds have a habit of building nests there. The nests could have become ignited without having come into direct contact with any flame.

2. Check all concealed spaces, particularly walls and under floors between the joists. If infrared heat detectors or thermal cameras are available, they can be used to help detect any hidden fires. If they are not available, fire in the walls can generally be located by running a hand over the wall. A hidden fire is indicated if a spot is found that is too hot for the hand. If such a spot is found, the wall should be opened and the fire extinguished. Continue to remove the plaster, wallboard, or other covering until all char to the wood is visible. After wetting down the charred area, it is generally best to use a small amount of water from a spray nozzle and direct it upward, allowing the water to cascade down the inside of the remaining wall.

   There are other potential indicators of a hidden fire that should be examined if they exist. Some of these are: peeling paint, rippled wallpaper, cracked plaster, and some emissions from cracks.

   If fire is suspected between floors, it is best to go below the suspicious spot and pull the ceiling. Ceilings are easier and cheaper to repair than floors. Of course, if fire is suspected under the floor and it is impossible to get below that floor, then do not hesitate to cut a hole in the floor.

Fire has a habit of getting under facings around windows, around doors, and at floor level. Baseboards and the facings around the doors and windows should be removed if fire is suspected in these locations. The bottoms of all vertical openings, including elevators, should be checked for burning debris whenever upper floors have been involved with fire.

3. Broken glass left in windows, doors, transoms, and so on presents a hazard to anyone entering the premises. Consequently, effective overhaul operations demand that all broken glass be removed from these areas. All broken glass on floors, sidewalks, streets, and the like should be swept up and safely discarded.

4. If not done during firefighting operations, the electricity should be shut off to the burned-out area if the insulation on any electrical wiring has been burned off. If this cannot be done by pulling a switch or circuit breaker at the main control station, then it may be necessary to have the power company cut the wires to the building. Gas and/or water lines should also be shut off if there is any evidence of leakage from these sources.

5. Be on the alert during the overhaul process for any material that is subject to spontaneous heating on becoming wet. If this material has become wet during the fire, it should be taken outside and laid out in thin layers to prevent the buildup of heat.

6. Take particular care when overhauling bales of cotton. Fire has a way of boring deep into a bale, making extinguishment difficult. Water will not normally penetrate to the seat of the fire due to its high surface tension. Wet water or Class A foam can be used to assist in extinguishment but it will reduce the salvage value of the cotton. The best way of making sure the fire is out is to tear open the bale. Cut the bale wires at the center first, then the outside wires. This will permit better control of the contents by limiting the amount of cotton strewn about.

7. Be cautious when overhauling material know to have a nitrocellulose base. These materials are capable of producing a flash fire when heated, due to the fact that they contain oxygen in their chemical makeup. It is best to immerse the material in a container of water.

8. Restrict the opening of packaged material to those packages that have been damaged by water or fire. Packaged material might be readily salvageable, which will help hold the loss to a minimum.

9. Fire in multiple-story buildings may make it necessary to throw materials out the window to the ground below. Before starting such operations, make sure that the space below is clear and that neither people nor property will be placed in jeopardy by such action. It is wise to place a firefighter near the place where the material will land to ensure that someone does not carelessly walk into the area. However, remember that it is good practice not to throw anything out the window that can be carried out.

10. Early in the overhaul process, any heavy objects such as baled cotton or similar material should be moved from the center of the room toward the wall or placed over supporting columns. Floors might have been weakened by the fire and the water absorbed by the material during the firefighting phase will have increased the weight. This combination could lead to floor collapse if corrective action is not taken. A window frame can be removed and the wall taken out to the floor level if it becomes necessary to remove some of these heavy objects from the building.

SECURING THE BUILDING

If possible, the building should be turned over to the owner when the department is ready to leave. If the owner is not present and cannot be contacted, it then becomes necessary to secure the building against unwanted visitors. All windows and doors that were broken during firefighting operations should be boarded up. Lumber can usually be found on the premises for this purpose. If not, interior doors can be removed and nailed over openings if needed.

The building should also be made as safe as possible from weather elements before leaving. If holes were cut in the roof or if the fire burned through the roof, the holes should be covered with tarpaper or plastic sheathing if there is a possibility of rain, snow, or other weather elements that might cause further damage.

At some fires it may be necessary to establish a fire watch. This can be done by leaving one or more firefighters on the scene with a portable radio or cellular phone. If this is impractical, then the first-in company can make a "drive by" a couple of times an hour to check for possible rekindles. It may be wise for the company to stop and reenter the building to check for any smoke developing.

◆ VENTILATION

ventilation The process of replacing a bad atmosphere with a good atmosphere.

Ventilation is a tactical task that is normally assigned to the truck company. (See Figures 2.17 and 2.18.) By definition, it is the process of replacing a bad atmosphere with a good atmosphere. As applied to the fire service, it means clearing a structure, vessel, or other area of hostile smoke, heat, or noxious gases through controlled channels and replacing the objectionable gases with fresh air. (See Figure 2.19.)

Proper ventilation has been called the key to successful firefighting. Stated in reverse, failure to ventilate in the proper place and at the proper time, or improper

FIGURE 2.17 ◆ A truck company member using a chain saw to ventilate the roof. *(Courtesy of Rick McClure, LAFD)*

FIGURE 2.18 ◆ Three truck company members preparing to break the glass window to provide horizontal ventilation. *(Courtesy of Rick McClure, LAFD)*

ventilation, has been termed one of the greatest and most common failures of fire-fighting operations. Some of the favorable results of proper ventilation are:

1. Firefighting and rescue operations are facilitated.
2. The danger of backdraft is reduced.
3. The area is made more comfortable for firefighting personnel.
4. The survival profile of people in the building is increased.
5. The spread of the fire is curtailed.
6. Toxic and explosive gases are removed from the area.
7. "Mushrooming" is eliminated.
8. The fire will burn more freely and the volume of smoke will be reduced.
9. Heat and smoke damage will be held to a minimum.
10. Fire, smoke, and gases can be properly channeled.

FIGURE 2.19 ◆ A well-vented industrial fire. Note how one of the master streams has been bent by the updrift from the fire. *(Courtesy of Rick McClure, LAFD)*

FIGURE 2.20 ◆ Truck company members making preparations to open the roof using a chain saw. *(Courtesy of Rick McClure, LAFD)*

To be most successful, ventilation operations should be given careful consideration during the size-up of the fire. The work should start immediately once the decision is made as to where ventilation should take place. Good firefighting techniques dictate that ventilation should not be started until sufficient hose lines are in place both to confine and to extinguish the fire. However, under normal conditions ventilation operations should not be held up waiting for this situation to develop, because it is almost impossible to ventilate before hose lines have been laid. Ventilating at the roof level, for example, requires that ladders be raised, firefighters ascend to the roof, the proper location determined as to where ventilation should take place, and a hole cut in the roof. These steps take much more time than does the laying of hose lines. Additionally, engine companies generally arrive on the scene prior to the truck companies and have a head start on operations.

There is no standard as to the type and amount of ventilation required at a fire. Each fire is different, and the conditions at each fire will dictate what is needed. Where possible, vertical ventilation is preferred to horizontal ventilation because smoke normally rises and seeks vertical outlets. This does not mean, however, that a hole should be cut in the roof at every fire. The height and construction of the building, along with the volume of smoke and the location of the fire, will dictate the type of ventilation required. (See Figure 2.20.)

ROOF VENTILATION

Roof ventilation should always be conducted with all firefighters in full protective gear including breathing apparatus and a charged line on the roof. (See Figure 2.21.) Roof ventilation will vary from a simple task to an almost impossible one. Operations might be one of simply removing a scuttle hole cover to a complex one of manually opening a large area. One of the first thoughts when preparing to ventilate a roof is to look for natural openings that can be used. Of course, these openings are of little value unless they are in the right location. They may pull fire to the opening, which could result in an expensive spread of the fire if they are not in the correct location.

FIGURE 2.21 ◆ A fire in the attic. A truck company member preparing to ventilate the roof using a chain saw. *(Courtesy of Rick McClure, LAFD)*

Some of the natural openings available for ventilation are scuttle hatches, skylights, roof monitors, vertical shafts, and stairwells. (See Figure 2.22.)

A **scuttle hatch** is an opening in the roof that extends directly into the attic. It is fitted with a cover that is sometimes locked. Scuttle hatches can normally be used to ventilate only the attic area, whereas some of the other natural openings can be used for additional ventilation. It is good practice to have the wind to your back so that the heat and gases will be released in a safe direction when lifting off scuttle hatch covers. A wooden cover can generally be removed by prying it open with an axe. Spring-loaded metal covers that are locked from the inside present more of a problem, and sometimes it is better to look for another location for ventilation rather than trying to open them.

scuttle hatch An opening in the roof that extends directly into the attic.

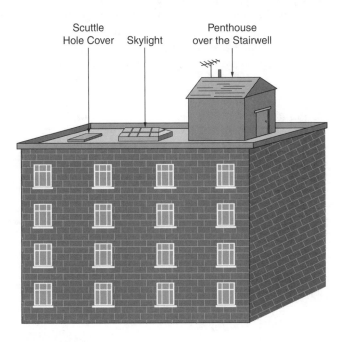

Scuttle Hole Cover   Skylight   Penthouse over the Stairwell

FIGURE 2.22 ◆ Look for natural ventilation openings.

skylights Used to provide
light to areas below the roof.

Skylights are used to provide light to areas below the roof. Their value for ventilation, as is so with other natural openings, depends on what area they will ventilate. In hotels and apartment houses they will generally ventilate the hallway of the top floor, but in some cases they will provide a vertical opening that extends from the first floor to the roof. Windows that open into this vertical shaft may be available at each floor. In commercial buildings they will generally ventilate the manufacturing area.

Skylights can generally be removed intact because the frames are normally only lightly secured at the corners. However, if any difficulty is encountered in removing them, and time is important, there should be no hesitation in breaking the glass. Some skylights are equipped with Plexiglas® rather than wire glass. The Plexiglas will melt at approximately 400° F, which normally provides self-ventilation.

The use of a stairway that terminates on the roof can be particularly valuable; however, it can also prove to be dangerous if improperly used. It is usually a simple operation as it requires only opening a door. However, it is generally necessary to prop the door open to maintain a continuous updraft. Previous knowledge of the building at this point can prove to be invaluable. The information gained can be used to ascertain what part of the building below the roof will be affected by opening the door. It should be remembered that extensive ventilation operations, particularly when they affect the life hazard, allow little room for error.

Two factors are of paramount importance when it becomes necessary to cut a hole in the roof. The first is that it be properly located. This means that the hole should be cut directly over the fire, *providing it is safe to do so.* Sometimes this spot is readily identified by simple observation. If not, look for discoloration of the roof or feel the roof for the "hot spot." (See Figure 2.23.)

Cutting a hole in the ideal location means that the place where the hole will be cut is usually the hottest and most dangerous place on the roof. This spot should be avoided and a safer location selected to make the cut if there is any doubt as to the safety of personnel. It is normally possible in a large building to cut the hole within 20 feet of the hottest spot without seriously pulling the fire through previously uninvolved portions of the building, Pulling the fire through uninvolved portions of the building should normally be avoided; however, there should be no hesitation to do so to protect life. There also should be no hesitation to do so when working on roofs of lightweight construction or other roofs that are a hazard to firefighters.

An evaluation of the roof should be made to ensure that it is safe before any firefighter is allowed on it. As firefighters enter the roof they should sound it out ahead

FIGURE 2.23 ◆ Be careful where the hole is cut.

of their movement to locate any weak areas. A rubbish hook or other long-handled tool can be used for the sounding. A good knowledge of roof construction is important in making this evaluation.

Some roofs by their very nature are hazardous and require extreme caution when working on them. An example is a roof covered with ½-inch plywood on a commercial building, which gives the appearance of being sturdy. Most roofs on buildings constructed prior to 1960 are solidly constructed and well supported. They are generally safe, although they might give the impression of being somewhat "springy." Extra care, however, should be taken when working on buildings constructed after 1960 that are classified as lightweight construction. Commercial buildings so classified generally have flat roofs with ½-inch plywood covering. Fire will weaken large sections of these roofs rapidly, exposing firefighters working on the roof to extreme danger. It is not safe to operate over any portion of the roof that is involved with fire. If the roof is obviously too "soft" or "spongy" to be safe, then all personnel should be removed from the roof. Generally, under these conditions, the fire is about to "self-ventilate." When it does so, no attempt should be made to extinguish it as it comes through the roof for it is doing exactly what would have been done with proper manual ventilation.

Working on a roof, directly over the fire, is a dangerous operation. A number of firefighters have gone through the roof and been killed during ventilation operations. Elevated platforms can be used as a safety factor when conditions are right. A line can be lowered from the platform and tied to the person opening the roof. The firefighter can be pulled to safety in the event the roof gives way. (See Figure 2.24.)

FIGURE 2.24 ◆ Think safety.

FIGURE 2.25 ◆ A typical truss construction. The metal gusset plates are pressed into place. Fire will cause the gusset plates to deform and pull away from the wood.

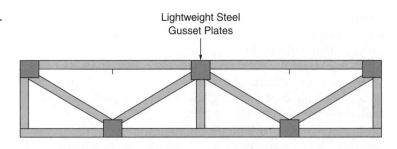

Lightweight Steel
Gusset Plates

A roof with truss construction is an excellent example of where this method of safety may be particularly important. These roofs are inherently dangerous. A large fire in the truss area is a good indication that the roof may not be safe for performing ventilation operations. Timber trusses can span areas as much as 100 feet and be spaced 20 feet on center. All members of the truss are dependent on each other to carry a portion of the load. Failure of one truss can cause a collapse of a large section of the roof with little or no warning. Figure 2.25 shows a typical truss construction.

The second factor to consider when cutting a hole in the roof is that the hole should be of sufficient size to vent the fire adequately. There is no set rule as to the size of hole to be cut. The intensity and size of the fire, the rapidity of its spread, the necessity for immediate ventilation, and the type of building construction all play an important part in making a decision. It is good practice, however, to adopt the principle of overventilating rather than following the natural tendency to underventilate. A 10' by 10' opening on the roof of a large building with a well-involved fire inside would not be considered out of line. Remember, a 10' by 10' opening will provide the same ventilation area as four 5' by 5' openings. Not only would the larger opening be more effective but it also would be easier and cheaper to repair than four smaller openings.

louvering An effective method of ventilating a roof.

Louvering is an effective method of ventilating a roof. (See Figure 2.26.) It is a system that helps reduce the exposure of personnel to smoke and heat as the roof is

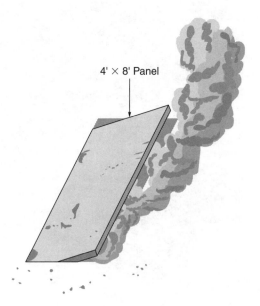

4' × 8' Panel

FIGURE 2.26 ◆ Louvering is an effective ventilation method.

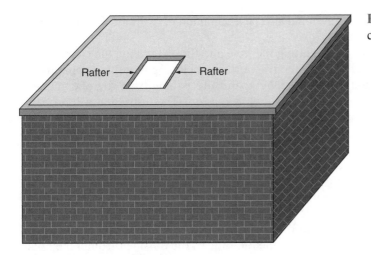

FIGURE 2.27 ◆ Make cuts along rafters.

vented. It can be used effectively on sheathing but is most effective on plywood pan-eled decking.

The operation consists of making two longitudinal cuts on either side of a single rafter with the cuts parallel to one another. The longitudinal cuts are then intersected with crosscuts. Once the cuts are made, one side of the cut panel can be pushed down, resulting in a louvering effect. This operation will ventilate the area below and direct most of the heat and smoke away from the firefighters conducting the operation. Cuts should be made so that the firefighters making the cut have the wind to their backs as the panel is louvered.

Some roofs are covered with tar paper or tar and gravel coatings. These materials should be removed before cutting a hole. The pick head of the axe will be effective for this operation.

When holes are cut in roofs, cuts should be made along the rafters to facilitate making repairs. (See Figure 2.27.) The rafters can usually be located by sounding with the axe. Rafters provide a dull sound as compared with the higher-pitch sound of the spacing between rafters. Cuts can be made with an axe or a chain saw. The holes should be oblong or square. All cutting should be made before any sheathing is removed. The firefighter removing the sheathing should stand to the windward side of the hole and start removing the sheathing from the leeward side, working toward the windward side. This will protect the crew from the heat and smoke. If the hole is of the proper size and the job of ventilating has been done correctly, the rafters will not be damaged as the sheathing is removed.

For reference, the **windward side** is the side from which the wind is blowing. The **leeward side** is the side toward which the wind is blowing. For example, if the wind is blowing from the north, then the north side of the building would be the windward side and the south side of the building would be the leeward side.

windward side The side from which the wind is blowing.

leeward side The side toward which the wind is blowing.

If plywood sheathing is used on the roof, the operation can be simplified by removing an entire panel intact. At least two adjacent panels should be removed and three would not be excessive when large areas are involved.

Heat, smoke, and fire should start issuing from the opening as the sheathing is removed if the hole has been properly located and there is nothing below to block its path. There should always be a charged line on the roof at this time to protect exposures from the resultant heat. There should also be a minimum of two ways for the firefighters to escape from the roof.

All firefighters, both those cutting the hole and those handling the charged line, should be fully equipped with protective clothing, including breathing apparatus. If there is a minimal amount of smoke, heat, and fire issuing from the opening, it is probably due to restrictions to the flow below the roof. The restriction will probably be a ceiling if the fire is not in the attic. This can usually be corrected by using a pike pole or similar object through the hole to open the ceiling below. The firefighter handling the pole should work from the windward side of the cut hole.

One last thought on roof ventilation. Timing is important. Although the ideal is seldom achieved, it should be strived for and the results of an early or late cutting should be visualized. Ideally it is best to open the roof a few seconds prior to the firefighters making entrance at the lower level and entering the area with charged lines. The opening of the roof will start a movement of the built-up heat, gases, and smoke from the interior of the building to the outside through the opening made. When the opening is made for the firefighters advancing the line to enter, fresh air will be pulled into the room, which will accelerate the outward movement of the heat, gases, and smoke. This will generally clear the interior atmosphere sufficiently to allow the hose handlers to make a rapid advance on the fire. (See Figure 2.28.)

If the roof is opened early, movement of the built-up heat, gases, and smoke will commence due to the natural tendency of heat to rise and the pressure buildup inside the building. However, the full effect of the opening will be somewhat restricted due to the lack of air entering the building at a lower level. The movement will be accelerated once an opening is provided at a lower level. Early opening provides more safety for those advancing hose lines than does a late opening.

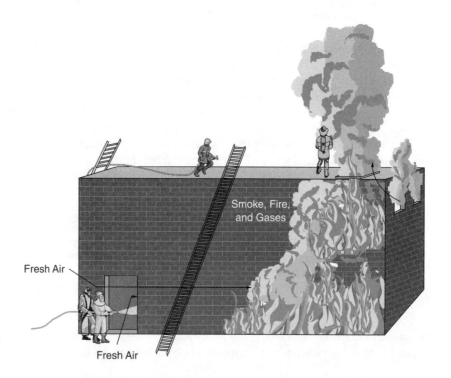

Smoke, Fire, and Gases

Fresh Air

Fresh Air

FIGURE 2.28 ◆ Proper timing helps the hosemen.

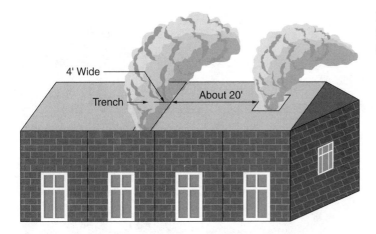

FIGURE 2.29 ◆ Trenching is a good defensive tactic. The trench should extend from one exterior wall to the opposite exterior wall.

A late opening is the least desirable of the three but probably occurs most often due to the time required for those responsible for providing the opening to ascend to the roof and make the opening. Under these circumstances, those advancing hose lines will be required to operate in an atmosphere with a heavy concentration of heat and smoke. However, their working conditions are immediately improved once the opening is made.

TRENCH CUT

A **trench** is a cut that is placed in the roof of a structure that divides a portion of the attic that is burning from a portion of the attic that is not burning. Trenching is not a ventilation method but rather a defensive operation. Its primary use is to stop the extension of an attic fire in a long, narrow structure. (See Figure 2.29.) A trench should never be started before the roof above the fire is properly vented. When properly used, trenching will confine a fire to a selected area of the building attic.

To be effective, the trench should be cut about 20 feet from the initial vent hole. The trench should be at least 4 feet wide and extend from one exterior wall to the opposite exterior wall. For maximum effectiveness in stopping the fire spread, a charged line should be extended into the uninvolved attic to stop any fire that may get past the trench.

trench A trench is a cut that is placed in the roof of a structure that divides a portion of the attic that is burning from a portion of the attic that is not burning.

USE OF VERTICAL OPENINGS

Because the natural tendency of fire and heated gases is to rise, built-in vertical openings provide an excellent means of venting the fire. However, care must be taken when selecting which vertical openings to use because a bad choice could increase the hazard to life or spread the fire. The three most common vertical openings that can be used for ventilation are stairwells, elevator shafts, and light wells.

Before using a stairwell for ventilation, make sure that all occupants of the building are below the fire floor and be certain that the stairwell will not be used for egress. The stairwell can be used most effectively if it extends through the roof. Operations only require that the door to the penthouse over the stairwell be opened at the roof; however, some action may be required to ensure that the door will remain open.

Additional steps must be taken if the stairwell terminates at the top floor. It will be necessary to cut a hole in the roof over the stairwell or to horizontally ventilate the top floor. It may be necessary also to provide horizontal ventilation on the floor below the

FIGURE 2.30 ◆ Ventilation
blowing up from the stairwell will
help facilitate the removal of smoke.

top floor if the smoke is extremely heavy. Smoke ejectors blowing up the stairwell from
intervening floors will help facilitate the removal of smoke. (See Figure 2.30.)

Enclosed elevator shafts can also be used effectively for vertical ventilation.
Make sure the cage is below the fire floor before attempting to use the shaft for ventilation. This will then provide a clear opening from the fire floor to the upper terminating point of the elevator. (See Figure 2.31.)

However, if the cage is secured at some point above the fire floor and cannot be
moved, and it is imperative or would be extremely beneficial to use the elevator shaft
for ventilation, then additional steps must be taken. The elevator doors at the floor below the secured cage must be forced open and horizontal ventilation provided at this
floor. Again, if the smoke is extremely heavy, horizontal ventilation should be repeated on the next floor below. (See Figure 2.32.)

HORIZONTAL VENTILATION

An evaluation should be made of the wind direction prior to commencing horizontal
ventilation operations whenever an entire floor or a large area is to be ventilated. The
operation will be most successful when the smoke and gases are channeled out the
leeward side of the building. Once the direction of the wind is determined, double-
hung windows on the leeward side should be opened from the top whereas sliding aluminum or sliding wooden windows should be fully opened. Remove all drapes,
shades, screens, and other obstructions that would impede the flow. Double-hung windows on the windward side should be opened from both the top and the bottom or

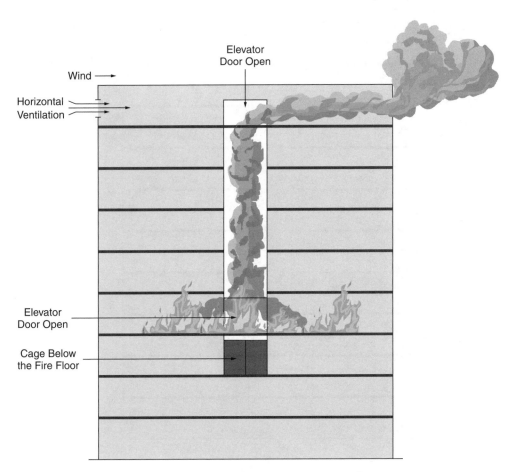

FIGURE 2.31 ◆ Elevators can be used effectively for ventilation.

from the top only after all the leeward windows have been opened. Sliding aluminum or sliding wooden windows on the windward side should be fully opened. All obstructions to the flow should be removed.

Occasionally, the air will be still and no horizontal movement can be established. In this case, double-hung windows should be opened two-thirds from the top and one-third from the bottom. Sliding aluminum or sliding wooden windows should be opened fully to provide for the maximum air movement. Smoke ejectors should be used to assist in moving the air. This method should also be used in small rooms where horizontal openings are restricted to one or two sides of the building.

It is good practice when providing horizontal ventilation to determine where the heated gases and fire will go once they leave the openings. It is possible that they could extend upward and through open windows on the floors above or through openings into adjoining buildings. If there is such a possibility, then all exposed unprotected openings should be closed or spray streams used at these openings to prevent any extension of the fire.

Sometimes it is desirable to provide horizontal ventilation where it is impossible to get to the windows that need to be opened. In this situation it may be necessary to break the windows. It is possible that this can be accomplished from the inside by the use of

FIGURE 2.32 ◆ Use of elevator shaft when the elevator is stuck above the fire floor.

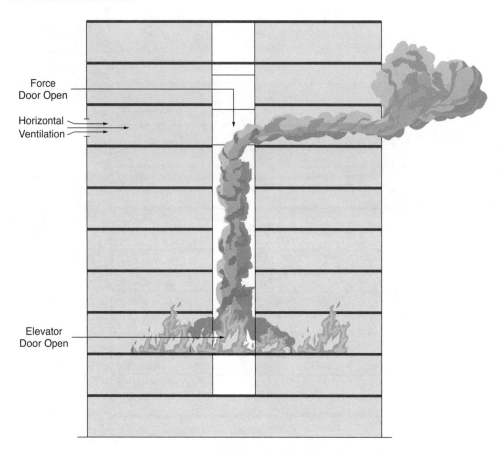

FIGURE 2.32 ◆ Use of elevator shaft when the elevator is stuck above the fire floor.

Force Door Open

Horizontal Ventilation

Elevator Door Open

hose streams. Hose streams may also be used outside for breaking windows, or the windows may be broken by the use of pike poles and ladders. If ladders are used, the ladder should be lowered into the window in such a manner that the glass will fall inside the building. Do not extend the ladder into the window. If this is done, panels of glass may slide down the ladder and seriously injure those on the ground. The ladder should be removed once the window is broken as fire is likely to shoot through the opening.

Smoke ejectors can be extremely useful as an aid in horizontal ventilation. They can be set up to create either a suction or blower effect. They will pull the smoke toward the blower and out the opening when used to create a suction effect. When used as a blower, they will move the smoke and gases toward openings. They can also be used in pairs with one providing a suction action and the other acting as a blower.

Whenever a blower is used to pull smoke and noxious gases out of a structure, it is referred to as **negative ventilation**.

negative ventilation
Whenever a blower is used to pull smoke and noxious gases out of a structure.

positive pressure ventilation
A method of clearing an area of smoke and gases for the purpose of gaining entry to extinguish the fire.

POSITIVE PRESSURE VENTILATION

Positive pressure ventilation is a method of clearing an area of smoke and gases for the purpose of gaining entry to extinguish the fire. It combines the process of making openings and using mechanical blowers. It requires time for maximum effectiveness. The theory of the positive pressure ventilation is that it is a method of creating a pressure difference between the inside of a structure and the outside air. When this condition is achieved, it forces the smoke and heat to move from a high-pressure zone to a lower pressure zone. There are a number of benefits claimed for the positive pressure ventilation method.

1. Positive pressure ventilation (PPV) quickly removes heat and smoke from the fire area, which reduces the ability of the fire to extend to other portions of the structure.
2. PPV may eliminate or reduce the need for and risk of roof ventilation.
3. The rapid removal of the smoke improves a search team's ability to conduct search and rescue operations.
4. The improved atmosphere reduces the heat stress to firefighters.
5. PPV reduces the smoke and fire damage.
6. The location of the fans does not interfere with ingress or egress.
7. The PPV can be set up without the necessity for firefighters to enter the smoke-filled environment.

The method to use for positive pressure ventilation can best be demonstrated by referring to Figure 2.33.

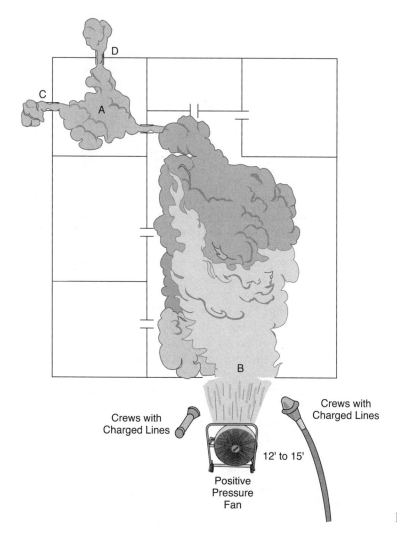

Crews with Charged Lines

Crews with Charged Lines

12' to 15'

Positive Pressure Fan

FIGURE 2.33 ◆ Positive pressure ventilation.

The fire is confined to room A and heavy black smoke has started to move to other areas of the house. Positive pressure fans have been set up about 12 to 15 feet from entry point B. The objective is to create a pressure cone from the fan that completely covers the door opening. Crews are waiting with charged lines at the door prepared to make their entry. When the proper conditions have been set up, the windows at points C and D are broken out, which creates a total exit area approximating that of the entry door. Fresh air will be forced into the fire area when the fans are started. The combination of the openings at points C and D and the forced air quickly clears the area of smoke, permitting the firefighters to enter for extinguishment. The blowers should continue to be operated after the fire is extinguished to aid in the removal of residual steam and smoke from the fire area. Following are several points that should be considered regarding positive pressure ventilation:

1. Do not attempt to use the positive pressure ventilation method for entry if the interior of the building is ripe for a backdraft.
2. Make the exhaust openings in the fire area as close to the fire as possible.
3. PPV may result in hidden fires being extended.

BASEMENT VENTILATION

Basement fires are not only one of the most difficult types to extinguish but also one of the most difficult to ventilate. The ventilation problem of basements is compounded by the fact that most basements have few openings directly to the outside and a relative small number from the basement to the ground floor. In fact, in some cases the only opening to the basement is a single entry from the ground floor. Additionally, the basements are typically used for storage. This creates a situation that makes it extremely difficult to anticipate what types of materials will be involved in the fire.

One of the objectives in fighting basement fires is to confine the fire, smoke, and heated gases to the basement area until means can be provided to control the path of their escape. Many times this presents a problem because some vertical openings originate in the basement, which permits the smoke and fire to rise through the walls if adequate fire stops are not provided. Because of the amount and different types of storage found in basements, it can be expected that the fire will produce large quantities of smoke.

Ventilation of the basement area might be simplified if there are openings to the outside at opposite ends of the basement. One of the openings can be used to rid the basement of unwanted gases while the other can be used to bring in fresh air. The opening used for exhaust of the gases should be opened first. Entering the other opening with blowers will help facilitate the movement of the gases, as will the use of spray or other streams.

If there is but a single opening into the basement, or only one of the openings available is suitable, it may be necessary to cut or breach a hole in the ground floor to provide ventilation. (See Figure 2.34.) This hole should be made next to a window or door opening so that the vented gases can be directed to the outside of the building.

Blowers or spray streams can be used effectively to assist the movement of the gases through the opened window or door. When using spray or fog nozzles to assist in ventilation, operate the nozzle at approximately a 60 degree angle and direct it in

FIGURE 2.34 ◆ One method of basement ventilation.

Fresh Air

such a manner that it will cover about 85 percent of the opening. (See Figure 2.35.) An adequate number of charged lines should be laid out prior to opening the hole to protect against the spread of fire once ventilation commences.

Positive pressure ventilation can be used effectively to remove the smoke from the basement. One or two fans can be used at the stairway entrance into the basement.

One last thought on ventilation of basements. Basement fires are often ripe for a backdraft due to the confined area and the tremendous amount of heat that can build up. It is good practice to consider that a backdraft will occur if adequate precautions are not taken to prevent it.

Cover 85% of Window Opening

60°

FIGURE 2.35 ◆ Spray streams can be used effectively as an aid in ventilation.

ADDITIONAL VENTILATION PRACTICES AND PRINCIPLES

Following are some general practices and principles that should be considered during ventilation operations.

1. Breathing apparatus should always be worn during ventilation operations. The types of materials used in building construction, as well as in storage, may produce toxic gases that can prove dangerous even when an atmosphere appears relatively clear.
2. It should be expected that the fire will increase in intensity once an opening is provided to the outside. Charged lines should always be in position to protect exposures and horizontal fire spread across the roof prior to providing the opening.
3. Firefighters should always work in pairs when ventilating a roof of considerable size or where large volumes of smoke are present. In fact, it is good practice to have firefighters work in pairs during all phases of ventilation operations, with a company officer present if possible.
4. A plan of escape should always be established prior to making an opening. At least two ladders should be raised to the roof if roof ventilation is to take place. This will provide a second means of escape in the event the planned route is cut off.
5. Unless life is in danger, or there is no alternative, do not provide openings that will pull the fire unnecessarily through uninvolved portions of the building.
6. Do not provide openings that will endanger adjacent property unless adequate lines are provided to protect these areas.
7. Do not ventilate a roof after the need has passed. Roof ventilation should serve a useful purpose and not be done automatically or as an afterthought.
8. Always expect and be prepared for a backdraft.
9. Always consider the direction and velocity of the wind when contemplating ventilation operations. The wind can greatly aid or hinder operations, depending on how plans are made for its use.
10. Delay ventilation of special types of fires, such as ship fires or fires involving electrical equipment, until overall fire strategy has been established.
11. Do not direct hose streams into openings provided for ventilation. Such operations not only eliminate the purpose of providing the opening but also will subject crews working in the interior of the building to unnecessary punishment.
12. Never cut into trusses or other structural supports.
13. Use good common sense when ventilating. Try to provide openings that will keep damage to a minimum. However, do not let the fear of damage be the determining factor as to where and how to ventilate. This is particularly true when life is at stake. Damage should be of little concern in such cases.

◆ SALVAGE

This function may be assigned to a company such as a squad company or a truck company. However, in some departments the function is not assigned to a particular company but rather given to a group or a company at a fire when its need is apparent.

Prior to 1900, the subject of salvage operations was not a particular concern of most fire departments. However, insurance companies, on the other hand, were quite concerned about the reckless use of water by some departments. This was understandable. In the final analysis, it was they who had to dig into their pockets and pay the bill for the water damage. Consequently, in order to protect their insured property, many of the larger insurance companies in the larger cities dispatched salvage companies to all fire alarms. The objective of these salvage companies was to protect the building and contents from water damage.

Fire officials' attention was drawn to salvage operations with the development of the Underwriters' Grading Schedule, following the Baltimore Conflagration of 1904. The schedule included the requirement that fire apparatus carry salvage covers. Failure to do so would possibly result in higher insurance rates for the citizens of individual cities. Essentially, this could cause fire chiefs to be blamed for higher rates if they failed to comply with the requirements.

Few fire officials will argue with the concept that the objective of a fire department is to extinguish a fire with a minimum of loss to life and property. Water damage is a part of that loss. In fact, at some fires the water damage is nine to ten times greater than the direct damage caused by the fire. It would seem to follow that salvage operations are a primary consideration of most fire officials. Unfortunately, this is not the case. In fact, salvage operations are almost nonexistent in some fire departments. There are several reasons for this.

One reason that salvage operations are the most neglected of those functions for which a fire department is responsible is that members of most departments are not totally salvage oriented. Most firefighters prefer throwing water or cutting a hole in a roof than they do to carrying a salvage cover into a building to cover up the furniture. Consequently, when a firefighter becomes an officer he or she generally gives little thought to salvage operations when in command of a company or the Incident Commander at an emergency. To overcome this deficiency, it is necessary that all officers be consciously aware of salvage and that salvage operations be so instilled in the minds of all new members during their rookie training that they carry the thinking into fire situations with them.

Another reason that salvage operations are neglected is that very few departments have a separate salvage company. This is not necessarily by choice but primarily a matter of economics. Unfortunately, fire departments throughout the country have been required over the years to do more and more with fewer firefighters and a reduced budget. The threat of terrorist activity has done nothing but compound the problem. Almost all of those departments that once had separate salvage companies have had to relegate this function to other companies or assign the function at a fire scene when it is apparent that it is needed. However, for the purpose of illustrating the importance of salvage operations to the overall coordination of fire operations, in this book it will be considered that the company or group performing salvage operations is salvage orientated and well trained in performing salvage operations.

In many departments, the responsibility for salvage operations is relegated to the truck company. Unfortunately, truck companies in many cities are understaffed. When salvage operations are relegated to understaffed companies that already have the responsibility for laddering, overhauling, ventilating, forcible entry, rescue, and controlling the utilities, something has to give. At some large fires it is desirable that all the functions for which the truck company has been given the responsibility, except the overhaul, be performed at or near the same time. At times this becomes an impossible task, and if the responsibility for salvage operations is also given to the truck company, it normally will become the most neglected factor.

There is a way, however, of overcoming this deficiency. If all members of a department are salvage oriented and thoroughly trained in salvage operations, the

Incident Commander can call for an extra engine company and assign it to the task of performing salvage operations only. Of course, if this type of operation is planned or anticipated, then it is important that all apparatus in the department carry a minimum number of salvage covers and some other tools used for salvage operations. It is also good practice for a department to maintain an apparatus in reserve, equipped with a minimum of 20 covers and salvage equipment, that can be dispatched on special calls. Two benefits would be derived from this action.

1. There would be a considerable reduction in the fire loss.
2. The public relations of the department would be enhanced.

It is probably worthwhile to expound on item 2. More letters of appreciation are received by fire departments who are salvage oriented and perform good salvage work at fires than are received for all the other combined functions performed by departments at a fire. An attic fire in a single-family dwelling might serve as a good example. The occupants see heavy black smoke rising from all parts of the house, they see the fire coming through the roof, and they watch the firefighters throwing water into the building from all sides. They envision all their furniture destroyed, their wall-to-wall carpeting ruined, and their house in shambles. It is not difficult to imagine their surprise and appreciation when they find the opposite. Everything in the house has been protected from water damage. The inside has been cleaned and all the furniture is back in place.

As previously mentioned, this section will discuss salvage operations as if a department operated with apparatus fully equipped with salvage equipment and under the command of an officer and crew who are salvage orientated. The principles of salvage operations are the same, regardless of who is given the responsibility for their execution.

Salvage work should be started immediately on the department's arrival at the fire if loss from water damage is to be held to a minimum. If a department is salvage oriented, some of the most effective work can be accomplished by the judicial use of water by those handling the hose lines. The problems of those doing salvage work are reduced considerably when the use of water is held to a minimum. The increasing use of Class A foam for attacking Class A fires should have a beneficial effect on this problem.

The three operations that should be performed by those doing the salvage work early during the fire operations are (1) spreading covers, (2) bagging floors, and (3) diverting and removing water. They are not necessarily performed in that order. (See Figure 2.36.)

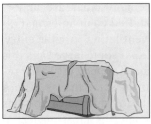 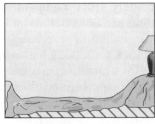 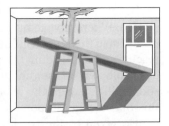

SPREADING COVERS BAGGING FLOORS DIVERTING AND REMOVING WATER

FIGURE 2.36 ◆ These operations should be performed early during the fire.

It is important that the officer responsible for the salvage operations make a size-up of the problems facing him or her prior to the commencement of operations. This officer should first determine the seat of the fire and make an estimate of the amount of water that will be used for extinguishment and where the excess water will most likely flow. With this knowledge, he or she will be able to establish a plan of action and estimate the number of firefighters it will take to carry out the plan. The officer should immediately inform the Incident Commander of the need for additional personnel if a sufficient number is not available on the scene to carry out the salvage plan effectively.

Estimating the amount of water that will be used will help the officer responsible for salvage operations to decide on whether the water can be held to the fire floor or whether it will be necessary to start operations on the floor below the fire and perhaps all floors below the fire floor. The type of construction and the type of occupancy will also play an important part in the officer's size-up.

At most fires, salvage work begins and ends on the fire floor. The quick removal of the excess water is particularly useful on hardwood floors. Hardwood floors warp easily and are expensive to replace.

There is another important operation that should be performed prior to lines being taken into the building if the fire is on the first floor of a dwelling that has wall-to-wall carpeting. Where possible, a floor runner should be laid from the entrance door to the room on fire. This will provide protection for the carpeting as the firefighters bring in lines and advance them to the fire area. This is particularly important if there happens to be mud or other elements outside that would be taken into the house and over the wall-to-wall carpeting on the boots of the firefighters. There are floor runners available from commercial companies that are made of vinyl-laminated nylon. These covers are lightweight, flexible, and tough; are both heat and water resistant; and are easy to maintain.

Although it may appear that the water will be held to the fire floor, it is important that a survey of the floor below the fire be made as soon as possible to see whether there is any possibility that the water will seep through the floor. It may be necessary to start operations simultaneously on both the fire floor and the floor below the fire.

Prior to starting the spreading of covers on the floor below the fire, an estimate should be made as to where the water will most likely come through the ceiling first. A logical spot is directly below the fire and through light fixtures in the ceiling.

Salvage operations will normally begin on the floor below the fire if the attic or the fire floor is well involved with fire on arrival. Timing is important. A good wooden floor and a plastered ceiling will hold water for 10 to 15 minutes. This will normally provide adequate time to cover items directly under the fire if salvage work is started early.

In dwelling fires, the furniture should be collected in compact piles. (See Figure 2.37.) It is best that the piles be made away from light fixtures or other ceiling openings as this will provide more time to do the necessary work. Each group of furniture should be held to a size that can be conveniently covered by a single salvage cover. One cover will generally cover all the objectives in a normal-size room in a single-family dwelling.

It is possible that once the pile is completed and covered that water may get on the floor and find its way to the covered furniture. This could result in a great deal of damage to the legs of some expensive furniture. One method of preventing this damage is to raise the furniture off the floor with some type of water-resistant material.

FIGURE 2.37 ◆ Collect furniture in a central pile and then cover it.

One possibility for accomplishing this is to use canned goods from the kitchen storage area. Another possibility is to ring the edge of the covering salvage cover with additional covers if they are available. Not many departments carry sawdust anymore, but those that do can use it to ring the covered pile.

All clothing should be removed from closets and placed in the pile prior to covering the furniture with a salvage cover. Additionally, pictures should be removed from the walls and costly draperies and curtains removed from their hangers. Anything of value lying around the room should, if possible, be placed in the furniture drawers. And last, any throw carpets on the floor should be rolled and placed on the top of the pile.

Particular care should be taken to protect fragile articles when preparing piles for covering. It is a good idea to protect these articles by covering them with clothing prior to covering the pile with a salvage cover. Articles of clothing placed between these valuables may also help protect them. The purpose of the salvage work could be rendered useless if valuables are broken during the process.

Most homes have wall-to-wall carpeting. The best way to protect the carpeting is to provide a method of diverting the water to the outside or capturing it in a catchall as it comes through the ceiling. Damp spots in the ceiling will provide a clue as to where the water is collecting overhead. There are two good methods of diverting the water to the outside: (a) make a drain using pike poles and a salvage cover or (2) make a chute using a ladder and a salvage cover.

To make a drain using a salvage cover and pike poles, lay the salvage cover out flat with the waterproof side up. Place a pike pole on each side of the cover and stick the end of each pike pole through a grommet. Tie it in place if the grommets are too small for the point of the pole. Fold the cover down until it clears the hooks and then roll each pike pole toward the center of the cover until the desired width is reach. Extend the handle end of the improvised drain over the windowsill and raise the hooked ends higher than the windowsill. Place them on an improvised stepladder or some other available support and secure them in place.

A ladder drain also can be quickly improvised. (See Figure 2.38.)

A ladder drain has an advantage in some instances because it can be longer than the length of a salvage cover. Place the end of the ladder over the windowsill and raise the other end higher than the windowsill and place it on an improvised stepladder

FIGURE 2.38 ◆ A ladder drain.

or some other suitable object. Lay a salvage cover in the accordion fold on the end of the ladder that is resting on the windowsill. Unfold the cover as it is spread upward on the ladder. If the cover is too short to reach the desired location, overlap the first cover approximately two feet with a second cover and continue to extend the cover up the ladder. Once the pike pole chute or the ladder chute has been completed, use a pike pole or another tool to poke a hole in the ceiling where the water is collecting. This will drain the water from the ceiling and allow it to flow to the outside of the building.

It is not always possible to set up a drain to divert the water to the outside because of the distance from where the water will come through the ceiling and a window. In this case it is best to make a catchall. (See Figure 2.39.) Catchalls can be made by

FIGURE 2.39 ◆ A catchall made from a salvage cover.

FIGURE 2.40 ◆ Volunteer firefighters using a salvage cover to make a water catchall.
(Courtesy of Eugene Mahoney)

stretching a salvage cover over chairs, benches, or other small pieces of furniture in such a manner that the cover will bag to the floor. This operation can be performed quickly and will provide for the containment of a considerable amount of waster. As with drains, a hole should be poked through the ceiling once the catchall is in place to allow the water to escape. The sooner the trapped water can be diverted to a drain or a catchall, the less it will spread out between floors.

In property other than residential, a priority system may have to be established for the order in which material is covered. (See Figure 2.40.) Early consideration should be given to computers and other sensitive electrical equipment that can be severely damaged by water. A general rule is that stock should be covered before machinery. The reason for this is that machinery is generally less susceptible to water damage. Little damage will occur to most machinery if it is wiped of all water as soon as possible. Of course, an exception to this is machinery that has to be completely dismantled and cleaned if it becomes wet.

The most valuable stock or that which is most susceptible to water damage should be covered first. Particular care should be given to stock in cardboard cartons. The cartons will generally collapse if they are wet, leaving the contents in one big pile. If cartons are stored on the floor and not skidded, it is best to lift them off the floor and place them on machinery, tables, shelves, boxes, and so on prior to covering. Water could get on the floor under the covered pile and collapse the entire pile that was covered if this is not done.

When covering large areas or extensive piles of stock it may be necessary to use more than one cover per pile. This requires that the covers be laid in such a manner as to provide an overlap. An overlap of two feet is generally sufficient to prevent water from getting to the stock. If a large amount of water is expected, it may be best to seal the overlap. This can be done by rolling together the edges of adjoining covers. Lay back the first cover approximately one foot. Place the second cover to where the first is laid back and roll the two together. If a third cover is required, repeat the process where the second and third covers meet.

Belt-driven machinery will occasionally present a problem in regard to covering. If possible, remove the belts. Cut the belts at the lacings if they cannot be removed. This will help solve the problem of covering and little damage will be done to the belts.

Many times vertical piping that extends from floor to floor will provide a pathway for water. The best method of alleviating this problem is to tie a salvage cover around the pipe and form a bag to catch the water.

It may not be possible to get everything on the floor below the fire covered prior to water coming through the ceiling at a fast-moving fire where a large amount of water is used. The officer responsible for salvage operations should personally supervise operations when faced with this problem. He or she should direct crews on what to cover and what to leave. It is best that firefighters work in pairs. Machinery and stock can be covered much quicker than if a single firefighter works alone. If sufficient personnel are available, a couple of firefighters should be detailed to bring additional covers to the needed area. This works better than having the firefighters who are doing the covering return to the apparatus for additional covers.

If it becomes necessary to bag the floor, it should be done after all covering has been completed. It will generally take a large number of covers to bag a floor in industrial and commercial buildings. The covers should be laid side by side and end to end with the edges overlapped and rolled to make a watertight seal. This will create a shallow basin that will hold a small amount of water. If more water than can be held by the covers is anticipated, then some means should be made for getting rid of the excess water. The simplest method is to use a shallow-draft electrical pump or some other mechanical device.

Stock is piled in rows with aisles between the rows in many large department stores, mercantile occupancies, and storage occupancies. Protection for this type of arrangement requires the combination of covering stock and bagging the floor. The floor should be bagged after the stock is covered. In these instances, the covers used for bagging should extend partway up over the covers used for covering. This will prevent water projecting from the aisles to the covered stock.

Occasionally it may be difficult to hold the water to the floor below the fire. In these cases it may be necessary to repeat the operation at the next lower floor. In some situations it may be necessary to cover all floors from the floor below the fire to the basement. Covering operations must stay ahead of the water flow to be successful. It is better practice to cover a floor unnecessarily than not to cover a floor that should have been covered.

WATER REMOVAL

It is possible that a building can become so overloaded with water that it will cause the collapse of floors or walls. This is most likely to happen at large fires where a number of heavy streams are projected into the building, at fires in sprinklered buildings where the fire has burned for some time undetected, from broken systems, or from broken water pipes in buildings. Every effort must be made to prevent this from happening.

It is generally necessary to remove the water quickly and successfully as soon as possible to prevent the floors from becoming overloaded. There are a number of methods that can be used to achieve this.

One method is to channel the water into permanently installed floor drains or scuppers. The water can be channeled by using salvage covers, or if sawdust is available it can be used. The tools best used for moving the water are squeegees and corn brooms. When using this method to remove water, screens should be placed over the drains or scuppers, if available, and should be checked periodically to make sure they are not clogged. The scuppers will normally divert the water directly to the outside of the building, and the floor drains will normally divert the water directly to the sewer system.

Stairways can be used effectively in multiple-story buildings for removing water. The task is simplified if the stairs are concrete with no covering. However, it they are wooden or carpeted it is best that they be covered with salvage covers prior to using them for water removal. Covers used for this operation should be opened up to half their width. It is best to start covering the stairway at the bottom step. Throw a cover from the bottom step toward the top step as far up as possible. Extend the cover over the banisters on both sides if the stairway is so equipped. The second cover should overlap the first by approximately two feet. Place the cover under the lip of the top step and secure it in place. The cover can be secured in place with a redwood lath if a bundle is carried on an apparatus. This operation should be repeated from floor to floor.

Salvage covers or sawdust should be used to control the flow of water when stairways are used. Channels should be made to divert the water to the stairway, and also to form dams on each landing to keep the water from being diverted to individual floors. (See Figure 2.41.)

Another way of clearing water from a floor is by the removal of toilets. (See Figure 2.42.) A three- or four-inch opening is provided directly to the sewer system once a toilet is removed. The valve at the bottom of the tank must be shut off if it is the tank type prior to removing the toilet. Salvage covers or sawdust can be used to channel the water to the opening provided by the removed toilet. If possible, some type of screen should be placed over the opening to prevent clogging.

Cast-iron or plastic sewer pipes run both vertically and horizontally through buildings. This piping can sometimes be used for water removal. It makes the task easier if cleanouts are available at floor level. If the cleanouts are above floor level and not usable, then it may be necessary to break the pipe. It may also be necessary to cut into a wall or floor to gain access to the pipe.

If there is no other practical means of removing water from a floor, then it may be necessary to cut a hole in the floor. The hole should be cut close to an outside window and be of sufficient size to handle a large amount of water. All stock and other materials located below the spot where the hole will be cut should be removed prior to performing this operation. A drain should be set up at this point that will divert the water to the outside of the building. A ladder and salvage cover can be used for this purpose.

FIGURE 2.41 ◆ Preparing a stairway chute.
(Courtesy of Chicago Fire Department)

FIGURE 2.42 ◆ Removal of a toilet provides a direct line to the sewer system.

Once the hole is cut, it is best to enlarge it from below so that the water will not be diverted to the area between floors.

It may be necessary to breach a wall in those instances where a large amount of water has collected on a floor and there are a limited number of methods for removing it. This operation can best be performed by selecting an outside window that has a windowsill relatively close to the floor. It is necessary to remove the glass and window frame. The wall below the windowsill can then be removed. This provides a large, clean opening directly to the outside.

Elevator shafts can also be used for water removal but should be low on the priority list. (See Figure 2.43.) If the shaft is to be used, the cage should be moved to a

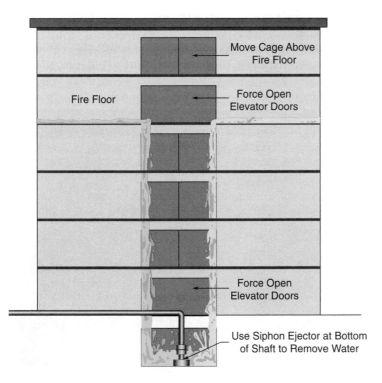

Move Cage Above Fire Floor

Fire Floor

Force Open Elevator Doors

Force Open Elevator Doors

Use Siphon Ejector at Bottom of Shaft to Remove Water

FIGURE 2.43 ◆ Elevator shafts can be used effectively for water removal.

floor above the upper floor from which water will be removed and the elevator taken out of service. It will be necessary to pry open the door at the upper floor and also at the first floor. If the elevator machinery is located at the bottom of the shaft, it should be covered before water enters the shaft. As early as possible, a siphon should be set up at the bottom of the shaft to remove the water from this location to the outside of the building. If the building has a basement, a check should be made to ensure that the water is not leaking out of the shaft and into the basement area.

If the amount of water to be removed is minimal, or water has soaked areas of carpeting, one of the fastest and easiest methods of removing the water is by the use of a water vacuum. Water vacuums are also capable of removing dirt and small debris from carpeting, tile, and other types of floor coverings. The water vacuum appliance consists of a nozzle and a tank that is normally worn on the back of a firefighter. Backpack-type tanks have a capacity of approximately five gallons. The tank can be emptied by simply pulling a lanyard that dumps the water out of the nozzle or through a separate drain hose.

FIRES IN SPRINKLERED BUILDINGS

Fires in sprinklered buildings almost always present a salvage problem. (See Figure 2.44.) Although the large majority of all fires in these buildings are extinguished or held in check by the sprinkler system, water continues to flow until the system is shut off.

Sprinkler system control, in such cases, is normally the responsibility of whichever company is given the responsibility for salvage operations. The Incident Commander must, however, keep in mind that the purpose of the sprinkler system is fire control and that a complete shutdown of the system should never be made until there is absolute assurance that the fire has been extinguished and that hose lines that have been laid are in position to ensure that there is no danger of fire spread. Prior to shutting down the system, salvage operations will consist of spreading covers and controlling the water being discharged by opened heads.

A check should be made of adjacent buildings if the fire is of sufficient size to cause an exposure problem. It is possible for the heat from a fire to enter an adjacent

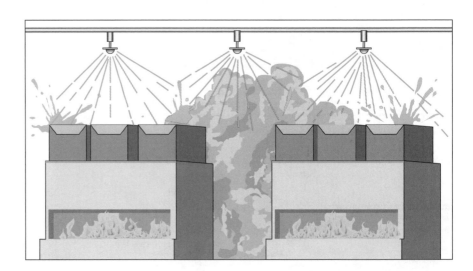

FIGURE 2.44 ◆ Fires in sprinklered buildings create a water control and removal problem.

building and activate one or more sprinkler heads. To prevent this, close the windows on the exposed side of the exposed building and open them on the unexposed side. This will result in restricting the entrance of additional heat into the building and reduce the internal heat caused by the fire. Shutoffs should be placed in the heads if the heads have been activated; however, the system should not be shut down until the exposure problem has been completely eliminated. It will be necessary to put this system back into operation prior to leaving the scene of the emergency.

Fire companies should carry devices for shutting off the flow to individual activated heads. These are normally inserted into the heads prior to the system being shut down. Most of the sprinkler shutoffs available are capable of completely shutting off the flow from a head, as long as the head is intact. Some departments use manufactured sprinkler shutoffs for this purpose; other departments use wooden wedges or similar devices. Occasionally, however, a head may be broken off in such a manner as to prohibit shutting off the flow of water at the head. Additionally with some flush-type heads it is difficult to stop the flow of water at the head by using available shutoff devices. A possible solution to this problem is to improvise a drain from the individual head. A 4- to 2½-inch reducing fitting can be attached to a section of 2½-inch hose and the hose extended to the outside of the building. The reducing fitting can then be secured over the ruptured head and held in place. This will divert the flow to the outside of the building. It will generally be necessary to hold the reducer in place until the system has been shut down and drained. (See Figure 2.45.)

The sprinkler system may be shut off once the officer conducting the salvage operations is assured that the fire is out or adequately controlled. However, prior to doing so he or she should notify the Incident Commander. The Incident Commander should also be notified when the system has been put back into operation.

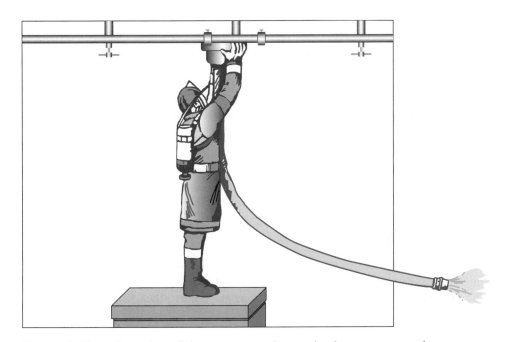

FIGURE 2.45 ◆ Sometimes it is necessary to improvise for water control.

To shut down the system completely, it is necessary that the main sprinkler shutoff valve be located. This value can normally be quickly located by noting the location of the fire department steamer connection or Fire Department Connection (FDC) on the outside of the building. The shutoff valve is usually inside of the building relatively close to the point where the steamer piping enters the building.

It is the practice of many departments for the first- or second-in engine company to lay into the sprinkler steamer connection whenever there is any possibility of a fire in a sprinklered building. Generally the pump operator will start flowing water to the system if there is any indication whatsoever that there is a fire in the building. The fire department steamer connection bypasses the main shutoff valve; consequently, water will flow into the system from a pumper even if the main shutoff valve is closed. It is therefore important that the officer responsible for salvage operations make sure that no pumper is pumping into the system prior to closing the main shutoff valve.

The sprinkler system can be shut down by first closing the main shutoff valve and then opening the drain valve. Most systems have a test valve at the far end of the system. This valve should be located and opened, and a firefighter should be stationed at the location. Opening this valve will allow air to enter the system and provide for a better drainage of the system. The ruptured heads can be removed and replaced once the system is drained.

In some communities the replacement of heads is the responsibility of the owner or sprinkler company. In other communities the fire department is responsible for replacing the heads and leaving the system in working order prior to leaving the premises. There normally will be an adequate supply of sprinkler heads available on the premises to replace the ruptured heads whenever the fire department is responsible for this function. If not, a sufficient number of heads is normally carried on the fire apparatus to make the necessary replacements. There are several points that should be remembered in replacing the heads:

1. Replacement heads should be of the proper temperature rating. The temperature rating of solder-type sprinklers is stamped on the solder link. For other types of heads it is stamped on one of the releasing parts. If the releasing parts cannot be found, it may be necessary to check the rating of one of the adjacent heads that has not been activated to determine the proper temperature rating for replacement.
2. Upright sprinkler heads should be installed with their framework parallel with the pipe so as to not disturb the spray pattern.
3. Upright and pendant-type heads should be replaced with heads of the same type.

The system is ready to be put back in service once the ruptured heads have been replaced. To put the system back in service, first close the drain valve. After the drain valve has been closed, open the test valve then the main shutoff valve. Water should be discharged from the test valve once the system is nearly filled. This valve should be closed when a steady flow of water is being discharged.

POSTSALVAGE OPERATIONS

After the main body of fire has been extinguished, salvage operations will consist of doing whatever is necessary to prevent additional water damage. The work includes removing water, as previously discussed; drying machinery, stock, furniture, and so on;

removing salvageable articles from the debris; and assisting the truck company in covering the roof, if necessary.

Some good salvage work can be done with the use of a sponge and chamois. These can be used to dry furniture and fixtures and thereby prevent the finish from being water spotted and ruined. They can also be used to dry machinery. Where possible, the machinery should be oiled after being dried to prevent rusting.

Some debris and water may have collected on top of some of the covers. The debris should be checked to ensure that it does not contain any hot embers. It may be necessary to wet down the debris prior to its removal. A hot ember can easily burn through a cover, possibly igniting the material being protected.

Care should be taken during the removal of a cover to ensure that the contents protected during the fire are not damaged by spilling water or debris on them. A cover should be folded from all sides in such a manner as to contain the water and debris inside. Care should also be taken so as not to damage any of the fragile articles that were protected. Once removed, the cover should be carried outside in such a manner that none of the debris or water trapped inside is spilled during transportation.

Stairway drains should be removed as soon as possible. This should be done so as to protect the covers from people walking over them. Floor runners can be placed over the stairs if it is necessary to protect finishing or carpeting from mud or debris.

Small holes in the roof are generally covered by the truck company if it becomes necessary to protect the interior of the building from rain or other weather elements. If the holes are large, or if the entire roof must be covered, it may be necessary to use salvage covers. Some departments carry rolls of plastic that can be used in lieu of covers. This operation is generally a joint effort of the company doing the salvage work and the truck company. If the entire roof is burned away, it may be necessary to bag the attic, the floor below the attic, or both. Again, if it is necessary to leave salvage covers on the premises, a signed acceptance by the owner for responsibility of the covers should be obtained.

◆ **SUMMARY**

Fire ground operations and an understanding of basic tactics and strategy will aid a firefighter when battling any situation. While maintaining a good attack on a fire, safety is always the most important factor for all involved. Safety of occupants and safety of fire service personnel must always be at the forefront of any Incident Commander's mind. There are rapid intervention teams that complete ventilation, provide rescue operations when needed, and must be able collectively to organize an assault on any given situation utilizing the skills and abilities of seasoned firefighters while not compromising safety for all on scene. Understanding the "how to" within the overarching scope of fire ground operations is where the rubber meets the road and where firefighters must be able to perform in any hazardous situation.

Review Questions

1. What is the term used to remember the seven tactical tasks used on the fire ground?
2. What are these seven tasks?
3. What are some of the questions that should be answered on arrival at a fire regarding rescue operations?
4. What is the general procedure for conducting rescue searches at a fire?
5. What is one of the primary contributing factors to loss of life when a large number of people are placed in danger at a fire situation?
6. What are some of the favorable results of proper ventilation?
7. What are the different types of internal and external exposures?
8. What are some of the natural openings available for roof ventilation?
9. Describe how and explain why salvage operations are conducted.
10. How are the windward and leeward sides of a building identified?
11. How can an elevator shaft be used for ventilation?
12. What method is used for setting up a trenching operation?
13. How is a positive pressure ventilation operation set up?
14. How should the windows be prepared when setting up a horizontal ventilation arrangement when no blowers are used?
15. How should the windows be prepared when smoke ejectors are used in a horizontal ventilation procedure?
16. What procedure should be used to break the glass when ladders are used to ventilate a window area?
17. What are some of the problems associated with ventilating a basement?
18. What are some of the various methods available for protecting a building and its contents from water damage?
19. What procedure should be used to protect the furniture from water damage at a residential fire?
20. What are some of the various methods available for removing water from the upper floors of multistory buildings?
21. What procedure should be used to place a sprinkler system back in operation when it has been shut down and drained at a fire?
22. Describe overhaul operations and list some examples of how overhaul is to be conducted.

The Model Incident Command System and Firefighters' Safety

3 CHAPTER

Key Terms

Objectives

The overall objective of this chapter is to introduce the reader to the Model Incident Command System and to firefighters' safety. Upon completing this chapter, the reader should be able to:

- Explain why and how the Incident Command System was developed.
- Define NIMS.
- Define common terms used in the explanation of the Model Incident Command System including, but not limited to, unity of command, unified command, and span of control.

♦ List the five major functional areas of the ICS.
♦ Explain what functions are performed in each of the major areas.
♦ Discuss the two major contributors to firefighter safety.
♦ Discuss the impact that civil disturbances have had on firefighter safety.
♦ Discuss terrorist operations.
♦ Explain the need for firefighters' physical fitness programs.

◆ INTRODUCTION

Understanding the basic principles of fire ground management and how this process correlates to firefighter safety is pertinent to all firefighters. With the ever-increasing changes to disaster response operations and with the need for all responders to ensure they know where they fit into the Incident Command System (ICS), the need for all personnel to be on the "same page" during a major event is not an option any longer; it is now the rule.

◆ FIRE GROUND MANAGEMENT

Incident Commander The individual in charge of an emergency.

incident action plan (IAP) The IAP is the basis for determining when and where resources will be assigned at the incident.

The management of a fire may be a relatively simple procedure involving a single company and a few firefighters. On the other hand, fire management can be extremely complicated and extensive when the incident is a major fire or an explosion extending over a wide area, using a large number of apparatus and personnel, and requiring the cooperation and coordination of a number of different agencies. However, a common thread extends throughout all incidents regardless of whether the incident is small and uncomplicated or extensive and complicated. This thread is the principle that someone must be in charge. This person is known as the **Incident Commander** (IC).

The Incident Commander is the individual at an emergency who is responsible for making the decisions that are essential for bringing the incident under control. He or she is also responsible for the results of the decisions made. The Incident Commander is the individual who develops the **incident action plan (IAP).** The IAP is the basis for determining when and where resources will be assigned at the incident.

It has often been said that decision making is the essence of management, and this is true. However, in making adequate decisions, the Incident Commander must have sufficient information on which to base the decisions. It is partly because of the need to ensure that this information is available and reliable that the Incident Command System was created.

◆ AN OVERVIEW OF THE INCIDENT COMMAND SYSTEM (ICS)

The ICS was developed as a result of a number of large wildland fires that occurred in southern California in 1970. These fires not only destroyed thousands of acres of wildland territory including hundreds of structures but also magnified the need for a system to provide for various agencies to work together in an effective and efficient manner in order to achieve a common goal. From this need a task force composed of a number of agencies was formed to develop a system that could be used not only for fires but also for emergencies such as tornadoes, riots, earthquakes, hazardous material incidents, floods, and other incidents caused by nature or human beings. It was important to all those involved that a system be developed that:

- Outlined procedures for controlling personnel, facilities, equipment, and communications.
- Provided for operations of an emergency from the time the emergency commenced until the need for management control and operations no longer existed.
- Could be utilized for any size of emergency, ranging from a small incident to a major one involving several agencies and the combining of a number of various resources.
- Contained procedures for expanding the organization to compensate for changing conditions.

◆ THE MODEL INCIDENT COMMAND SYSTEM

Out of the meetings of the combined agencies, came a system that has been referred to as the Incident Command System. Some municipalities in the United States have adopted and utilize the system in its entirety. Others use only a portion of the system based on the needs within their individual communities. Some departments merely use the concept as a portion of their standard operating procedures. Regardless of the extent and usage of the system by individual municipalities and related agencies, the overall concept has been widely accepted by the fire service.

The title of the material used in the concept also varies. Some agencies refer to it as the Incident Management System (IMS). Others use the more commonly accepted term Incident Command System (ICS). The National Fire Academy (NFA) uses the term ICS. The information in this text is based on the material that is taught at the NFA. Although the system was designed to be used at all types of human or natural caused incidents, for the purpose of this book the explanation and discussion of the system will be limited to fire emergencies.

Although most emergency situations are handled at a local level, when there is a major incident, other jurisdictions may need to get involved to help mitigate the event. State and federal government type of support may be called in to assist. National Incident Management System (NIMS) was developed so responders from different jurisdictions and disciplines can work together better to respond to natural disasters and emergencies, including but not limited to acts of terrorism. NIMS includes a unified approach to incident management; standard command and management structures; and emphasis on preparedness, mutual aid, and resource management. For additional information on NIMS, one can search the Internet as organizations throughout the United Sates adopt NIMS standards and protocols for emergency management. One website is www.fema.gov/emergency/nims/index.shtm.

There are a number of components included in the ICS that need to work together for the system to function effectively. These components are: common terminology, modular organization, integrated communications, unified command structure, consolidated action plans, manageable span of control, predesignated incident facilities, and comprehensive resource management.

COMMON TERMINOLOGY

The preceding list contains the components that are included within the ICS. For effective operation of the system, one of the first factors that should be understood is

FIGURE 3.1 ◆ Fire
command chart.

LOCAL UNIFIED COMMAND

FIRE	POLICE	WATER

MULTI-AGENCY UNIFIED COMMAND

U.S. FOREST SERVICE	CALIF. DEPT. OF FORESTRY	FEMA

the requirement of a common terminology. Following are some definitions that need to be understood to comprehend the system fully.

Unity of Command. **Unity of command** is the principle that an individual can work effectively for only one boss and that everyone within an organization has a designated supervisor to whom he or she reports. It is the principle that, except in an emergency, every order is processed through the proper chain of command. An officer should not be giving orders to an individual from a different company or unit than his or hers, nor should an individual accept orders from an officer directing another segment of the system. However, if an individual receives an order that is different from that issued by his or her superior officer, he or she should inform the officer giving the conflicting order and be guided by his or her reply. The unity of command principle does not mean, however, that a member should not advise members of another unit of the system of situational changes that may affect their safety. When an organization provides for the implementation of the unity of command principle at fire emergencies, it helps assure that an effective accountability system is in use and that provisions are made for the location and tracking of all personnel at the emergency.

unity of command The principle that an individual can work effectively for only one boss and that everyone within an organization has a designated supervisor to whom he or she reports.

Unified Command. Care must be taken not to confuse the term unified command with unity of command. **Unified command** refers to a situation in which agencies from several jurisdictions or in which several departments from the same jurisdiction work together for a common cause. In a unified command structure, individuals who have been designated by their jurisdictions or by a department within the same jurisdiction, work together to determine the objectives, strategies, and priorities to be used to control a large-scale emergency. Figure 3.1 shows examples of a local unified command that might be used for a large fire in a large city and a multi-agency unified command that might be used at a wildfire.

unified command A situation in which agencies from several jurisdictions or in which several departments from the same jurisdiction work together for a common cause.

Span of Control. **Span of control** is defined as the number of personnel that one supervisor can manage effectively. An effective span of control is particularly important at those incidents where firefighter safety and personnel accountability have top priority. In the ICS the span of control is considered to range from three to seven with five being the optimum amount. It is recommended that five be the maximum number in any fast-moving complex incident. In general, the span of control at a fire emergency can be influenced by:

span of control The number of personnel that one supervisor can manage effectively.

- ◆ The availability and capabilities of personnel or resources assigned to a supervisor.
- ◆ The number and difficulty of the assignments needed to control the incident effectively.
- ◆ The physical distance between subordinates being supervised.
- ◆ The availability of resources.

Examples of an acceptable and an unacceptable span of control follow in Figure 3.2.

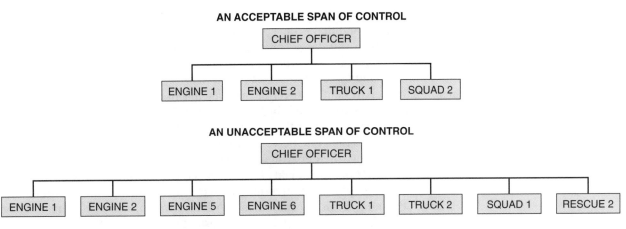

FIGURE 3.2 ◆ Span of control chart.

◆ ORGANIZATION AND OPERATIONS OF THE INCIDENT COMMAND SYSTEM

Figure 3.3 is an organization chart of the Model ICS. At first glance the organization appears to be overwhelming. However, there are several factors that an individual being exposed to the organization for the first time should take into account. First, there

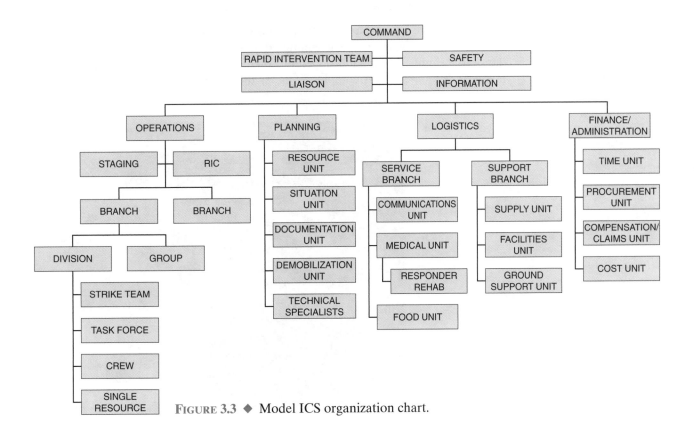

FIGURE 3.3 ◆ Model ICS organization chart.

is only one position on the entire chart that is always placed into operation at an emergency incident, and that is the command function. Second, with the exception of those departments in very large communities or those individuals working within a large organization involved in firefighting operations, such as the California Department of Forestry and the United States Forest Service, it is very unlikely that a firefighter or a fire department will ever be involved in a fire situation in which every box in the organization chart is filled. In fact, with most fire departments, the chance of working at an incident in which the full magnitude of the organization is placed into effect is probably less than one percent. However, it is still extremely important that every member of every fire department be completely familiar with the organization and understand all of the functional areas contained in the organization.

The model ICS system has five major functional areas. These are: Command, Operations, Planning, Logistics, and Finance/Administration. When referring to all or a portion of the organization chart, it very important to remember that the individual blocks in the chart are functions. They are not positions that must be staffed. In other words, this is a functional chart, not a position chart. This point will be stressed throughout the explanation of the chart.

Another point that should be kept in mind is that the functional organization as shown in the chart does not manifest itself suddenly at an emergency, regardless of the size and complexity of the emergency. Generally it starts small with nothing more than an Incident Commander and a few companies. It grows in size as the demands and extent of the emergency increase and the need for more personnel, equipment, and functional support grows.

THE COMMAND FUNCTION

command function The command function refers to the Incident Commander or to the unified command, depending on the incident and the number of agencies involved.

The command function shown on the chart is the only function that is staffed at every emergency. (See Figure 3.4.) The **command function** refers to the Incident Commander or to the unified command, depending on the incident and the number of agencies involved. If an incident is completely under the control of a single fire department,

FIGURE 3.4 ◆ Keeping records at a structure fire command post. The IC is wearing the white helmet with the identification D3. *(Courtesy of Rick McClure, LAFD)*

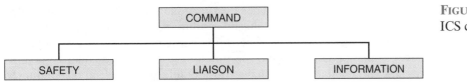

FIGURE 3.5 ◆ Functions of ICS chart.

then it refers to the Incident Commander. If multiple agencies are involved, it refers to the unified command. Regardless of who is in charge, the command function is responsible for:

- Determining the overall strategy to be used.
- Selecting the tactics to be used.
- Putting the plan into action.
- Developing the organization to be used and coordinating the resource activities.
- Providing for the safety of those working at the incident.
- Releasing information about the incident to the media.
- Coordinating with agencies not included in the Unified Command.

The individual or group staffing Command is responsible for all the functional activities within the organization. While some of the functions can be delegated to other major functional areas within the organization, certain functions are considered staff functions to Command and cannot be delegated. These functions are safety, liaison, and information. They are indicated on the chart in Figure 3.5.

Safety.   The overall responsibility for safety at a fire is inherent in the position of Incident Commander (IC). However, because of the complexity of operations at some fires, the IC generally delegates this task to a responsible individual known as the **safety officer.** (See Figure 3.6.) This individual's function occupies the box on the ICS organization chart identified by the word Safety.

safety officer   An officer at the scene of an emergency who has the authority to take corrective action at the scene, if possible, and to monitor the actions of those at the scene to ensure that safety procedures are followed.

FIGURE 3.6 ◆ A jacket similar to this is recommended for use by all safety officers. *(Courtesy of Las Vegas Fire & Rescue PIO)*

The officer occupying this position has the functional responsibility of identifying all present and potential hazardous or unsafe conditions, and ensuring that all safety procedures and practices are followed. This individual should be very knowledgeable in fire behavior, building construction, the potential for structure collapse, department safety rules and regulations, and have considerable experience in responding to fire incidents.

The first step that the safety officer should take is to communicate with Command to find out what specific concerns that position may have regarding the emergency. The next step is for the safety officer to make a survey of the incident scene to check for unsafe conditions. This walk-around is known as making a risk analysis of the fire scene. The risk analysis should determine where a high degree of risk exists in regards to operations or physical conditions. During this assessment, if the safety officer discovers an operation that involves an unsafe act that needs immediate attention, he or she has the responsibility and authority to correct the unsafe situation. However, the normal procedure is for the safety officer to confer with the individual commander responsible for the unsafe act and suggest methods for eliminating it. This suggestion is generally considered as an order.

liaison In the Incident Command System, the functional responsibility for identifying, contacting, assisting, and cooperating with outside agencies.

Liaison. **Liaison** is the functional responsibility for identifying, contacting, assisting, and cooperating with outside agencies. The individual occupying this functional area is known as the liaison officer. (See Figure 3.7.) He or she is the contact person for all agencies working at the incident. This position maximizes the effort to prevent duplication at the emergency. The appointment of a liaison officer reduces the load on the Incident Commander and assures the proper utilization of all the various agencies assisting in the operations at the fire.

Information. A large fire always attracts the attention of many reporters and representatives from interested agencies. These individuals are seeking as much information as possible about the size and extent of the fire, and where and how fast the fire is anticipated to travel. Because of the scope of the operations, the Incident

FIGURE 3.7 ◆ The incident Liaison Officer keeps records for the purpose of assisting outside agencies.
(Courtesy of Rick McClure, LAFD)

FIGURE 3.8 ◆ A special jacket worn by the public information officer helps identify him to those needing assistance or information.
(Courtesy of Las Vegas Fire & Rescue PIO)

Commander is not in a position to answer the questions. He or she will delegate this functional responsibility to the information officer. (See Figure 3.8.)

The **information officer** is responsible for the production and release of all information about the incident to the media and other appropriate agencies. (See Figure 3.9.) He or she provides an interface with the media. However, this officer is also responsible to be sure any sensitive information is first approved by Command prior to its release to the media.

In addition to the staff functions reporting to Command, Operations, Planning, Logistics, and Finance/Administration also report to the Command position. Each of these major subdivisions of the organization is examined individually. (See Figure 3.10.)

information officer   The individual responsible for the production and release of all information about the incident to the media and other appropriate agencies.

FIGURE 3.9 ◆ A public information officer providing the press with information on the fire. This takes a big burden off the Incident Commander.
(Courtesy of Las Vegas Fire & Rescue PIO)

FIGURE 3.10 ◆
Subdivisions of ICS chart.

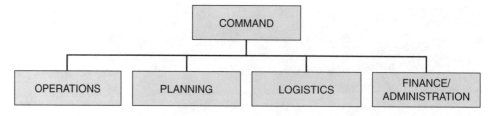

THE OPERATIONS FUNCTION

operations function The objective of the operations function is to provide a system for dividing up the incident into more manageable sectors or areas.

The objective of the **operations function** is to provide a system for dividing up the incident into more manageable sectors or areas. This relieves the Incident Commander or Command of the increasing burden created by the development of the size and complexity of the emergency. Such division can be by geographical area or by functional sectoring, depending on the type and extent of the emergency. The model chart shows this division of management responsibilities by branches, divisions, and groups. The divisions are made whenever it is apparent that the span of control at any level is becoming excessive.

When the emergency is divided up geographically, it places a well-defined area of the emergency under the control of an individual who has complete authority and the responsibility for controlling all operations within the area as if it were an individual emergency. The Model Incident Command System refers to these geographical areas as divisions. The division might be identified for communication and control purposes by number or letter. The individual in command of a division is referred to as the division commander.

At large structure fires, the division of the fire is by location inside or outside of the structure. Each section of the division may be identified numerically or alphabetically. If identified numerically, the front of the building would be number 1, the left side of the building from the front would be number 2, the rear number 3, and the right side number 4. (See Figure 3.11.)

If identified alphabetically, the front of the building would be A, the left side B, the rear C, and the right side D.

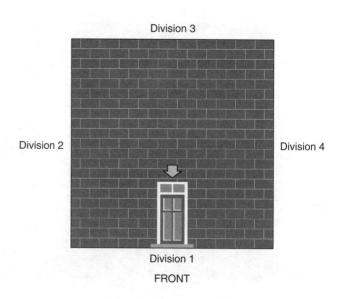

FIGURE 3.11 ◆ Sides of a
structure (numerically).

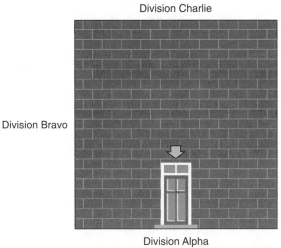

Division Charlie

Division Bravo

Division Delta

Division Alpha

FRONT

FIGURE 3.12 ◆
Sides of a structure
(alphabetically).

To avoid confusion, some departments use the standard terms of Alpha, Bravo, Charlie, and Delta for all communications. (See Figure 3.12.)

Exposures can also be identified in the same manner with radio transmission identifying Exposure Charlie, Exposure 2, and so on.

In multistory buildings, for simplicity the divisions are identified by floor. For example, the ninth floor is Division 9, and the twelfth floor is Division 12. Below-ground-level floors are identified as Basement 1, Basement 2, and so on. (See Figure 3.13.)

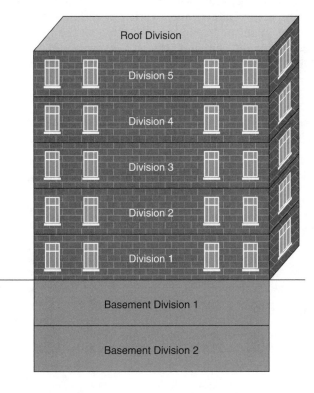

Roof Division

Division 5

Division 4

Division 3

Division 2

Division 1

Basement Division 1

Basement Division 2

FIGURE 3.13 ◆
Multistory buildings and
divisions.

FIGURE 3.14 ◆ Command function under ICS chart.

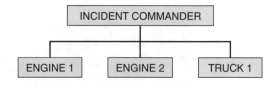

The individual assigned to the operations function is known as the operations officer. When the Incident Commander appoints an individual to occupy this position, he or she discusses with the individual the strategy and tactics that have been approved and are in place. The IC also discusses his or her expectations with the operations officer for the progress of the emergency and any anticipated problems or immediate concerns. Once appointed, the operations officer has the responsibility for the management of all operations that are directly connected with the primary mission of the organization. If the incident is a fire incident, the operations officer is responsible for all activities directly connected to the extinguishment and control of the fire. Management of operations is conducted in accordance with the strategic goals developed by Command. The operations officer also has the responsibility for allocating and assigning the resources necessary for the control of the incident.

The operations function at an incident may be relatively simple or extremely complex. For example, in a typical city the initial response to a single-family structure fire may be two engine companies, a truck company, and a Chief Officer. If a working fire is encountered, the chief officer would be the Incident Commander and have full responsibility for the command function. He or she would not need to assign someone to this position. In this incident, the command function would appear as in Figure 3.14.

However, the incident may be a large-scale wildfire involving numerous strike teams, task forces, individual units, major staffing, and a variety of different agencies. At such an emergency, operations might be expanded to include all the functional areas that are part of the Model Incident Command System, as shown in Figure 3.15.

It should be noted that, with the exception of staging, all the functional boxes on the organization chart in Figure 3.15 are involved directly in tne extinguishment and control of the fire. Staging, however, is a staff function reporting directly to the operations officer.

staging The staging area is a location to which incoming units report and can be rapidly deployed.

Staging. While **staging** is introduced in the organization for large-scale emergencies, it is often used at smaller fires in individual communities or in areas protected by a volunteer department. (See Figure 3.16.) Staging is responsible for the coordination, support, and distribution of all incoming resources. It is a location within the incident area that is used for temporarily locating resources that are available for immediate use. Staging provides the dual benefits of establishing accountability and eliminating freelancing.

freelancing The problem of company commanders committing the company to a certain task without orders from the Incident Commander.

Freelancing is the problem of company commanders committing the company to a certain task without orders from the Incident Commander. Many times the Incident Commander has to call for or use another unit at a location where he or she had planned to use the freelancing company.

The staging officer at a large-scale emergency is given the responsibility for a much larger operation. He or she is responsible for all activities that occur within the staging area. The staging officer does not need to be a chief officer. At large-scale emergencies it is best, however, that this officer at least hold the rank of company

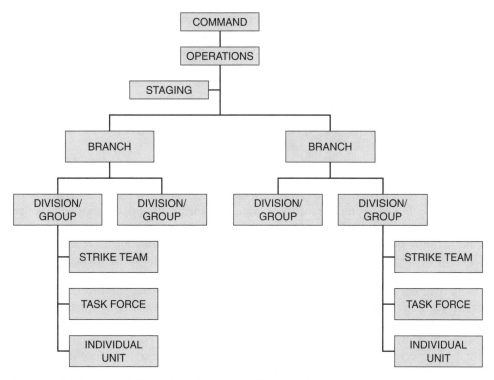

FIGURE 3.15 ◆ Model Incident Command System chart.

officer. All units responding to the emergency will report automatically to the staging officer and await orders. It is good practice to rotate companies from the fire area to the staging areas to give tired firefighters a rest. The staging officer will request food, fuel, sanitation, and so on from the logistics section when needed.

Staging for the initial response units of a community can be included in the department's standard operating procedures (SOPs). For example, the SOP may state that only the first arriving engine company and the chief officer should proceed

FIGURE 3.16 ◆ A large collection of apparatus in the staging area waiting for an assignment.
(Courtesy of Rick McClure, LAFD)

directly to the reported address of the emergency. Second and later arriving companies should hold at the last available hydrant or, if in an area lacking hydrants, at least a block away headed toward the reported fire area until instructed by the Incident Commander to do otherwise.

The staging for second or greater alarm fires or mutual-aid companies is different. For these situations, a staging area should be established and its location announced. Flat areas such as school yards are ideal for this purpose. Any resources beyond the initial alarm assignment should report directly to the staging area and await instructions.

rapid intervention team A standby team that is available for the rescuing of firefighters who become lost, incapacitated, or trapped in a fire due to a backdraft, a flashover, a collapse, an SCBA malfunction, an injury, or some other similar event.

Rapid Intervention Team (RIT). The **rapid intervention team** (sometimes referred to as RIT, RIC, or rapid intervention crew) is a standby team that is available for the rescuing of firefighters who become lost, incapacitated, or trapped in a fire due to a backdraft, a flashover, a collapse, an SCBA malfunction, an injury, or some other similar event. A RIT may be assigned as a staff function to Command, to Operations, or to both, depending on the anticipated needs at the emergency. (See Figure 3.17.)

It is important that that the RIT be set up early during the fire because the events leading to the need to rescue a firefighter usually come in the early stages of the fire. A good procedure is to assign the entire complement of a truck, engine, or rescue company to the RIT. If that type of assignment is not possible due to a limited number of personnel at the fire, the team can consist of two or three firefighters. The positioning of the RIT can be flexible, but in a multicompany operation it is best to locate the unit near the command post. This gives the person in charge of the unit the opportunity to review the situation sheets that track the location of all units at the fire.

At least one RIT should be established at all fires of any size with the exception that RITs are not generally used at wildland and brush fires.

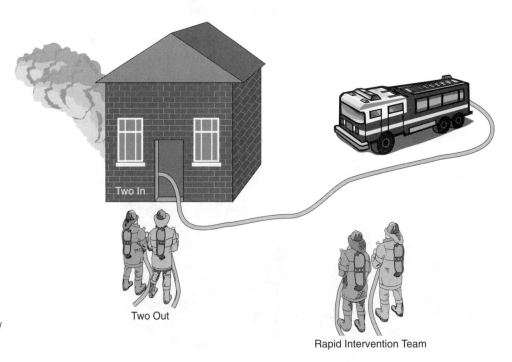

FIGURE 3.17 ◆ Two-in/two-out concept.

Branches. **Branches** are portions of an incident used for management purposes. Branches are the Incident Commander's or operations chief's primary method for correcting a span-of-control problem that had been caused by the creation of too many divisions and/or groups. At wildland and widespread brush fires, branches may be assigned by geographical area or by function. Branches are not normally used at large structure fires such as those in high-rise buildings as a portion of the firefighting forces; however, a branch may be assigned to perform a function such as rescue, salvage, and so on. The maximum number of branches that should be assigned to any individual at a fire is five.

branches Branches are portions of an incident used for management purposes.

Divisions. As shown in the organization chart, the geographical area of a large-scale emergency can be further divided into areas designated as divisions. On the chart, a **division** is an organizational level responsible to the branch chief for operations within the assigned geographical area. The chief officer assigned to this position is designated as a division commander.

division A portion of the command function that can be allocated by geographical area or by functional sectoring depending on the type and extent of the emergency.

A division may be a floor or one side of a building at a structure fire or an individual geographical area responsible directly to Operations at a brush or wildland fire. This type of organization is not identified on the Model Incident Command System chart; however, it was examined earlier under the operations function.

Groups. **Groups** are units of personnel generally assigned to a function such as ventilation, search and rescue, salvage, and so on. Groups are responsible for the entire job to which they are assigned and therefore frequently will work across geographical areas assigned to a division. When working across geographical areas, a group always should consult with the division supervisor prior to commencing operations.

groups Groups are units of personnel generally assigned to a function such as ventilation, search and rescue, salvage, and so on.

Task Force. A **task force** is a combination of any type of resources assembled to carry out a specific assignment. A task force should not exceed five units. The individual in charge of the task force is identified as a task force leader. A common formation of a task force for a large-scale emergency consists of two engine companies and one or two truck companies under the command of a chief officer.

task force A combination of resources assembled to carry out a specific assignment.

Some departments dispatch task forces as a portion of their initial response assignments and also to greater alarms. An example is Los Angeles City. The Los Angeles task forces consist of two engine companies and a truck company. The commanding officer of one of the companies is designed as the task force leader. All communications from the dispatch office are directed to the task force leader. When additional companies are needed at a greater alarm fire, it is much easier and more efficient for the Incident Commander to call for one or two more task forces than it is to place a call for two to four more engine companies and two or more truck companies.

Strike Team. A **strike team** is a combination of five identical units under the command of an individual designated as the strike team leader. The strike team leader reports to the Incident Commander or operations officer for an assignment. A common arrangement for a strike team is five fully equipped and staffed pumpers with a chief officer as the strike team leader. (See Figure 3.18.)

strike team A combination of five identical units under the command of an individual designated as the strike team leader.

The most prevalent use of strike teams is at wildland fires. Sometimes as many as 200 engine companies are used at these emergencies. The use of strike teams reduces the amount of radio communications by limiting all exchanges between dispatch and the strike team leaders rather than trying to communicate with 200 individual companies.

FIGURE 3.18 ◆ A strike team of five engine companies with a chief officer as the strike team leader waiting for an assignment.
(Courtesy of Rick McClure, LAFD)

Strike teams and task forces are often used by division and group supervisors to correct a span-of-control problem.

PLANNING

Planning is responsible for a number of functions that are important to the Incident Commander. One of the functions for which Planning is responsible is the reviewing and evaluation of the organization structure. Information regarding the organizational structure is maintained on an organization chart that provides the Incident Commander the opportunity to see immediately which of his functional responsibilities have been delegated and which he or she has maintained. It also gives the Incident Commander the information needed to make additional assignments or adjustments to operations as needed.

Other areas for which the planning officer has responsibility are:

- Interacting with any technical specialists assigned to the emergency.
- Predicting the need for any additional resources.
- Keeping records of incident scenes.
- Developing a demobilization plan.

Planning is also responsible for maintaining the resource and situation sheets. These sheets contain information that should be shown on a rough drawing or a plot plan of the fire area and any exposures. This drawing keeps the Incident Commander apprised of the distribution of resources and indicates where additional resources may be needed.

The planning function portion of the Model Incident Command System organization chart is repeated in Figure 3.19. The individual assigned to Planning is referred to as the planning chief. The individual assigned to this position is normally a senior chief officer.

It should be noted that five units are included within the responsibility of the Planning Chief. They are the resource unit, situation unit, documentation unit, demobilization unit, and technical specialists.

resource unit This unit is responsible for tracking all of the resources that have been requested, dispatched, or are on the scene at an incident.

Resource Unit. The **resource unit** is also referred to as the resources status unit (RESTAT). This unit is responsible for tracking all of the resources that have been requested, dispatched, or are on the scene at an incident.

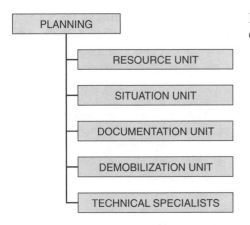

FIGURE 3.19 ◆ The planning function chart.

Depending on the type of fire and the part of the country where the fire occurs, the resources will vary considerably. Engine companies are almost always part of the resources at the fire. Truck companies and other companies such as rescue companies may also be included. There is considerable variation in the number of personnel assigned to a company. The unit may have only one firefighter or may have an officer and five firefighters. Where the company can be assigned at the fire will depend on both the staffing and the capabilities of the unit.

In addition to companies and personnel, a large number of support items may be required at the fire. These needs include water, foam, medical supplies, air for breathing apparatus, and specialized equipment.

In addition to the resource unit, at complex incidents, operations, divisions, and groups will also be responsible for tracking resources assigned to them.

Situation Unit. The **situation unit** is also referred to as the situation status unit (SITSTAT). This unit is responsible for the collection, tracking, and displacing of all information relative to the status of the incident. (See Figure 3.20.) The unit prepares situation displays and summaries, and develops maps and projections.

situation unit This unit is responsible for the collection, tracking, and displacing of all information relative to the status of the incident.

FIGURE 3.20 ◆ The situation unit keeps a record of the entire incident situation. *(Courtesy of Rick McClure, LAFD)*

Documentation Unit (DOC). This unit is responsible for all records and protection of all documents relevant to the incident. The unit documents the incident action plan, maintains all incident-related documentation, and provides duplication services. The unit's responsibility includes reports such as incident reports, injury reports, and claims for overtime compensation.

demobilization unit The demobilization unit is responsible for the orderly, safe, and efficient stand-down of an incident.

Demobilization Unit (DEMOB). The **demobilization unit** is responsible for the orderly, safe, and efficient stand-down of an incident. Costs may be involved in the stand-down of the emergency as a large number of varied units may be involved. When personnel are no longer required at the incident, this unit ensures that an orderly, safe, and cost-effective movement of personnel is made.

Technical Specialists. The technical specialists unit is used primarily at situations other than fire incidents. An example is some hazardous materials incidents where the advice of technical specialists is sought.

LOGISTICS

logistics The Logistics section is responsible for ordering the personnel, equipment, and resources required to control the incident and support the responding personnel.

Logistics is responsible for ordering the personnel, equipment, and resources required to control the incident and support the responding personnel. At small incidents these functions are the responsibility of the individual assigned to the unit. At larger incidents, the responsibilities are shared in accordance to the organization chart. The logistics portion of the organization chart is repeated in Figure 3.21.

The demands placed on Logistics at a large-scale emergency such as wildland fires can be extremely demanding and extensive. These fires normally extend over several days and operations may proceed for a week or more. This requires that large camps be set up with facilities to provide for sleeping, eating, and bathing. Additionally, vehicles will be needed to provide for filling air bottles, refueling apparatus, and providing water for both ground and air operations.

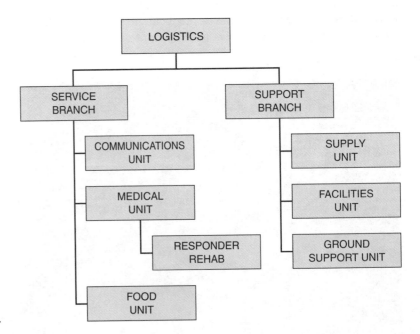

FIGURE 3.21 ◆ Logistics function chart.

One of the first operations of Logistics is the setting up of a base for operating. The base setup should be located a safe distance from the emergency. The base serves an entirely different function than does the staging function. Whereas the staging area is a location to which incoming units report and can be rapidly deployed, the base is an area where units will not be readily deployable.

It should be noted that, at major incidents, logistics is divided into two branches: the service branch and the support branch. Each of these divisions is further divided into units.

Service Branch.   The service branch is responsible for all the service activities involved in the incident. These activities are handled by a communications unit, a medical unit, and a food unit. The individual assigned to service branch is known as the service branch director.

Communications Unit.   The communications unit is responsible for planning for the effective use of all the communication equipment used at the incident, the distribution of the equipment to personnel, the supervision of the communication network, and the maintenance and repair of communication equipment. This unit also manages the incident communication center.

The use of cellular phones at emergencies is increasing. A cellular phone will allow a field unit to communicate with dispatch through a relatively unmonitored line. It allows the Incident Commander to communicate with a number of different units without interfering with radio communications. Providing that transmissions are readable, cellular phones can be used by two separate rescue teams to communicate with each other without the necessity of tying up the open communication channel.

Medical Unit.   The **medical unit** is responsible for the emergency treatment and transportation of any individual injured or taken sick at the incident. Although the primary objective of this unit is to treat those personnel assigned duties at the emergency, it also may be used to assist civilians on the scene.

Responder Rehab.   Under the control of the medical unit is the **responder rehab section.** This section is responsible for the rehabilitation of firefighters. When a firefighter is sent to rehab, the emergency medical services (EMS) will evaluate him or her and take vital signs. If the individual has an elevated blood pressure or other negative vital signs, the EMS will monitor the signs to see if they return to normal. Regardless of whether the firefighter wants to return to duty, he or she is not permitted to do so until it is approved by the EMS.

Food Unit.   The **food unit** is responsible for feeding those at the emergency. This unit is particularly valuable at those incidents where operations extend over several days. Firefighting units at the emergency are rotated between the fire line and the staging area or base where they have the opportunity to rest. The food unit is responsible for determining the food and water requirements at the incident. It may also prepare menus and food at the emergency, provide the food through catering services, or use a combination of the two methods.

Support Branch.   The **support branch** is a staff function responsible for the development and implementation of logistics plans designed to support the incident's action plans. The individual in charge of this branch is referred to as the support

medical unit   The medical unit is responsible for the emergency treatment and transportation of any individual injured or taken sick at the emergency.

responder rehab section   The section is responsible for the rehabilitation of firefighters.

food unit   The food unit is responsible for feeding those at the emergency.

support branch   The support branch is a staff function responsible for the development and implementation of logistics plans designed to support the incident's action plans.

branch director. The director's responsibilities are shared by a supply unit, a facilities unit, and a ground support unit.

Supply Unit. The **supply unit** is responsible for the ordering of all personnel, equipment, and supplies required at the incident. It receives and stores all supplies and provides the services for nonexpendable supplies and equipment.

Facilities Unit. The **facilities unit** is responsible for providing the layout and activation of those fixed facilities required at the incident. Such facilities might include a command post, sanitation facilities, a base camp, and/or staging areas for supplies. The unit is also responsible for providing managers for the incident base and camp.

Ground Support Unit. The ground support unit is responsible for the refueling of vehicles, the transportation of personnel and supplies, and the maintenance and repair of vehicles and equipment. The security support for the facilities is also supplied by this unit.

FINANCE/ADMINISTRATION

The Finance/Administration officer position can be set up at a large-scale incident such as an extended wildfire where it is evident that a major cost recovery will be involved. If possible, it is a good idea to staff this function with a finance-oriented individual such as a city or county financial officer. This section is responsible for monitoring incident related costs and administering the necessary procurement contracts. The finance/administration unit is divided into the time unit, the procurement unit, the compensations/claims unit, and the cost unit. (See Figure 3.22.)

Time Unit. The **time unit** is responsible for tracking and recording the on- and off-duty time for all personnel working at the incident.

Procurement Unit. The **procurement unit** is responsible for all financial matters pertaining to vendor contracts. The unit processes all the paperwork associated with equipment rental and supply contracts.

Compensation/Claims Unit. The **compensation/claims unit** is responsible for the financial concerns over serious injuries and deaths that occur as a result of operations at the incident. The unit is responsible for seeing that all documentation related to

supply unit The supply unit is responsible for the ordering of all personnel, equipment, and supplies required at the incident.

facilities unit The facilities unit is responsible for providing the layout and activation of those fixed facilities required at the incident.

time unit The time unit is responsible for tracking and recording the on- and off-duty time for all personnel working at the incident.

procurement unit The procurement unit is responsible for all financial matters pertaining to vendor contracts.

compensation/claims unit The compensation/claims unit is responsible for the financial concerns over serious injuries and deaths that occur as a result of operations at the incident.

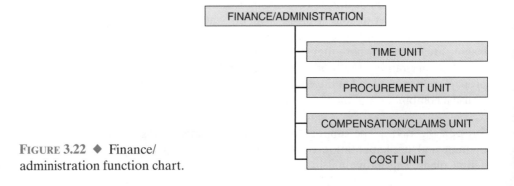

FIGURE 3.22 ◆ Finance/administration function chart.

workers' compensation is correctly recorded and completed. The unit maintains all the files related to injuries or illnesses associated with the incident.

Claims Unit.   "Claims" is responsible for investigating all claims involving damaged or lost property associated with or involved in the incident.

Cost Unit.   The **cost unit** is responsible for tracking cost data, analyzing the data, making cost estimates, and recommending cost-saving measures.

cost unit   The cost unit is responsible for tracking cost data, analyzing the data, making cost estimates, and recommending cost-saving measures.

◆   IMPLEMENTATION OF THE ICS

The ICS is implemented by the first officer arriving at the scene of a fire. This individual will size up the situation in order to develop a plan of operation. During the size up, he or she will evaluate various factors affecting functions performed at a fire such as:

1. Is anyone in immediate danger who needs to be rescued?
2. What is the present situation regarding the fire?
3. Where is the fire likely to go?
4. What resources will be dispatched on the initial assignment?
5. What resources will be need to stop the progress of the fire and eventually extinguish it?

If there are no life-threatening situations that require immediate attention, the Incident Commander should develop an incident action plan. This plan should include such factors as:

1. Safety to personnel.
2. Rescuing or evacuating occupants.
3. Stopping the spread of the fire and protecting exposures.
4. Reducing the loss to a minimum.
5. Overhauling and securing the occupancy.

As soon as the size-up has been made and the determination made as to what additional help might be needed, where arriving units should be placed, and so forth, a report should be made to the dispatch officer. Such a report should include:

1. Address of the fire.
2. Size and type of building.
3. The fire conditions.
4. The location of exposures.
5. What action is being taken by the first company on the scene.
6. Instructions for other arriving units.
7. Request for any additional help that might be needed.
8. The announcement of the location of the command post and the staging area.

The original report to the dispatch officer may go something like this:

> Dispatch from Engine 2. At 733 North Beach Street we have a 2,000 square foot single-family dwelling with one room on the Charlie side well-involved with fire. No exposures exist. Engine 2 is taking a ½-inch line inside to attack the fire. Have the second engine company lay us a supply line. Truck 2 ventilate the roof. The command post will be in front of the building.

◆ PASSING THE COMMAND

In most cases a chief officer will arrive on the scene a short time after the arrival of the first company. The Incident Commander will inform the chief of the action that has been taken and plans that have been placed into operation. When the chief officer is ready to take command, he or she will so inform the dispatch officer. The message may go like this:

> Dispatch from Battalion 3. On the scene at 733 North Beach Street. I am taking command of this fire.

At greater alarm fires, more than one chief officer will most likely respond to the fire. Some of these may be of a higher rank than the present Incident Commander. This does not mean that the arrival of a superior officer automatically means that there will be a change of command. The superior officer may have responded to the emergency because the standard operation procedures of the department required it, or he or she may have responded out of curiosity. Regardless, if he or she does not intend to take command, the dispatch office should be so informed. The radio message might go something like this:

> Dispatch from Division 1. I am on the scene at 733 North Beach Street. Battalion 3 will continue as the Incident Commander.

At times, the first arriving officer on the scene may be in charge of a company with a limited number of personnel. The action he or she decides to take may require that he or she remain as an active member of the company. If this is the case, the officer has the option of passing the command of the fire to the second arriving officer. However, passing of the command cannot take place unless the second arriving officer is actually on the scene. The dispatch office should be notified of the action taken. The report may go something like this:

> Dispatch from Engine 1. At 513 West 9th Street. We have a three-story brick building with smoke and fire showing on the second floor. There are people showing in several windows on the third floor. Give me a second alarm assignment. I am taking my company inside to start rescue operations. Have Truck 2 raise a ladder to the third floor. Engine 4 has just arrived at the fire. I am passing command of the fire to Engine 4.

◆ EXPANDING THE ORGANIZATION

As a fire grows in size, there is a greater requirement for additional resources and staff assistance. Although the general principle is that the ICS should be expanded only to the point where the fire can be handled safely and efficiently, there should never be a hesitation to order more personnel and equipment if the Incident Commander decides that it MIGHT be needed. As the incident grows in size, the position of Incident Commander may be changed several times. This requires that an efficient system be used for the tracking of resources. The tracking of resources is one of the most important functions of the ICS. This responsibility belongs to the resource unit.

◆ **TERMINATING THE INCIDENT**

Once the fire has been brought under control, the need for the size and complexity of the in-place organization diminishes. At this time there must be a plan in place for the releasing of personnel and resources that are no longer needed. The demobilization must take place methodically and efficiently. The responsibility for accomplishing this belongs to the demobilization unit.

◆ **FIREFIGHTER SAFETY AND SURVIVAL**

BACKGROUND

For many years, neither the fire service nor government agencies had given much thought to firefighter safety. In fact, as late as in the first half of the twentieth century firefighters were still proud of the amount of smoke they could endure and held their heads high when referred to as leather-lunged smoke eaters. To many, being a firefighter was a childhood dream come true.

Traditionally, firefighters rode the tailboard of apparatus or hung on to the sides of ladder apparatus while responding to emergencies. They waved to the citizens and small children as they responded proudly to alarms, and both the grownups and the children waved back.

While they were aware that riding the tailboard contained a number of risks, most firefighters enjoyed it. They considered that riding the tailboard provided a certain amount of romance and excitement. It was difficult to explain the thrill, the increased heart rate, and the flow of adrenaline that resulted when they turned their heads in the direction of the reported fire to see a large column of black smoke arising, indicating they had a working fire on their hands.

The hanging on to the crossbar on the back of a pumper continued for many years despite the fact that a number of firefighters had been killed due to falling off an apparatus while the apparatus was underway. However, this did not seem to bother either the officers or the firefighters. The issue of firefighter safety was seldom raised.

Firefighters generally arrived at fires wearing their uniform clothing, a turnout coat, and a helmet. Turnout pants and boots replaced the uniform clothes on night responses. This was the total amount of protective gear firefighters used in fighting fires. Not only was this the limit of their protective gear, but also in many departments rookie firefighters had to purchase their own equipment when hired. Any damaged equipment that had to be replaced over the years was also at the firefighters' expense.

For many years both firefighters and officers fought the use of some of the equipment that is considered as standard today. Breathing apparatus was available to fire departments in the 1960s but was seldom used. Any member choosing to do so was considered a sissy. Some firefighters objected to the use of gloves as they complained that the gloves were cumbersome and restricted the effectiveness of operations while fighting fires.

Neither the firefighters nor the officers liked to have their ears covered while firefighting, as they said the ears were very sensitive to heat and provided them with an early warning when they were involved in situations that were too hazardous.

During the latter part of the twentieth century, a number of incidents occurred that resulted in the fire service, fire service organizations, and government agencies focusing on safety for firefighters. This focus on safety was a tremendous change in an organization based strongly on long tradition. The focus began with the issuance in 1987 of NFPA 1500. This standard is referred to as the *Standard on Fire Department Occupation Safety and Health Program.*

NFPA 1500 is recognized as the umbrella standard for a fire service occupational safety and health program. The purpose of an occupational safety and health program for the fire service is for firefighters to leave the profession in the same condition as when they entered it. The result is that today operations and equipment supporting firefighter safety are a major part of every fire department.

Two primary types of incidents that have had an impact on this change are civil disturbances and terrorism.

CIVIL DISTURBANCES

civil disturbance An intentional act of disobedient behavior by a group of people in violation of public policy or established laws or regulations that results in acts of violence directed at persons or property.

A **civil disturbance** is an intentional act of disobedient behavior by a group of people in violation of public policy or established laws or regulations that results in acts of violence directed at persons or property.

Today, civil disturbances that were so prevalent during the 1960 s have become a way of life for U.S. citizens. Fortunately, the lessons learned from the civil disturbances and riots of the 1960s have had a positive influence on the need to provide for the safety of firefighters working in a civil disturbance environment. The perils of evil visually demonstrated by participants in the civil disturbance activities also strongly indicated that there was a positive need to develop nontraditional methods of fighting fires when working in this type of environment.

One factor of importance that was recognized early when responding to a civil disturbance incident was the need to protect firefighters riding the tailboard and sides of fire apparatus. Figure 3.23 shows the type of open cab apparatus that was used by most fire departments prior to 1960. It is apparent to anyone who may choose to do so that the opportunity to do physical harm to the responding firefighters is readily available. Today, firefighters no longer ride the tailboard of a pumper or hang on to the sides of a ladder truck. Instead they ride in closed cabs, which provide protection from thrown objects and eliminate the possibility of a firefighter being seriously injured or killed by falling off the back of a moving vehicle. Figure 3.24 shows the type of closed cab apparatus that is almost standard on all departments today.

riots Riots can generally be defined as a spontaneous outburst of group violence characterized by extreme excitement mixed with rage.

Riots can generally be defined as a spontaneous outburst of group violence characterized by extreme excitement mixed with rage. The rage is normally directed at law enforcement personnel but may expand to acts of violence directed at fire and EMS personnel who have been dispatched for the purpose of bringing aid. During the August 1965 riot in Los Angeles, firefighters were intimidated with threats of harm; spat on; injured by bricks, bottles, and other items thrown at them, and in several instances wounded by bullets. The riot itself lasted for five days

FIGURE 3.23   ◆   This is a 1923 Seagrave. It was still in service in the 1950s. *(Courtesy of Frank Borden, LAFD Historical Society)*

and ultimately required the assistance of 13,000 National Guard troops to gain control.

What appears to be a peaceful demonstration can change rapidly into a violent disturbance by some small, unexpected incident. Because of this possibility, whenever a civil disturbance is taking place within their community, fire officials should be prepared to become involved with little or no warning.

FIGURE 3.24   ◆   Today, firefighters ride inside an enclosed cab that protects them from weather, thrown rocks and bottles, and the possibility of falling off the apparatus. *(Courtesy of Eugene Mahoney)*

Most of the major civil disturbances involving large-scale fire operations that have taken place in the United States have occurred in larger cities. However, many smaller cities have experienced a number of disturbances that could have developed into widespread violence. The cities most affected have been those containing colleges or universities and those having structures within their boundaries that house activities having a highly emotional and controversial basis. Although many fire officials believe that civil disturbances will never occur in their community, no harm can be done by examining the principles involved and developing some logical plan just in case they are wrong. Such an approach is highly recommended.

The Anatomy of a Riot. A riot may begin as a civil disturbance consisting of a group of people demonstrating for or against a subject or issue, as a civil disturbance created by a small group of angry people who have gathered to express their displeasure with their law enforcement agency, or as a result of a few unrelated incidents in a particular part of the community that eventually come together as a group with a common cause. Regardless of its initial purpose, any civil disturbance can develop into a riot situation if the crowd increases in size, the anger grows more intense, and the area of operation increases in size.

Angry mobs generally first express their unhappiness by throwing rocks and bottles at law enforcement and/or fire department units or personnel. Their actions will most likely include the destruction of property if the intensity of the crowd continues to grow. The primary tool used for the destruction of property is usually fire.

The first fires normally set by a violent mob are generally confined to vehicles or bonfires. It is not unusual for the crowd to pull drivers out of vehicles and set the vehicle on fire. The mob arsonists will most likely direct their attention to structures as the violence and size of the crowd increases.

The tool most often used by riot arsonists in igniting structures is the Molotov cocktail. A **Molotov cocktail** is a thin-skinned bottle filled with gasoline with a rag in the neck of the bottle that is used as a wick. The method used to start a structure fire is to break out a large window, light the rag that is used as a wick, and throw the bottle a long distance into the building. The bottle breaks, scattering the gasoline that is immediately ignited by the wick. This results in the entire area rapidly becoming well involved with fire.

As a result of these tactics, a small fire in a structure is seldom encountered when the fire department first arrives on the scene. A number of fires scattered over a large area with angry mobs roaming the streets results in a scenario that is uncommon in the day-to-day operations of the fire department. Firefighters are required to undertake the extinguishment of a large number of simultaneous large structural fires while under physical attack by hostile mobs.

Molotov cocktail A thin-skinned bottle filled with gasoline and equipped with a rag in the neck of the bottle that is used as a wick.

General Guidelines for Firefighters

1. The safety of firefighters should be the number-one consideration at all times.
2. With the exception of SCBA, firefighters should be equipped at all times with full personal protective equipment. If available, consideration should be given to the use of ballistic vests when conditions warrant it.
3. SCBA or gas masks should be worn or immediately available by personnel when responding into an area where tear gas or other chemical crowd control agents are being used or are anticipated to be used.

4. Task forces with law enforcement protection should be used for all responses in an involved area.

5. Every effort should be made to complete the task as quickly as possible when performing a firefighting operation. The crew should be made available immediately for reassignment when an assignment is completed.

6. All crew members should remain with their apparatus and be prepared to vacate the area immediately when not engaged in performing a task.

7. Crews should be deployed as a unit and maintain a high level of personnel accountability.

8. Every effort should be made to avoid responding into a dead-end street or a street blocked with only one way in and out.

Fire Suppression.   Firefighters routinely work in environments fraught with hazards and calculated risks. Standard operating procedures can be developed that provide for extinguishing a fire with a minimum of loss to life and property. However, when a civil disturbance exists with excited and aggressive crowds gathered in large numbers, unique operational challenges develop that present a level of threat to firefighters that is not present in normal day-to-day operations. Consequently, operational changes are required in order to provide for the safety of personnel. The primary goal is changed from confining a fire to a room, floor, or building of origin to one of stopping conflagrations. The tactics and procedures recommended herein are based on providing the maximum level of safety for personnel and preventing conflagrations.

1. In order to minimize the potential for a conflagration, exposure protection should be the primary consideration of the fire suppression tactics employed.

2. The second priority of the fire suppression tactics should be to knock down the fire as quickly as possible using deck guns mounted on the apparatus. Portable monitors should be used if deck guns are not available.

3. Operations should be conducted in a manner that will permit the quick withdrawal of personnel, apparatus, and equipment.

4. The water in the tank should be used for small fires rather than the laying of a supply line. Tanks shall be refilled after the apparatus has withdrawn from the area of operation.

5. When available, Class A foam should be used to enhance extinguishment and increase water penetration.

6. The pump panel of the attack engine should be oriented toward the fire so that firefighting personnel are kept within sight of one another and the apparatus is between the firefighting personnel and the crowd. If law enforcement officers are available to assist, an officer should be placed on the crowd side of the apparatus to provide protection for the firefighters.

7. A minimum of two personnel should remain with the pumping apparatus at all times. One of the members should be assigned to watch the crowd.

8. SCBA should be available to firefighters if needed; however, operations requiring its use should be kept to an absolute minimum.

9. Crew members should be kept together at all times and operate as close to the apparatus as possible. No crew member should be permitted to operate alone.

10. The length of hose lines should be kept to a minimum.

11. No salvage or overhaul operations should be performed.

12. No structures should be laddered.

13. No personnel should be permitted on any roof.
14. Hit-and-run tactics should be employed. The fire should be knocked down as quickly as possible, and the task force made available for another assignment.
15. No interior fire attacks should be made.

On Scene Scenario

On April 10, 2001, the Cincinnati, Ohio, fire department participated in a civil disturbance. During the disturbance

1. The rioters pulled motorists from vehicles and physically attacked them.
2. Cars were overturned and windows of businesses were shattered.
3. Task force operations were instituted early. The task forces consisted of two engine companies, two truck companies, and a chief officer.
4. All responses by the task forces were made with a police escort.
5. The "hit and run" type of fire attack was employed. A quick knockdown with a minimum amount of overhaul.
6. All fire companies were kept informed of areas where sniper shots were reported.
7. Fires that posed no exposure or life-threatening problem were allowed to burn themselves out.
8. Body armor to protect firefighters was used when available.
9. Apparatus was parked at emergencies in the position best providing for a rapid retreat.

TERRORISM

terrorism A type of act that is designed to kill innocent people or disrupt the lives of the survivors of those killed and create panic in a community or nation.

Terrorism is a type of violent act that is designed to intimidate or coerce a government or its people for the purpose of furtherance of a political or social objective. A secondary objective of a terrorist or a terrorist organization is to disrupt the lives of the survivors of those killed and to create fear or panic in a community or nation. Terrorism is an act that can be designed and executed by a single individual working alone or by an individual or group of individuals working as part of a terrorist organization. Terrorist acts are not confined to any particular country, nationality, or religion. Such acts can be committed by anyone, regardless of nationality or religion.

For many years, terrorism was confined to countries outside of the United States. Today terrorism or fear of terrorism is a part of the life pattern of all citizens in the country. It may be committed in any city, at any time. Terrorism in the United States may be thought of as being divided into two types—domestic attacks and attacks from organized foreign terrorist organizations. In most cases, an explosion initiates a terrorist act; however, fire departments and other responders should take the position that a secondary bomb or other device is always a possibility.

Domestic Terrorism. The history of domestic terrorist acts in the United States is not extensive but has left a lasting imprint on the American people. Most people think

Figure 3.25 ◆ April 19, 1995. The Alfred P. Murrah Federal Building in Oklahoma City after being bombed by domestic terrorists. Firefighters are shown performing rescue operations. *(Courtesy of AP Wide World Photos)*

that domestic terrorism is new to the United States; however, it goes back to 1920. On September 16, 1920, a TNT bomb was planted in an unattended horse-drawn wagon on Wall Street in New York City. The exploding bomb killed 35 people and injured hundreds more. It was believed that the bombing was done by Bolshevist or anarchist terrorists; however, the crime was never solved.

The second incident also occurred in New York City, this one on January 24, 1975. A bomb exploded in the historical Fraunces Tavern killing four and injuring more than 50. This time a Puerto Rican nationalist group claimed responsibility.

However, it wasn't until 9:03 A.M. on April 19, 1995, that the attention of the entire United States was grabbed. This time the attack was on the federal building in the heartland of the country—Oklahoma City. This act was committed by a hate-mongering ex-soldier who used a massive bomb inside a rental truck. His motivation for the attack was a strong hatred for the United States government. Not only were the lives of 168 people, including 19 children, taken but the lives of thousands more were violently affected. Over 220 buildings sustained damage. (See Figure 3.25.)

International Terrorism.   Although most international terrorist acts have occurred in countries other than the United States, the most drastic and damaging attacks occurred in the United States on September 11, 2001, with attacks on the Twin Towers of the World Trade Center and the Pentagon. No one doubted that these well-planned and well-executed attacks were carried out by a professional, organized foreign terrorist group. The planning and execution closely resembled the Japanese attack on Pearl Harbor in 1941. The 9/11 attacks not only killed 2,726 people, including 343 firefighters, but also changed the lifestyle of every American man, woman, and child to such a degree that the United States will never again be the same.

The Twin Towers. September 10, 2001.

FIGURE 3.26 ◆ The Twin Towers.
September 10, 2001.
*(Courtesy of District Chief Chris E.
Mickal, New Orleans Fire Department—
Photo Unit)*

The Twin Towers. September 11, 2001.

FIGURE 3.27 ◆ The North Tower has
been hit and is on fire. The second
plane, hijacked United Airlines Flight
175, is headed toward the South Tower.
(Courtesy of Corbis/Bettman)

The Twin Towers. September 11, 2001. *(Continued)*

FIGURE 3.28 ◆ The second plane hits the South Tower in a ball of flame, sending pieces of shattered glass and debris to the ground below. *(Courtesy of AP Wide World Photos)*

A look at Ground Zero

FIGURE 3.29 ◆ September 27, 2001. The remaining section of the World Trade Center is surrounded by a mountain of rubble following the September 11 terrorist attack. *(Courtesy of FEMA)*

(Continued)

A look at Ground Zero *(Continued)*

FIGURE 3.30 ◆ New York, New York. October 5, 2001.
Rescue workers continue their efforts at the World
Trade Center.
(Courtesy of FEMA)

FIGURE 3.31 ◆ Ground Zero. February
21, 2002.
*(Courtesy of District Chief Chris E.
Mickal, New Orleans Fire Department—
Photo Unit)*

With the exception of the attack on the World Trade Center, the terrorist acts both overseas and in the United States have not created any serious firefighting problems. Fire department participation in these attacks has consisted primarily of providing emergency medical care and assisting other agencies in rescue efforts. However, future attacks may change this pattern.

One type of attack by organized international organizations that has not yet manifested itself in the United States, but eventually will, is attacks by suicide bombers. Although it has nothing to do with operational procedures, it is a good idea that all firefighters have some idea of how these individuals who volunteer to kill themselves think.

It is extremely difficult for people who enjoy freedom and love life to understand why anyone would be willing to give up his or her life for a radical cause. However, it is also difficult for recruited suicide bombers to understand why anyone would not be willing to exchange the ultimate sacrifice for the benefits to be received.

Fire Department Responsibilities.   Regardless of whether they like it, it is imperative that fire departments prepare for the possibility of responding to a terrorist act. Fire administrators must accept the fact that a terrorist attack will come completely by surprise and that when it comes law enforcement officers, firefighters, and emergency medical teams will be the first responders. As demonstrated on 9/11, it is the first responders who are placed at the greatest risk. Unfortunately, at this time even the best prepared communities do not have either the equipment or the resources to cope with all types of terrorist attacks adequately.

It is recognized from the experience with terrorist attacks that have occurred throughout the world that an initial attack will most likely be initiated by an explosion. However, the greatest fear of those planning for or anticipating attacks is the realistic concern that future attacks may be initiated by a nuclear device, a chemical weapon, or a biological weapon. An attack utilizing any one of these devices could very likely cause more deaths and widespread panic than any previous attack, including 9/11. Of these three, a nuclear attack is most unlikely but not an impossibility. However, the use of a "dirty bomb" is a strong possibility. A **dirty bomb** is an explosive device that spreads radioactive materials over a specific area. The bomb will most likely kill a number of people in the initial blast and will develop radiation sickness in many more. The contamination in buildings will depend on how much radioactive material is used in the bomb. All fire departments should be prepared to operate in a radioactive environment.

dirty bomb   An explosive device that spreads radioactive materials over a specific area.

It is chemical weapons and biological weapons that throw the most fear into the general public and require extensive protective equipment for fire department members. **Chemical weapons** are designed to attack an individual's nervous system, eyes, skin, intestinal tract, and mucous membranes. The weapons generally contain either blister agents, nerve agents, or blood agents. Almost all chemical agents are heavier than air and like gasoline vapors will likely collect in the lowest levels available. This makes basements, subways, and other similar locations particularly hazardous.

chemical weapons   Weapons designed to attack an individual's nervous system, eyes, skin, intestinal tract and mucous membranes.

Biological weapons contain toxins that can enter the body through inhalation or skin absorption. They contain material such as anthrax, plague, and viruses. Once absorbed into the body, the result can be fatal.

biological weapons   Weapons that contain toxins that can enter the body through inhalation or skin absorpotion.

There are a number of actions a fire department can take in preparation for a potential terrorist attack.

1. Accept the fact that while most terrorist attacks take place in larger cities, they can occur in any city or community at any time.

2. Pre-attack planning should include identifying and prioritizing all sites likely to be targets. These include utilities, places of public assembly or any location such as shopping malls where a large number of people gather, historical locations, chemical plants or storage areas, government buildings, transportation facilities, and telecommunication facilities.
3. Plan on the fact that regular water supplies may not be available. Make plans for the use of auxiliary supply sources.
4. Be aware that the terrorist may make plans to slow down the response of emergency vehicles or perhaps prevent vehicles from ever reaching ground zero.
5. Review the present lines of communication with law enforcement agencies and military organizations if located within a reasonable distance. If the present lines of communication have flaws, do whatever is possible to correct the deficiencies.
6. Be aware that the initial response to a terrorist attack may be a trap. This is most likely to occur if the initial explosion is started by a timing device or by remote control. The overall plan of the terrorist is to bring all emergency forces—firefighters, emergency medical care units, and law enforcement units—to ground zero. When all units are on the scene, the terrorist sets off a second, and usually larger and more powerful, bomb, killing many of the first responders. A suicide bomber does not initiate this type of attack.
7. Review the standard operating procedures for responding to hazardous chemical incidents.
8. Prepare and conduct training sessions. If possible, include law enforcement agents, emergency medical teams, and mutual-aid companies in the training sessions. Try to plan and conduct the training sessions as realistically as possible.

Protection of Firefighters. In order to protect both firefighting personnel and the general population adequately, it is imperative that fire officials accept the fact that they are operating in a "here and now" era. The **"here and now" era** refers to the fact that a terrorist attack can occur at any time, at any place, and possibly involve weapons of mass destruction (WMD). WMD include both chemical and biological weapons. Operating in the here and now era requires that fire officials recognize that protective clothing provided for the prior normal operations is not sufficient for adequate protection. Many of the responses that firefighters will make might result in situations that are life threatening or potentially fatal. Without a doubt, firefighters should be equipped with appropriate specialized protective equipment and gas masks. Gas masks should be issued to every firefighter and kept readily available in individual turnout coats.

"here and now" era The here and now era refers to the fact that a terrorist attack can occur at any time, at any place, and possibly involve weapons of mass destruction (WMD).

FIGURE 3.32 ◆ Firefighters are decontaminated after a hazardous incident. *(Courtesy of Rick McClure, LAFD)*

Guidelines to assist fire officials in providing adequate protection are included in NFPA 1994, which was adopted in 2001. This standard is entitled *Protective Ensembles for Chemical/Biological Terrorism Incidents.*

In preparing for responding to reported terrorist acts, it is important to recognize that providing for the safety of personnel is critical. Firefighters should always wear full protective gear including breathing apparatus whenever they are operating in an area identified as a hot zone. Firefighting personnel should be decontaminated after leaving the hot zone and before removing their breathing apparatus. (See Figure 3.32.)

On Scene Scenario

New York City—September 11, 2001

8:45 A.M. (all times EDT) Hijacked America Airlines Flight 11 with 81 passengers and 11 crew members was deliberately flown into the North Tower of the World Trade Center. The impact tore a huge gap in the building and set it on fire.

9:00 A.M. A second hijacked airliner, United Airlines Flight 175, with 56 passengers and 9 crew members was flown into the South Tower of the World Trade Center. The airliner exploded on contact and set the building on fire.

9:43 A.M. Hijacked American Airlines Flight 77 with 58 passengers and 9 crew members crashed into the Pentagon, sending up a huge column of smoke.

10:00 A.M. Hijacked United Airlines Flight 93 crashed in Somerset County, Pennsylvania. It is believed that it was headed for the Capitol, the White House, or Camp David.

10:05 A.M. One hour and 5 minutes after being hit, the South Tower of the World Trade Center collapsed.

10:10 A.M. Twenty-seven minutes after being hit, a portion of the Pentagon collapsed.

10:28 A.M. The North Tower of the World Trade Center collapsed.

4:10 P.M. The 47-story Building 7 of the World Trade Center was on fire.

5:20 P.M. Building 7 collapsed.

FIRE GROUND SAFETY

According to Chapter 19, Section 7, of the nineteenth edition of the *NFPA Handbook of Fire Protection,* a number of key facets can minimize firefighter injuries and deaths. These factors are:

- Thorough training of officers and response personnel.
- Considerable medical and physical requirements and health maintenance.
- A detailed and well-functioning incident management system.
- Adequate staffing.
- Closely supervised operating teams or crews.
- State-of-the-art personal protective equipment (PPE), including protective ensembles, clothing, and equipment that are consistently and properly used.

Firefighter safety is a primary concern of every firefighter and every fire officer on every fire department. Many concepts have been developed to assist in preventing injury and death to firefighters on the fire ground. Some of these concepts have been codified and established as standard operating procedures for fire operations. However, every firefighter should remember that concepts are not infallible, and every firefighter should remain alert and flexible at all times in order to adapt to any changing situation.

Fire Protective Clothing and Equipment. Even after the rash of civil disturbances during the latter part of the 1960s and the early years of the 1970s, the terms "safety" and "health" within the fire service were not heard to any degree until the mid-to-late 1970s. A good portion of the advances in protective equipment for firefighters came about as a result of a research program conducted in the late 1970s and early 1980s under the name Project FIRES. This research was sponsored by the U.S. Fire Administration. Results of this research form a part of the NFPA (National Fire Protection Association) and OSHA (Occupational Safety and Health Administration) standards.

Today, safety protocols have been established by NFPA, OSHA, a number of state agencies, various fire departments, and a number of labor organizations. In general, these protocols recommend on-scene safety officers, rapid intervention teams, incident command teams, and a good personnel accountability system.

The NFPA and OSHA standards provide guidelines on the type and specifications of protective clothing that should be worn by firefighters. NFPA 1971, *Standard on Protective Ensemble for Structure Fire Fighting*, establishes the protective clothing that should be worn by structure firefighters. Such protective clothing includes a protective hood, a turnout coat and pants designed for short-term steam and thermal protection, eye protection, boots or safety shoes, gloves, a helmet, and a personal alert safety system (PASS). OSHA requires that the head protection must consist of a protective head device with earflaps and a chin strap. An SCBA is a standard part of a firefighter's protective clothing and equipment. A PASS device is also considered standard. (See Figure 3.33.)

Protective equipment for wildland firefighters differs considerably from that of structural firefighters. Wildland firefighter's protective clothing includes gloves, goggles, brush jackets and pants or a one-piece jumpsuit, head and neck protection, and footwear. Wildland firefighters also use different forms of respiratory protection. Specifications for this personal protective clothing and equipment are contained in NFPA 1977, *Standard on Protective Clothing and Equipment for Wildland Fire Fighting*.

The trousers and shirts are flame-resistant and do not absorb moisture. This permits air to pass through the fabric and allows free movement during periods of long hours when working at high temperatures. The cuffs of the sleeves and the pants legs are closed snugly around the wrists and ankles. The hard hat is lightweight in design and ventilated to protect the firefighter against heat stress. The safety goggles are also ventilated to minimize fogging. The goggles are equipped with impact-resistant lenses.

FIGURE 3.33 ◆ Special protection is required for members of the hazmat squad.
(Courtesy of Las Vegas Fire & Rescue PIO)

The Personnel Alert Safety System (PASS). **PASS** is a personnel alert system that can assist rescuers in locating a missing or trapped firefighter. The use of a PASS device by all firefighters and rescuers is mandatory under NFPA 1500. The standards for the device are established in NFPA 1982. However, it must be kept in mind that NFPA standards are only advisory; they have no power of enforcement unless adopted by a jurisdiction.

These devices are about the size of a portable transistor radio and are worn on a firefighter's SCBA or turnout coat. They are activated by a lack of movement of a firefighter for 30 seconds and emit a loud, pulsating shriek, or they can be activated manually by the wearer. The shrieking will continue for at least one hour.

Some of the earlier models required the system to be manually activated; however, current standards require a PASS device in all self-contained breathing apparatus (SCBA). The requirement is that the PASS is activated automatically whenever the SCBA cylinder is opened.

One of the problems with a manually activated PASS is that it issues many false alarms, which quickly become irritating to firefighters. Consequently, many firefighters fail to activate their units or turn them off prior to entering a structure. This action eliminates the protection that was designed into the unit for the firefighter. A number of firefighters have died at fires as a result of being trapped or lost. No one was aware of their predicament because they had failed to activate their PASS devices.

There are some PASS devices that detect heat, and some even send a signal to the Incident Commander by a remote transmitter that lets the Incident Commander know that a firefighter is in trouble.

PASS A personnel alert safety system that can assist rescuers in locating a missing or trapped firefighter.

The Two-In/Two-Out Concept OSHA and the NFPA both issue standards designed to protect firefighters operating at fires. OSHA describes structure fires that are beyond the incipient stage as **IDLH** (immediately dangerous to life or health) atmospheres. It mandates that all personnel working in these atmospheres use self-contained breathing apparatus (SCBA).

The two-in/two-out concept is mandated in the OSHA 29 CFR 1910.134, Respiratory Protection Standard. This standard dictates that two firefighters must enter the structure as a team when commencing interior firefighting operations in IDLH atmospheres. OSHA regulations further state that "once firefighters begin the attack on an interior structure fire, the atmosphere is assumed to be IDLH and the two-in/two-out rule applies." The standard also mandates that these two individuals remain in direct voice or visual contact at all times while in the structure. Radio contact is not acceptable. As a backup, at least two other fully equipped and trained firefighters must be outside the IDLH atmosphere monitoring those inside and be prepared to rescue the two members inside in the event the need arises. One of the outside members can be the driver/operator or the Incident Commander.

The two firefighters working in an IDLH atmosphere are referred to as the inside team. The two standby firefighters are referred to as the outside team. Outside of the IDLH atmosphere does not necessarily mean outside the building, and inside the IDLH atmosphere does not necessarily mean anywhere inside the structure. As an example, the fire, and consequently the IDLH atmosphere, could exist and be contained on the third floor of a three-story building. Firefighters inside the structure on the first and second floors would not be considered as inside the IDLH atmosphere. When the inside team enters the IDLH atmosphere, the outside team could be located on the second floor. This arrangement would be in conformity to the two-in/two-out rule.

IDLH Refers to firefighting atmospheres that are considered immediately dangerous to life and health.

The standard further identifies what the outside members must wear, what action they can take, and under what conditions a fire department can deviate from the standard. The two-in/two-out deviation is permitted in those situations in which firefighters find that immediate action is necessary to save a life. This type of deviation must be an exception to a department's standard operating procedures and not considered as standard operations

The OSHA standards are not mandatory for all fire departments in the United States. Some states are designated as OSHA states and there the standards are mandatory. Other states that have not adopted the OSHA standards are not considered OSHA states. Approximately half of the states fall into this category. However, fire departments in non-OSHA states are fully aware that the provisions of the OSHA regulations establishes a recognizable "standard of care." It is possible that this standard of care could make a fire department liable for the death or injury of a firefighter if the death or injury could have been prevented, or the injury could have been minimized, had the department followed the provisions of the OSHA standard.

The Rapid Intervention Team (RIT). A RIT is a standby rescue team trained and prepared to provide immediate assistance to any firefighter in distress and to effect a quick removal of the victim to a secure area outside of the structure. The training of individuals assigned to a RIT includes training in both individual and team skills. Individual skills include such factors as the use of the buddy breathing system and the use of thermal imaging devices.

buddy breathing A means for two firefighters to share the air supply from a single air chamber.

Buddy breathing allows two firefighters to share the air supply from a single air chamber. Team training includes organized search procedures in large areas and the team removal of a trapped or downed firefighter.

While a RIT may consist of a team of only two firefighters, ideally a full company should be assigned. Teams on some departments that perform the same functions are identified by such terms as RIC (rapid intervention crew) and FAST (firefighter assist and search team).

Several factors should be considered when organizing an RIT. First, it is difficult for two firefighters to remove a downed firefighter from a heavy smoke-filled environment. Dragging an unconscious firefighter over piles of debris and out through an unfamiliar area is both difficult and time-consuming. The strenuous effort results in rescuing members using up their oxygen much quicker than normal.

While the location of the RIT may be flexible at a fire, at multicompany operations it is best that the team be positioned near the command post. This gives members of the team a better chance to review the situation sheets visually that track the location of all units at the emergency. However, at high-rise fires, at least one RIT should be assigned to the staging area.

The RIT should be adequately equipped with the tools and equipment necessary to carry out the assigned functions effectively.

Equipment should include a thermal imaging camera, if one is available, spare oxygen bottles, a portable radio, and a hand light together with tools for prying and forcible entry. (See Figure 3.34.)

thermal imaging unit A unit designed to detect variations in temperature.

A **thermal imaging unit** is designed to detect variations in temperature. It should be kept in mind, however, that the unit reads surface temperature and not air temperature. It is capable of penetrating smoke, structure features, and darkness. The unit can provide a RIT team with a visual image of reflected temperatures. It is valuable for locating downed and lost firefighters. Additionally, at a fire scene it is useful when searching for victims, identifying fire locations, and detecting fire in concealed spaces. When considering the purchase of a thermal imaging device, the question facing a fire

FIGURE 3.34 ◆ Firefighters use a thermal camera to see into an obstructed area. *(Courtesy of International Safety Instruments)*

chief should not be whether the department should have one but rather how soon one should be purchased. With the strong emphasis placed on firefighter safety, a thermal imaging unit should be part of the standard equipment of every RIT team.

The Personnel Accountability System.   The objective of a **personnel accountability system** is to ensure that the Incident Commander has consistent and up-to-date information on the location of every member and unit at the emergency. The system should be designed so that the Incident Commander knows where every section officer is and what he or she is doing, that every section officer knows where every company is and what it is doing, and that every company officer knows the location of every member of his or her team and what each one is doing. Such a system assists the Incident Commander in knowing where everyone is working and what is being done. It provides a means for the Incident Commander to initiate an immediate search for any firefighter or unit trapped or caught in a dangerous situation. Too many firefighters have died at fires because they were trapped in a fire, were injured, or had their breathing apparatus run out of air. The unfortunate part is that many of these firefighters were not discovered as being missing until it was too late.

> **personnel accountability system**   A system designed to ensure that the Incident Commander has consistent and up-to-date information on the location of every member and unit at an emergency.

The development of a personnel accountability system is the responsibility of the fire chief of every department. The system developed by a department should be understood by every member on the department, and every member of the department should be trained in its use. The system should include the accountability for personnel who respond to the emergency in vehicles other than fire department apparatus. This is most likely to occur on volunteer departments.

The developed system depends on those working at the emergency to keep the staff member responsible for the implementation and maintenance of a personnel accountability report (PAR) informed of their approximate location at the incident and the task or function they are performing. The PAR provides written information on the location of all individuals and units at the emergency. The staff officer generally assigned the responsibility for this report is either the planning officer or the safety officer. The PAR should be implemented early and be consistently accurate.

The Incident Safety Officer (ISO).   There should be at least one member on every department designed as a safety officer with the rank and authority to carry out the duties required of the position. The requirements for this officer are contained in NFPA 1521, *Standard for Fire Department Safety Officer.* On large departments, it is best that the individual identified as the department primary safety officer be a chief officer. A higher-ranking officer generally provides more respect to the position and his or her recommendations normally carry more weight and receive better consideration. On smaller departments, this position can be allocated to a company officer.

While some departments can get along with a single safety officer, it is preferred that at least one safety officer be on duty at all times. In these incidents, the on-duty safety officer should be available to respond to any incident. Those designated as safety officers should be properly trained and motivated.

Where possible, a safety officer should be dispatched to every fire of any size and to any incident such as an excavation where the safety of members may be a problem. If the department does not dispatch a safety officer routinely to an incident, the Incident Commander should assign an individual to this position early in the development of a working fire. The safety officer must have the authority to take corrective action at the scene, if possible, and to monitor the actions of those at the scene to ensure that safety procedures are followed. At the minimum, a safety officer should take note of safety rules that are violated for the use at debriefing discussions and at future training sessions.

When the two-in/two-out concept is employed, it is valuable for the safety officer to be at the entrance to the IDLH area to monitor those entering the atmosphere. He or she should monitor those prepared to enter to ensure that each member is properly equipped and that the PASS device has been activated.

It is best that the safety officer be easily identified at emergency situations. This can be accomplished by the safety officer wearing a particular colored vest with the words "Safety Officer" written on the back. If a department safety officer is routinely dispatched to every fire, then the safety officer should maintain his or her own vest. However, if the SOP of a department is to appoint a safety officer at the scene of every fire, then it is a good procedure for every chief officer to carry a vest in his or her vehicle.

At the scene of an emergency, the safety officer should take the responsibility for risk management. This means that he or she should evaluate the building and the operations of various units continuously to determine whether any conditions present an unusual risk to the safety of members. This continuous assessment can prove valuable to the safety officer in monitoring operations and taking corrective action when safety practices are violated.

The Ultimate Safety Factor. The organization and implementation of a department's outside safety factors are extremely important. Having a safety officer on the scene, a well-organized accountability system, the use of the two-in/two-out procedure, and always having a RIT in place and ready to move in should all be part of a department's standard operating procedure. However, the survival of a firefighter in serious trouble inside a building depends more on what takes place inside rather than what is organized outside. The actions taken inside are referred to as the ultimate safety factor.

An aggressive attack on the fire is laudable; however, more important is the safety of the firefighters making the attack. No firefighter should be taught, encouraged, or allowed to operate inside an IDLH atmosphere on his or her own. Inside operations should always be conducted on a team basis whereby every firefighter on the team is constantly aware of exactly where every other team member is located and what each of them is doing. Each firefighter should be in position and have the necessary resources to aid any of his or her companions who get into trouble. Teamwork and watching out for one another is the key to ultimate safety. Few firefighters have been killed at fires when working within sight and reach of fellow team members.

Of course a situation such as a structure failure or a flashover may result, which could cause every firefighter in harm's way individually to seek any means of exiting the area. When such a situation develops, it should be keep in mind that hose lines should not be removed from the building until positive assurance has been made that every firefighter is out of the structure. Hose lines not only provide a quick and positive pathway to the outside but also provide a pathway to where a firefighter in trouble inside who was working on the fire is most likely to be found.

PERSONAL SAFETY RESPONSIBILITY

A good portion of the safety factors initiated or implemented during the last decade of the twentieth century was directly concerned with operations at emergencies. However, statistics gathered by the NFPA strongly indicate that two important factors related to firefighter safety have not been given the attention they deserve. The importance of these factors to firefighter safety can best be expressed by analyzing the cause of on-duty deaths of firefighters.

According to the NFPA, in 2002 a total of 97 firefighters died on duty. Fires were not the primary killer of these individuals. Heart attacks and motor vehicle crashes caused more on-duty deaths than did smoke, heat, flames, or collapsing buildings.

Although fires were not the number-one cause of on-duty deaths, they did contribute to the overall loss and indicate a strong need for training in personal safety. Several factors should be considered when developing a personal safety program.

1. Every firefighter who is a part of a crew assigned to advance and utilize a hose line should remember never to leave the hose line. The line extends to the outside the same way it came in and is a round-trip ticket to safety. It is the basis of the second consideration.
2. Always have an escape plan. The hose line is the best escape plan; however, the overall escape plan should include the positive monitoring of the each individual's air supply. It is poor practice to wait until the warning bell rings before starting a retreat.
3. Constantly be on the alert for a crack or other signs that could indicate the possibility of a structure collapse.
4. Keep in mind that the best source of help if you get into trouble is the firefighters who are inside with you. Base your operation on the neighborhood watch program whereby every firefighter watches out for each of his or her team members.

Heart attacks continued to be the number-one cause of on-duty deaths, claiming the lives of 37 firefighters. Thirteen of these occurred on the fire ground, eight while traveling to or from a fire or other emergency, seven while engaged in normal administrative activities, six at nonfire emergencies, two during training activities, and one while cleaning up after a tornado.

On-duty heart attacks generally involve older firefighters, those 38 years of age or older. They are more likely to occur in volunteer departments than in full-paid departments. One of the reasons for this is that older firefighters (60 years and older) are more likely to serve on volunteer departments.

The second cause of on-duty deaths was motor vehicle accidents, claiming 29 lives. Twenty-two firefighters were killed in crashes and seven were struck by vehicles.

Statistics drawn from the years previous to 2002 point out the difference between the on-duty deaths of younger firefighters and those of older firefighters. The younger firefighters are more likely to die as a result of a vehicle accident or from trauma or asphyxiation resulting from becoming trapped or overcome during firefighting operations. However, as a firefighter gets older, his or her chance of dying on duty from a heart attack increases. An analysis of on-duty fire deaths over the years strongly indicates a greater need to concentrate on training and physical fitness.

On Scene Scenerio

On Friday, August 20, 2004, thousands of firefighters, some from as far away as New York, filled the Cathedral of Our Lady of the Angeles in downtown Los Angeles to pay tribute to Jaime L. Foster, the first female Los Angeles firefighter to die in the line of duty.

Training, Monitoring and Enforcement.   It is almost a crime to read about a firefighter who is killed at a fire due to his or her failure to follow a simple safety procedure. Almost every year a firefighter is lost or trapped in a fire while wearing a manually activated PASS device with the device set in the "off" position.

Failure to set the device in the "on" position is due to simple carelessness and strongly emphasizes the need to increase training; the need for firefighters working under the buddy system to monitor each other; and the need for officers, particularly the safety officer, to enforce safety procedures at fires.

The same kind of carelessness exists in firefighter deaths due to vehicle accidents. Nearly every year a firefighter is needlessly killed in a vehicle accident who may have been saved if he or she had been wearing a seat belt. Others are killed because of rollovers or collisions resulting from the driver driving too fast for the road and weather conditions. Again this indicates the need for company officers to enforce safe driving conditions and to conduct safety training sessions in the station continuously.

Physical Fitness.   One of the most important parts of any occupational health and safety program is physical fitness. This is particularly true for the fire service. The fire service is the only occupation in the world in which an individual may be awakened from a sound sleep and three to four minutes later be required to perform using his or her maximum strength and endurance. It takes little imagination to recognize what this does to a person. It also is not difficult to visualize what effect this would have on an individual who is not in good physical condition.

Newly hired firefighters are normally in good to excellent physical condition. Unfortunately, however, some of the new firefighters in certain communities are not in as good a physical condition as those hired in other communities. The primary reason for this is that some communities have a more stringent physical ability testing program for firefighter candidates than others. For the safety of firefighters, there should be a standard test for all candidates throughout the country. Fortunately, this issue has been addressed by a joint task force of the IAFF (the International Association of Fire Fighters), the IAFC (the International Association of Fire Chiefs), 10 cities, and a number of local firefighter unions. The results of the task force have been formalized in a recommended system called The Fire Service Joint Labor Management Wellness/Fitness Initiative—Candidate Physical Ability Test (CPAT). Information on this recommended program is available from the IAFF.

Regardless of the physical condition of a newly hired firefighter, as the years pass his or her physical condition tends to deteriorate and by the time he or she reaches age 40, he or she may be a prime risk for a heart attack. Although it cannot be proven in each case, those suffering a heart attack on duty are probably not in top physical condition. Their cholesterol and triglycerides are likely to be too high, they probably weigh too much, and most likely they are trapped by the smoking habit.

Many fire administrators have recognized the problem and have tried to improve the physical condition of their firefighters by instituting a physical training program on duty. NFPA 1583, *Standard for Health Related Fitness Program for Fire Fighters*, establishes certain guidelines. Unfortunately, few of the programs instituted by fire administrators have been fully successful. Most firefighters feel they are in good physical condition and that they get a sufficient amount of exercise while off duty. For many, this analogy is true; however, one shoe does not fit all.

Many years ago, U.S. Army General George C. Marshall made a statement that might be applicable to the situation. He said, "Fix the problem, not the blame." Fire administrators would be wise to heed this advice and fix the problem. They should recognize that the physical fitness of a firefighter is primarily an individual problem. Although physical fitness programs in the engine house are difficult to enforce, it is wise and inexpensive for the fire administration to provide exercise equipment for those who recognize the need to keep in good physical condition. On the other hand, each individual should accept the fact that physical fitness for firefighters is an individual choice. It is up to every firefighter to choose between the possibility of completing an exciting and challenging career and the fringe benefit of a long and healthy retirement or of becoming a statistical figure in the chart of firefighters' deaths.

Obtaining and maintaining good health and physical condition is primarily a matter of eating and exercise. It may be necessary for an individual to change his or her lifestyle, concentrating on the items that are essential for his or her success. Serious thought should be given to losing weight, stopping smoking, decreasing alcohol intake, decreasing sugar and carbohydrates intake, and increasing physical activity.

For maximum benefit, an eating pattern should include fruits, vegetables, and a low-fat protein source such as poultry, fish, and meats. The responsibility for achieving this in the fire station rests primarily with the individual preparing the meals. However, the responsibility while off duty is 100 percent that of each individual firefighter.

A general principle is that it is better to eat five or six smaller meals each day rather than three large meals. This would be difficult to achieve while on duty; however, it should be given serious consideration while off duty.

It is helpful for an individual to obtain a blood test to determine cholesterol and triglycerides levels. The results of this test may provide a road path for the development of good eating patterns.

Unfortunately, firehouse cooking will probably not provide the best selection of foods for a healthy lifestyle. However, most firefighters look forward to the excellent meals prepared in the station, and especially to the deserts and ice cream that follow the evening meal. This is part of the club atmosphere enjoyed by firefighters. It just means that individuals will have to make up for the firehouse meals while off duty.

The importance of regular exercise along with a satisfactory eating pattern cannot be overemphasized. Unfortunately, as an individual ages, his or her tendency to exercise normally declines. Without a doubt, this has a bearing on the statistics that most heart attacks that occur on duty are experienced by those over age 40. It is fairly easy for a firefighter to choose the comfort of the lounge chair in front of the TV over the choice to exert a little energy in an exercise program. It is easy, but not intelligent, to make the excuse that a rest may be beneficial because the alarm may sound and the remainder of the shift may be spent at a working fire.

Many fire stations do not provide exercise equipment for those who wish to devote part of the shift to an exercise program. However, firefighters have traditionally pooled their resources to buy a TV for the use of all shifts, and each member has contributed regardless of whether he or she watches TV. The same type of cooperation can be used to purchase exercise equipment in the event the department does not provide it. Lacking such equipment, a simple walk several times a day around the engine house or outside yard is much more productive than staring at the TV.

A final word of caution. It is a good idea to have a complete medical examination including the recommended blood test prior to embarking on a vigorous physical fitness program. A stress test to determine the present state of the firefighter's cardiovascular system should be part of the examination.

Physical Fitness Assessment Programs.   Almost all fire departments use a physical agility test as part of the hiring process for new firefighters. However, once an individual is hired and completes a probationary period, he or she is seldom required ever again to prove he or she has the strength or endurance to perform the duties that may be required. This trend continues despite the strong emphasis that has been placed on firefighter safety and the fact that heart attacks on duty continue year after year to be the number-one cause of firefighters' deaths. It appears wise for management and department firefighter labor organizations to work together to provide a system to reduce the number of on-duty deaths of members. Whatever system is designed, it should be one that the membership considers to be fair and objective. There are two programs that might be considered for a part of the total package.

One is an annual physical fitness assessment program. This program should not in any way be considered as a punitive measure but more as a means for each firefighter continuously to be aware of his or her ability to do the job when the time arrives. A suggestion is that the assessment test should closely parallel the test required to be passed by newly hired firefighters. The ability to perform the physical duties of a firefighter does not change over time. There should be some way for a red flag to be raised if the time arrives when a firefighter is no longer able to perform the physical functions required. Failing to do so is the same as recognizing that a firefighter is trapped in a dangerous situation and not taking any action to save oneself.

A second part of the total package should be an annual assessment of a member's cardiovascular system. Such assessment should be sponsored and paid for by the department. The results of this assessment need not be given to management but should be provided to the member for his or her benefit. In the final analysis, the overall reduction of on-duty heart attack deaths will be achieved only through the recognition by all members that the problem is an individual one that can only be solved by individual effort. There is probably no mandatory program that can be instituted by management that will achieve the desired results without the full cooperation of all members.

On Scene Scenario

The following article appeared in the November–December 2003 issue of the *Los Angeles Firefighter.*

16107 Victory Boulevard, Van Nuys area. Four Fighters suffered minor injuries and were transported to area hospitals in stable condition. One sustained first-degree burns to his forehead, another sustained first-degree burns to his knees, a third Firefighter suffered lower back strain while a fourth had a laceration to his hand. Loss from the fire, which destroyed the business, was estimated at 3.8 million dollars. The cause of the fire is categorized as electrical. Injuries can and do happen. Firefighters must be as prepared as possible physically to deal with hazards of the job.

◆ **SUMMARY**

It would be very difficult to operate a cruise ship if 10 people wanted to be the captain. There is only one captain on a ship who makes all the decisions while in command. This is to ensure that all decisions are made in a timely manner and that everyone who works on the ship understands where he or she is in the hierarchy of the ship's operations. The same holds true on a fire incident. Someone is in charge and that person needs to hold the authority and have the responsibility to make good timely decisions. All personnel must also know their individual assignments and where they are located on the organization chart. The Incident Command System and now the National Incident Command System allow responders to identify with a specific role and placement within the response organization. These systems allow for a flow of information and provide responders with a process from which to work and complete assigned tasks.

■■

Review Questions

1. What is the definition of the term unity of command?
2. What is the definition of unified command?
3. What is the definition of span of control?
4. What is the maximum span of control that is acceptable for use in the ICS?
5. How was the ICS developed?
6. What are the five major functional areas in the ICS?
7. Which of the major areas is directly responsible for the extinguishment of the fire?
8. In the ICS, to which of the major functional areas has the responsibility for directing firefighting operations been delegated?
9. At a large wildland fire, which of the major functional areas would be responsible for setting up a camp?
10. What improvements in firefighter safety came about as a result of civil disturbances?
11. What are some of the occupancies that might be locations for civil disturbances?
12. What type of occupancy would be ideal for the location of a staging area for a civil disturbance?
13. What should be done to prepare apparatus for a possible response into a civil disturbance area?
14. What was the most disastrous domestic terrorist attack that has occurred in the United States?
15. What is a dirty bomb?
16. What is the primary cause of on-duty deaths of firefighters?
17. Why is it important for firefighters to maintain good physical fitness?
18. What is the ultimate safety factor for firefighters?
19. Define the National Incident Management System.

CHAPTER 4

Fire Ground Planning and Tactics

Key Terms

Objectives

The objective of this chapter is to introduce the reader to the pre-incident planning program, the pre-incident planning inspection, size-up, engine company operations, and truck company operations. Upon completing this chapter, the reader should be able to:

- Explain the objectives of a pre-alarm size-up.
- List some of the information that should be gathered during a pre-incident planning inspection of an occupancy.
- List the factors that should be considered during the response size-up.
- Discuss the preliminary and continuous size-ups.
- List the three basic functions for which an engine company is responsible.
- List the factors to consider regarding the locating of the fire.
- Explain the methods of preventing external and interior spreading of the fire.
- Explain the objectives and methods used for both a direct and an indirect attack on the fire.
- List the tasks for which a truck company is responsible.
- Discuss ladder operations, forcible entry, and controlling the utilities.

Between June 3 and June 7, 1942, land- and carrier-based planes attacked the Japanese fleet northwest of the island of Midway. In the opinion of many military experts, the U.S. victory in this battle was the turning point in the Pacific campaign of World War II. The Battle of Midway was the first Japanese defeat. Not only did it end Japan's thoughts of seizing Midway Island and using it as a base to attack the Hawaiian Islands, but it also crippled Japan's naval air power. However, the battle was not completely one-sided. U.S. naval forces also had their share of casualties. One of the situations that occurred during the battle will probably go down in history as the most disastrous defeat of a naval aircraft squadron. Every plane from Torpedo Squadron Eight was shot out of the air, and every pilot and crewman—except one—was killed. The destruction of Torpedo Squadron Eight while making an attack on a Japanese carrier will remain a prime example to tactical commanders of all emergency forces of the tremendous losses that can occur when a planned attack lacks coordination.

During World War II, a torpedo squadron was one of three squadrons required to make a successful coordinated attack on an enemy carrier. The other two squadrons were the fighter squadron and the dive bomber squadron. Coordination required that planes from all three squadrons commence their attack at the same time. The fighter planes would come in at an angle and strafe the deck, hoping to clear it of firepower. The dive bombers would approach from overhead and release their bombs while the torpedo planes would launch their attack at water level to drop torpedoes. This attack was generally successful, if properly coordinated. Unfortunately, on the day of the conflict, planes from Torpedo Squadron Eight arrived on the battle scene prior to the arrival of the fighter and dive bomber squadrons. The planes were low on fuel, having just barely enough to make an attack and return to their own carrier. Although it would have been better to revert to a defensive mode, the commander of Torpedo Squadron Eight decided that he could not wait for the arrival of the other two squadrons and ordered the attack. The outcome is history.

The necessity for planning the strategy to be used on the fire ground is just as important as it is for a military operation. Although the elements required to be considered in the development of the strategy are different, the result of failure is the same. Either the battle will be lost or the losses during the battle will be greater than desired.

To understand the importance of planning the strategy and the manner in which objectives will be achieved, it is first necessary to be familiar with what operations are required in order to extinguish the fire with the least possible loss of life and property. It is also crucial to understand what is required to carry out an operation successfully and who is responsible for seeing that the operation is properly completed.

It has often been said that no two fires are exactly alike. However, it has also been said that all fires are somewhat alike. Because there is a degree of similarity in all fires, thought must be given to certain factors regardless of what type or size of occupancy in which the fire occurs. Consideration must also be given to the tasks to be performed on the fire ground, and the manner in which these tasks can best be accomplished. Preparing the strategy to be used on the fire ground requires knowledge on the part of every individual involved as to the exact part he or she plays. It is much better when the assignment of the tasks to be performed can be made in advance and not left to the snap judgment of an Incident Commander at the scene of an emergency. Consequently, the tasks to be performed should be tentatively divided and assigned long

FIGURE 4.1 ◆
Coordination of
functions is important.

before the bell rings. To do this, it is first necessary to review the tasks and ensure that those individuals responsible for their performance are properly trained. The importance of this might be related to an article written by a famous football coach many years ago and published in the *Saturday Evening Post*. The essence of the message in the article was summarized in the title: "Football Games Are Not Won on Saturday." (See Figure 4.1.)

Most of the tactical tasks to be performed were examined in Chapter Two. Some others that are important will be examined in this chapter. However, to do an efficient job of planning strategy for the fire ground, the first thing that is required is information: information on the building in which the fire occurs and information regarding the equipment available to perform the tasks. There is a distinct advantage in gathering information on the building prior to a fire occurring. This can be done by making a **pre-incident planning inspection.** A pre-incident planning inspection is a portion of a larger process referred to as pre-incident planning.

pre-incident planning inspection An inspection of a building to gather as much information as possible about the building for the purpose of fighting a fire before it occurs.

◆ PRE-INCIDENT PLANNING

pre-incident planning Planning to ensure that personnel responding to a fire incident know as much as possible about the building or group of buildings to which they are responding.

Pre-incident planning is extremely important to successful operations in large or complex structures or group of structures and in buildings that are protected by built-in fire protection systems. The objective of pre-incident planning is to ensure that personnel responding to a fire incident know as much as possible about the building or group of buildings to which they are responding. In addition to the information contained in this chapter, those interested in developing pre-incident plans should consult NFPA 1620, *Recommended Practice for Pre-Incident Planning*.

Although it would be ideal, most departments do not have the personnel and resources to pre-incident plan every major building or group of buildings within the community. Some company officers that are assigned to a primary residential first-in district may possibly complete pre-incident plans for the major buildings in their district. However, an officer assigned to a principal commercial/industrial first-in district, or to a district in the downtown area of a large city with its multitude of high-rise structures, would be hard pressed to accomplish the desired task. Consequently, in these districts it is necessary to establish a priority system for pre-incident planning.

The generally accepted principle is that pre-incident planning should be done on those occupancies identified as **target hazards.** Target hazards include those buildings or occupancies that:

> **target hazard**   An occupancy that constitutes a large collection of burnable valuables or an occupancy in which the life hazard is severe.

- Have the potential for a large loss of life.
- Could create a conflagration.
- Have a record of having an above average number of fires.
- Could create safety problems for firefighters responding to a fire in the occupancy. (Example: uses or stores hazardous materials, or has structural weaknesses such as lightweight building components)

◆ THE SIZE-UP

It would be very difficult, if not impossible, to conduct operations at the scene of an emergency intelligently and successfully without first making an estimate of existing conditions. By necessity, existing conditions include available personnel, equipment, water supply, the life hazard, the time of day, weather, type and size of occupancy, what is burning, and the extent of the fire, just to name a few. This estimate of existing conditions is referred to as the **size-up.**

> **size-up**   An estimate of existing conditions.

Although the tactical size-up for overall operations at the fire is the responsibility of the Incident Commander, every firefighter at the incident, from the newest rookie to the most senior fire officer, consciously or unconsciously participates to some degree in some portion of the size-up process. Consequently, the more knowledgeable a firefighter is regarding the size-up process, the more likely he or she is to make a good decision regarding his or her actions on the fire ground.

The size-up commences long before the alarm sounds and continues throughout the duration of the emergency. For practical purposes, it can be divided into three parts—the pre-alarm size-up, the response size-up, and the fire ground size-up.

THE PRE-ALARM SIZE-UP

The **pre-alarm size-up** first manifests itself when any information on the fire building or its exposures that could affect operations on the fire ground is initially gathered. This might have taken place on a previous response to the occupancy; however, it generally occurs during an inspection of the potential fire building that is made prior to the fire occurring by companies that normally will arrive first to a fire in the building. This inspection is a portion of an overall process referred to as a *pre-incident planning inspection.*

> **pre-alarm size-up**   A size-up made prior to the alarm sounding. It normally first manifests itself when any information on the fire building or its exposures that could affect operations on the fire ground is obtained.

The pre-incident planning program for a department is the responsibility of the fire chief. The fire chief should establish a policy requiring that company officers conduct pre-incident planning inspections of target hazards on a regular basis and establish the procedure on how the inspection should be conducted. Department forms for recording the information gathered should be designed and standardized. It is important that the department's forms include a checkoff list that can be used by a company officer to ensure that nothing of importance is forgotten during the inspection. The pre-incident planning program should include the requirement that engine company drills be conducted periodically by company officers to inform the engine company members of the plans made for extinguishing fires in various sections of the inspected building.

A chief officer should be delegated the responsibility of ensuring that the policy for the pre-incident planning inspections established by the fire chief are carried out. When possible, this officer should sit in on company training sessions on the occupancy.

Pre-incident Planning Inspections. The objective of a pre-incident planning inspection is to gather information to fight the fire before it occurs. The pre-incident planning inspection should not be confused with a fire inspection (or target inspection) of the building. Code inspections and pre-incident planning inspections are two distinctively different functions with two different purposes. No notices of fire violations found should be written during this inspection and the purpose of the inspection should be made clear to the occupancy management. It is important to explain to the building management the nature of the information that the department wants to gather as it may be necessary for management to arrange to have those employees most knowledgeable about certain processes available when the department arrives.

It is important that fire company officers and other fire officials recognize that the occupancy to be inspected is conducting a business, and that any interruption of the business cycle can have a cost impact on operations. It is therefore good policy to call the manager of the occupancy at least a week prior to the intended inspection and make an appointment for the inspection on a day and at a time most convenient for the occupancy management. It is good public relations to inform management at that time of the purpose for the inspection and emphasize that the inspection is for the mutual benefit of both parties. It is also suggested that the manager be informed that no notices for fire violations will be issued during the inspection. It is suggested that the company officer who will make the inspection invite the chief officer to whom he or she reports to join the company in the inspection. This chief officer will probably be the individual who will assume the position of Incident Commander (or operation chief) in the event of a fire in the building.

The actual inspection involves collecting information and using the information to plan ahead of time how fires will be fought if they occur in various parts of the building. The pre-incident planning inspection can provide a majority of the information required in order to make a successful analysis of the situation on the fire ground. During the inspection, the information gathered should be placed on a form developed by the department that indicates the information the department wishes to record as a result of the inspection. If the department has computers in the apparatus, the information should be accessible to the company commander. If the department does not wish to develop a form for its own use, the National Fire Academy offers one referred to as a quick action plan (QAP), which might provide spaces for the information the department wants to record. In addition to its use for recording information, the developed form can be used for future training sessions in the engine house.

The unit or group making the inspection should arrive at the occupancy location a few minutes prior to the appointment time. A drive-around or walk-around of the building should be completed prior to entering the building. At this time, information can be gathered as to potential entry points and in particular the location of hydrants that will

FIGURE 4.2 ◆ Consider the reach of streams.

most likely be used in the event of a fire. On return to quarters, the hydrant records should be examined to help determine the minimum amount of water that can be expected from each of the available hydrants. Both the information on the location of the hydrants and the amount of water available should be entered on the department form that will be used as a database.

Some of the information that should be gathered once the occupancy is entered is:

1. *The size and construction of the building.* Knowing this information will help determine whether hand lines can be used or whether it may be necessary to resort to heavy stream appliances. It should be remembered that the reach of a handheld line is about 50 feet. It is possible for the fire to burn unchecked in the center of an open area in the building if it is attacked from both sides with hand lines if the width of a building exceeds 100 feet. (See Figure 4.2.)

The hazard of this was brought to the attention of the entire fire service during the early 1950s with the destruction of the General Motors plant in Livonia, Michigan. This fire occurred during the daytime when the plant was in full operation. Attacks were made from both sides of the structure; however, the plant was lost partly because the streams did not have the reach to knock down the fire in all parts of the building.

The construction of the building also plays a major part in making a decision as to how the fire should be attacked and what paths it is most likely to take in spreading. Although a building of concrete construction may prove to be beneficial in limiting external fire spread, it may also make access to the fire much more difficult due to the large amount of gases, smoke, and heat that will be retained within the building.

In some cases, such as in high-rise buildings, the interior of the building may become like a heating oven due to the entrapment of the fire, heat, and smoke. It is estimated that the temperature inside the First Interstate Bank Building in May of 1988 reached approximately 2000°F, due to the construction of the building.

Buildings constructed of unprotected steel may prove to be even more of a problem. The steel will begin to lose its strength in as short a period of time as 10 minutes when subjected to severe fire conditions. An estimate should be made of the possibility of early collapse of the building if the fire has burned unchecked for any period of time. (See Figure 4.3.)

Additionally, the possible adverse effects that the building construction could have on various factors such as access, ventilation, salvage, and so on should be evaluated.

While gathering information on the building construction, it is good practice to visualize fires mentally in various parts of the building and estimate the fire flow that will

FIGURE 4.3 ◆ Metal buildings can collapse quickly when the metal is exposed to extreme heat.
(Courtesy of District Chief Chris E. Mickal, New Orleans Fire Department—Photo Unit)

be needed if the level of involvement of the fire is 25 percent, 50 percent, 75 percent, or 100 percent. An examination of methods for estimating the fire flow needed is included in Chapter Five. A space for the estimated fire flow required should be provided on the inspection form. (See Figure 4.4.)

2. *The life hazard.* Saving lives takes first priority at every fire. An aggressive attack on the fire may be the best method of saving lives under practical conditions, but thought should always be given to the possibility of people being trapped. Consequently, every possible condition that may contribute to loss of life during a fire should be considered during the pre-incident planning inspection. How many people are likely to be in the building at the time of the alarm? Will all of them be able to leave the building unassisted? Is it likely that people might become trapped above the fire floor? Is there an adequate number of exits in the building? Are the exits likely to be blocked by fire during the emergency? What will be the best way to get to people in various parts of the building, and what will be the best way to get them out?

3. *Fire and smoke travel.* Consideration must be given during the inspection as to how the fire, heat, and smoke will most likely travel to various parts of the building. Evaluation of conditions should include thought given to heat traveling by radiation, convection, and conduction, and the possibility of fire traveling from room to room, from floor to floor, and from the fire building to other structures or material outside

FIGURE 4.4 ◆ The size and construction of a building play an important part in fire extinguishment.
(Courtesy of District Chief Chris E. Mickal, New Orleans Fire Department—Photo Unit)

of the building. When considering this factor, it should be remembered that fire can travel from room to room on the same floor:

   a. Through unprotected horizontal openings such as doorways, hallways, interior windows, transoms, breaks in walls, and so on.

   b. By convection of heated air, smoke, and gases.

   c. As a result of backdrafts or flashovers.

   d. By the conduction of heat through metal pipes, other metal objects, or possibly through walls.

   e. Through concealed spaces, air-conditioning systems, and the like.

Fire can travel from floor to floor:

   a. Through unprotected vertical openings such as open stairwells, elevator shafts, light wells, and so on. The fire itself may travel through these openings or the extension of the fire may be caused by convection heat.

   b. Through windows, by fire extending up the outside of the building and through the windows of upper floors. This is a particular problem in high-rise buildings.

   c. By sparks or burning material falling to lower floors.

   d. By roof or floors collapsing onto lower floors.

   e. Through air-conditioning and similar systems.

Fire may extend from building to building, or from the fire building to material outside of the building:

   a. By conduction, convection, or radiation.

   b. Through unprotected wall openings.

   c. Through combustible walls or roofs.

   d. By failure of walls through explosions or structural weaknesses.

   e. By flying brands landing on combustible roofs of buildings.

   4. *The contents of the building.* Knowledge of the contents of the building is extremely important when assessing the speed with which the fire can be expected to travel. If flammable liquids are stored in the building, it can be expected that the fire will spread rapidly if they become involved. Rapid expansion of the fire can also be expected if material such as cotton is involved. More than having an effect on the fire spread, what is stored inside the building may have an adverse effect on floor collapse. Thought should be given not only to the weight of the material in its stored condition but also to its potential weight when it absorbs water.

5. *On-site fire protection.* The location of all on-site fire protection systems should be identified. All shutoffs and fire department inlets should be located if the building has a sprinkler system. The areas controlled by shutoffs and inlets should be recorded. Particular attention should be paid to any parts of the building that are not covered by the sprinkler system. If the occupancy is of sufficient size to warrant an annunciator or central control station that will identify the location within the building where a sprinkler head is flowing water or an alarm box has been pulled, then this location should be noted for future reference. The time required to find the seat of the fire will be shortened if responding companies are familiar with the location of these central controls. Controls for carbon dioxide and other fire protection systems should also be identified and their areas of protection noted.

If the building is five or more stories in height, it will probably be equipped with a dry standpipe system for fire department use. It may also be so equipped in buildings less than five stories in height. This system is particularly helpful when relaying water to upper floors. The inlets to this system will most likely be found alongside those for the sprinkler system; if they are not, however, they should be located during the preincident planning inspection. Of course, it is important that an assessment be made of the length of lines that will be required to be laid into all the inlets and the length of lines that will be required from the dry pipe outlets to the location of the fire in various sections of the building.

6. *Ventilation problems.* The pre-incident planning inspection is the ideal time to make an analysis of the best methods for ventilating the building during the fire and the ventilation problems that are likely to arise. It is generally best to start the inspection of the building on the roof. Not only does this location provide an opportunity to size up the external exposure problem, but it also gives firefighters an opportunity to determine the best means of providing roof ventilation. While on the roof, the inspection party should look for natural openings such as scuttle holes and skylights. The inspectors should also take note where the elevator shaft and stairwells come through the roof. It is a good idea to determine the construction of the roof during the inspection.

7. *Windows.* The lower floors will provide information that may later be helpful in horizontal ventilation. Note should be taken of the location and types of windows found in the building. Can they be opened easily, or might it be necessary to break them? Can they be broken from the outside if it is desired to use them for ventilation? Plate glass, wire glass, and tempered glass windows can prove to be a difficult problem.

8. *Building access.* What problems might arise while trying to gain access to the building during the fire? Does the building have a Knox box? Are means of entering available on four sides, or might access be limited to the front and back—or maybe even just the front? Will it be possible to use all natural entry points? Plate glass or rolling steel doors will present serious problems. Are there gates (or guard dogs) on the premises that might slow down access to the rear and sides of the building? What length ladders will be required to gain ingress to the building on upper floors? (See Figures 4.5 and 4.6.)

9. *Hazardous materials.* It is import to determine where any hazardous materials are located on the premises. The names of all materials should be written down during the inspection and their hazards determined after returning to quarters. Where are the material safety data sheets located? An analysis should be made as to how these materials can best be protected from becoming involved during fire operations, and a

FIGURE 4.5 ◆ Fences and similar objects can hamper access to the fire. *(Courtesy of District Chief Chris E. Mickal, New Orleans Fire Department— Photo Unit)*

determination made as to what reactions are likely if they do become involved. Which of the materials are water reactive, heat reactive, poisonous, and so on, and what is the best method of extinguishing a fire involving these materials? Are incompatible materials stored on the premises? If so, it is imperative that these materials be kept from mixing during fire operations. A careless hose line could cause the death of a number of firefighters.

   10. *Hazards to firefighters.* In addition to hazardous materials, there might be a number of structural defects that could be a hazard to firefighters during fire operations. The roof is a logical place to start checking for these hazards. When analyzing potential hazards, it is best to consider what condition will most likely exist during the emergency. A television antenna located five feet off the roof does not look hazardous during an inspection, but it could cause a firefighter to lose an eye if the roof is covered with heavy black smoke at the time of the fire. Evaluate what it will be like when a firefighter comes onto the roof in smoky conditions. How will a firefighter ventilate the roof (will special tools be needed)? This concept should be carried out when evaluating

FIGURE 4.6 ◆ Sometimes it is necessary to breach a fence to lay an attack line to the fire. *(Courtesy of Las Vegas Fire & Rescue PIO)*

conditions elsewhere in the building. Are there any pits or unprotected vertical openings that might present a hazard to firefighters crawling into the building with a hose line? It is best not to overlook any area of the building when considering this factor.

11. *Utility controls.* It might be imperative that the gas, electricity, or water be shut off during the early stages of the emergency. The moment of need is not the time when a search for these controls should be made. The location of the shutoffs and the areas they control should be determined during the pre-incident planning inspection.

12. *Salvage.* Thought should be given during the inspection to the salvage problem. Can the floors be expected to hold the water for any length of time? Can the elevator shaft and/or the stairwells be used for water removal? How susceptible is the material stored on the premises to water damage? Will it be necessary to cover the floors to protect them? What are some of the alternative methods that can be used for water removal?

13. *Hose requirements.* How much hose is it going to take to reach various parts of the first floor from the front door? How much will be required if the hose is brought in through other access points? What if lines are advanced up stairwells or fire escapes to upper floors? What lengths will be required if the hose is taken aloft and worked off the dry standpipe system? It is much better to determine the answer to some of these questions prior to the fire than trying to make a snap judgment during the stress of the emergency.

14. *Water supply.* Where are the nearest hydrants located and how much water can be expected from the system? Are there any auxiliary supplies such as water tanks, cisterns, or swimming pools available? If so, how much water can be expected from these sources? Is the water available from these sources adequate to cope with the expected size of the emergency, or might it be necessary to relay water from another source? If so, what source will provide the quickest and best supply?

Gathering information during a pre-incident planning inspection is just the beginning of the pre-incident planning process. The information is of little use if it is not available or known to the Incident Commander at the emergency. Although the information may be known by a company officer, the odds are that the officer will not be on duty at the time of the fire. Consequently, the information must be considered as a database and incorporated into the department's response plan for the building.

The department's response plan must be one that provides the first-in company officer and the first-in chief officer the capability of quickly and easily accessing the information, preferably prior to responding or on the way to the emergency. The fastest and most efficient method of achieving this is to store the data in a computer at the dispatch office and relay the information to the apparatus computer screens of all responding units. (See Figure 4.7.)

If computer screens are not available on apparatus, the information can be sent to computer screens in the stations or be printed out on a "tear and run" printer in the station. This will provide officers with the information prior to leaving quarters. The information formatted should be user friendly and be limited to that most valuable to the Incident Commander. It is important that it include a floor plan of the occupancy indicating life hazards, location of dangerous chemicals and known hazards to firefighters. For maximum effectiveness, not only should the information appear on the computer screen but it also should be quickly printed out so that the receiving officer can take a copy with him or her.

Unfortunately, not all departments are presently equipped to provide this capability. As an alternative, three-ring binders carried on responding apparatus and in command cars can be used to handle a number of pre-incident plans. The information

FIGURE 4.7 ◆ A computer in a command vehicle can be extremely valuable to an Incident Commander.
(Courtesy of Las Vegas Fire & Rescue PIO)

in these binders should be limited to what is most valuable to the Incident Commander yet include floor plans of the occupancy indicating the location of hazards.

Another method that should be considered is for the dispatch office to provide responding units with information on the occupancy by radio. This will provide all officers responding to the emergency with important but limited information on the occupancy. Regardless of how limited, any information is better than none.

THE RESPONSE SIZE-UP

Many thoughts go through the minds of every firefighter when the bell sounds. Will this one be routine or is this the big one? Is this the one where I might have to crawl on my hands and knees to save some child? Will people be trapped, which might require the use of special equipment?

The most tangled thoughts, however, will flash through the mind of the officer who is expected to be on scene first. That individual knows he or she will be responsible for operations until relieved by a superior. That particular officer will have to make the initial determination as to the number of companies needed, where lines are to be placed, and the thousands of little details involving the factors of rescue, extinguishment, exposures, ladder, ingress, ventilation, salvage, and so on. He or she hopes that a pre-incident planning inspection of the building has been made, but more than likely this will not be the case. If this is so, there are a number of factors the officer will have to consider until he or she arrives at the scene where some of the information will then be able to be determined.

The Response. Knowing the location of the emergency will provide knowledge as to the number and types of companies that can be expected on the initial assignment. If the address is in a residential area, at least two engine companies and a truck company will normally be dispatched. However, in larger cities this number can increase sufficiently, depending on the area of response. As many as four engine companies, two truck companies, and additional equipment such as rescue companies, squad companies, boat companies, and special equipment might be dispatched if the department is so equipped and the area has unique problems. The officer who is first on the scene should quickly review in his or her mind the staffing of these companies, the types of engine companies that will be responding, whether the truck companies are equipped with aerial ladders or elevated platforms, and the directions from which all companies will be arriving. He or she should also consider the number of additional companies that will arrive on the scene if the fire is beyond the control of the first alarm assignment and it becomes necessary to call for a second or third alarm.

The District. The district in which the emergency is located will have an influence on the size-up. The life hazard in a residential area will normally be considerably less than one in a hotel or apartment house district. The size of the fire will be different in an industrial zone than in a limited commercial area. The speed with which the fire is likely to spread will vary according to the construction in the area and the spacing of the buildings. A response into a brush or wildfire area will present many different problems than one in a residential area. The factors influencing the area of response should be analyzed as the apparatus moves from the engine house toward the fire.

The Time of Day. Generally the life hazard and the probable size of the fire is much greater between midnight and 4:00 A.M. than it is at 2:00 in the afternoon; however, this may not be the case. The type of occupancy matched with the time of the alarm plays a vital part in the development of a critical situation. The life hazard in a school is much greater at 10:00 in the morning than at 10:00 at night, whereas the life hazard in an apartment house is much greater at 3:00 A.M. than at 3:00 P.M. Thought must be given to the life hazard during the response size-up. Is it likely that people will be in the building or is it more likely that the building will be empty? Will the occupants most likely be awake or asleep? Will they be able to get out of the building unassisted or will it be necessary to remove most of them physically?

The time of response will also have an effect on the speed of the response itself. What influence will the time of response have on the traffic problem? A considerable delay can be expected when the workforce is leaving a factory or a large building. It can also be expected that crowds will be larger during certain hours than at others. Parked cars may also hamper access to hydrants during some hours, whereas the same hydrants will be freely accessible at other times. All the factors influenced by the time of response have to be taken into consideration during the size-up.

The Weather. The weather will have an impact on the response, the fire spread, and fire tactics. For example, weather conditions such as rain, snow, and fog will have an adverse effect on response time and, in some cases, in locating the fire. Certain areas of the city might become flooded during heavy rains, which could cause a change in response routes. This not only slows down response time but also might bring the company into the fire from a completely different direction than is expected by the Incident Commander. (See Figure 4.8.)

There have been cases where a company could respond only in one direction from the engine house due to deep flooding in the other directions. Heavy fog not only will slow down response but also may make it extremely difficult to locate the fire. Companies have pulled up in front of well-involved buildings during heavy fog conditions before seeing the fire. This reduces the time available for planning hose lays to zero.

The weather may also have a tremendous impact on fire spread. The mere delay in response will give the fire additional time to progress before hose lines can be brought into position to restrict its spread; however, more important is the direct

FIGURE 4.8 ◆ Cold weather can hamper firefighting operations.
(Courtesy of Minneapolis, Minnesota, Fire Department)

impact it has on contributing to the fire spread. The weather factor most feared is high winds. Winds can lift roof shingles high in the air and drop them on other combustible roofs some distance from the original fire, and they can bend high-pressure hose streams so badly that little of the water reaches the fire. For a number of years, winds in excess of 30 miles per hour have been the leading contributor to conflagrations. The problem is particularly acute in wildfires.

Weather also plays an important part in fire operations. It is much more difficult to attack a fire during periods of heavy snow than it is on warm clear days. It is also different fighting a fire when the outside temperature is 15 degrees than it is when the temperature is 120 degrees. (See Figure 4.9.)

FIGURE 4.9 ◆ Cold weather also has an effect on firefighters.
(Courtesy of Minneapolis, Minnesota, Fire Department)

In fact, it is possible that what might have been a greater alarm fire might not even be reported because of the weather. This was strongly illustrated a number of years ago in the western part of the United States. A fire occurred in a large industrial building sometime around midnight. The first report the fire department received on the fire was the next morning when employees reported to work to find the building burned to the ground. During the night, the area had been blanketed by thick fog. No one had seen the fire, despite the fact that the building was located near a well-traveled street.

THE FIRE GROUND SIZE-UP

The fire ground size-up can be divided into two distinct phases—the preliminary size-up and the continuous size-up.

preliminary size-up The immediate estimate of the situation made by the fire officer who arrives on the scene first.

The Preliminary Size-Up. The **preliminary size-up** is the immediate estimate of the situation made by the fire officer who arrives on the scene first. It is made in accordance to his or her best judgment prior to the commitment of forces and prior to reporting to the dispatch office. It is at this point that all of the knowledge gained by the pre-alarm size-up, the response size-up, and the conditions observed on arrival are brought into focus. The preliminary size-up forms the basis for the initial deployment of personnel and equipment and for the calling for additional help if it is determined that it *might* be needed. (See Figure 4.10.)

FIGURE 4.10 ◆ The first arriving officer would report this fire as "coming through the roof."
(Courtesy of District Chief Chris E. Mickal, New Orleans Fire Department—Photo Unit)

FIGURE 4.11 ◆ Where is it going to go?

The preliminary size-up is the responsibility of the first arriving officer, regardless of rank or seniority. (See Figure 4.11.) It must be done with initiative and aggressiveness; however, it should not be made in a haphazard or indiscriminate manner. The size-up must be based on adequate knowledge and an intelligent survey of existing conditions. There is no place in aggressive firefighting for a first arriving officer who neglects the responsibility for the size-up and instead commits his or her company to the fire and leaves this responsibility for the arrival of a chief officer.

It is good practice never to underestimate the size and number of lines needed or the amount of equipment and personnel that might be necessary. There is every excuse for overestimation, but none for underestimation.

The Continuous Size-Up. The **continuous size-up,** as the name implies, is the continuous and comprehensive estimate of the situation as firefighting operations proceed. Things can change rapidly at a fire. Where it appeared that it might go in one direction, the fire may suddenly reverse itself and proceed in an entirely unexpected direction. Fires might suddenly appear in previously uninvolved portions of the building. Explosions, backdrafts, or flashovers can quickly change what appeared to be a controlled situation into a nightmare. The Incident Commander must constantly expect the unexpected and be prepared to change tactics to combat it.

continuous size-up The continuous and comprehensive estimate of the situation as firefighting operations proceed.

◆ **FIRE COMPANIES**

To do the job of planning strategy on the fire ground adequately, the Incident Commander must know the types of companies at the fire or due to arrive at the fire and what tactical tasks each of these companies can be expected to perform.

FIGURE 4.12 ◆ Engine companies require apparatus that carries hose and water, and is equipped with a pump.
(Courtesy of District Chief Chris E. Mickal, New Orleans Fire Department— Photo Unit)

FIGURE 4.13 ◆ Truck companies require apparatus that carries a complete assortment of ladders and sufficient tools and equipment to carry out the responsibilities assigned to the company effectively.
(Courtesy of District Chief Chris E. Mickal, New Orleans Fire Department— Photo Unit)

The type of company that is available at all fire emergencies is the engine company. (See Figure 4.12.) The second basic company provided on larger departments is the truck company (referred to as a ladder company in some parts of the country). (See Figure 4.13.) It is worthwhile to explore the tactical functions that are normally performed by each of these companies.

ENGINE COMPANY OPERATIONS

The basic objectives of an engine company are quite simple. They are to locate, confine, and extinguish the fire. (See Figures 4.14, 4.15, and 4.16.)

Locating the Fire. Most people envision a fire as lots of flames and smoke. That is not always true. From a firefighting standpoint, locating the fire encompasses more

FIGURE 4.14 ◆ The apparatus operator on an engine company is responsible for supplying firefighters at the end of a working line with water. *(Courtesy of Rick McClure, LAFD)*

FIGURE 4.15 ◆ Engine company members using both hand lines and heavy stream to make a direct attack on a fire. *(Courtesy of Rick McClure, LAFD)*

FIGURE 4.16 ◆ Engine company members using a portable monitor and a wagon battery on the fire. *(Courtesy of Rick McClure, LAFD)*

FIGURE 4.17 ◆ The location of
the fire is easy to determine,
but what is burning and where
is it likely to go?
*(Courtesy of District Chief
Chris E. Mickal, New Orleans
Fire Department—Photo Unit)*

than just searching for and finding the flames. Tactically, it includes (1) locating the
seat of the fire, (2) determining what is burning, (3) determining the extent of the
main body of the fire, and (4) evaluating where and how fast it is likely to travel.
(See Figure 4.17.)

Locating the Seat of the Fire.   The seat of the fire is the hottest part of the fire. It
is usually located at the point of origin, but not always. Because the seat of the fire is
producing heat faster than at any other location, maximum return on the water used
normally can be obtained if water is directed at this point. However, it is necessary to
find and identify this location.

The identification of the seat of the fire is usually not difficult if the so-called
burning materials are producing flame and the fire is relatively small. It becomes a
problem, however, when flames cannot be found. (See Figure 4.18.)

There is an old saying that where there is smoke there is fire. Don't you believe it.
Fire was previously defined as rapid oxidation accompanied by heat and flame (or
heat and light). It is possible to have a smoldering fire with some glowing embers, with
no flame but lots of smoke. It is also possible to receive an alarm of fire and on arrival
find a building or portion of a building filled with smoke but no visible signs of a fire.
In these cases it is probable that the fire has gone out or there has been a source of
heat that has produced a lot of smoke followed by the elimination of the heat source.
Finding the cause of smoke falls under the category of locating the fire.

FIGURE 4.18 ◆ It is easy to determine that the fire is in the attic, but how far has it extended and where is it going? *(Courtesy of District Chief Chris E. Mickal, New Orleans Fire Department— Photo Unit)*

The sense of smell will many times provide a clue as to the possible source of the smoke. The alarm classified as "food on the stove" is a typical example. At times the smell is so strong that the first arriving company officer can determine the source of the smoke when the apparatus first pulls up in front of the dwelling.

The sense of smell could also be useful in extremely light smoke conditions. Of course, if there is any amount of smoke whatsoever, breathing apparatus will be worn that will render the sense of smell useless.

Some burning materials such as cloth, rags, or insulation on electrical wires give off a distinct odor. If there is no smoke present, using the sense of smell may guide the search party immediately to the source; however, at other times it is extremely difficult to locate the source by using the sense of smell. This is particularly true if the source is not found immediately. The reason for this is that the sense of smell goes dormant after a period of time. A means of compensating for this is to bring in someone who has not previously been exposed to the smell and give him or her a try.

The appearance of the smoke will sometimes give a clue as to whether the source is still active. Heat will cause the smoke to be buoyant. It will rise and swirl about, the degree of which depends on the amount of heat providing the stimulus. Smoke in this condition is referred to as **live smoke. Dead smoke,** on the other hand, is not buoyant. It will remain relatively still and may collect in pockets. Dead smoke normally indicates that there is no active fire and that the source of the heat has been eliminated.

Regardless, it is necessary to locate the source of the smoke. This can be done only through a systematic search. The search party should be organized and each firefighter given a definite assignment. A group of firefighters wandering aimlessly about normally produces nothing.

Check all possible sources. One of the first things to check is electrical wiring, equipment, and appliances. Search out all electrical motors. Remember—motors can be found in refrigerators, air-conditioning systems, heating systems, escalators, elevators, washers, dryers, and so on. The best method for a firefighter to determine whether a motor could have produced the smoke is to place the back of his or her hand on the motor to see if it is hot. When doing this, the person should make it a rapid touch so as not to burn his or her hand if the motor has been overheated. Carefully check neon lights and fluorescent fixtures. Light ballasts in fluorescent fixtures

live smoke Heat will cause the smoke to be buoyant. It will rise and swirl about, the degree of which depends on the amount of heat providing the stimulus. Smoke in this condition is referred to as live smoke.

dead smoke Nonbuoyant smoke that remains relatively still and may collect in pockets.

are a common cause of trouble. Check electrical wires that are plugged into outlets to see if there is any smoke around the outlets, and also all electrical appliances for the possibility of recent heat. All of the walls should be observed to see if there is any discoloration of the wallpaper. This would most likely indicate a partition fire. The other walls should be felt with the hand to make sure that there is no fire hidden there. Check all locations, including storage closets and hallways. A thermal camera is very useful when making this search.

Sometimes the source of the smoke may be some distance from the body of smoke itself. A good example is a multiple-story building with the top floor filled with smoke. It is good practice to check out the top floor and the basement simultaneously when arriving at a scene where this condition exists. Fires in basements of multiple-story buildings have a habit of sending large quantities of smoke up unprotected vertical openings to fill the top floor with smoke.

Another type of occupancy where the location of the smoke can be deceiving is a string of mercantile occupancies with a common attic. The determination that the attic is filled with smoke should initiate a search of not only the attic but also all occupancies below the attic.

Do not be too quick to leave the premises if the source of the smoke cannot be found. Make sure that nothing is overlooked. Particular care should be taken in hotels and apartment houses. If necessary, every room in the building should be opened and checked. Sometimes an occupant will have a fire or an extinguished fire in his or her room and try to hide the fact from the firefighters. This is particularly true when there is something in the room that the occupant does not want outsiders to see. The individual will extinguish the fire himself or herself, not call the fire department, and when the firefighters knock on the door the occupant will respond with the answer "There's no fire in here." No harm can generally be done to accept this type of response when it is first given; however, do not fail to return to each of these locations and demand a "look-see" if the source of the smoke cannot be found elsewhere. Another reason to check every room is that the occupant may have had a fire, put it out, and was overcome by the smoke. There is at least one case on record where this occurred and only a diligent and thorough search by the firefighters saved the person's life.

Determining What Is Burning.   Determining the type and amount of material involved in the fire is extremely important in developing the tactics to be used for extinguishment. This knowledge is required for the selection of the extinguishing agent and, if water is to be used, the number and size of lines that will be required. However, obtaining this information is not always easy. The task is somewhat simplified if a previous inspection has been made of the building and the responding officers are familiar with what was stored in the area involved. Otherwise, if the fire is of any size, it might be necessary to estimate what is burning by what is seen and the type of occupancy involved. For example, from past experience firefighters know that in dwellings, office buildings, and small mercantile establishments the fire will normally be fed by various types of ordinary combustible materials. However, it is good practice not to become complacent because the materials involved may not be what is commonly expected in that type of occupancy. A hotter fire than normal or one that is moving faster than expected could indicate that an accelerant had been used to start the fire or that flammable liquids are stored on the premises. (See Figure 4.19.)

Another clue as to what is burning is the color, amount, density, and odor of the smoke. Materials such as roofing paper, asphalt, rubber, and petroleum products most often produce clouds of heavy black smoke. A white, grayish smoke usually indicates

FIGURE 4.19 ◆ Some fires extend very rapidly and are well involved when the first company arrives. Is this what was expected in this occupancy? *(Courtesy of District Chief Chris E. Mickal, New Orleans Fire Department— Photo Unit)*

a grass fire. Abnormal colors in the smoke, such as yellow and red, generally indicate that chemicals are involved. A normal room fire containing furnishings and painted walls will generally produce a dark gray smoke, whereas a structure such as a house or garage gives off a medium-brown smoke. If the structure and the contents are both well involved, the smoke produced may be almost black. Some products have characteristics all of their own. Magnesium burns with a brilliant white light, and hydrogen burns with no visible flame or smoke.

Determining the Extent of the Fire.   After determining what is burning, it is necessary to find out the extent of the fire. Is it confined to a single room or are several rooms involved? Has the fire spread to adjacent exposures? Is an entire floor of a multiple-story building involved or has the fire extended to upper floors? Answers to these questions will help determine the personnel, equipment, and number and size of hose lines required for confinement and extinguishment.

When evaluating the extent of the fire, it is important that all fire be located. There have been cases where firefighters have shut down their lines in preparation for picking up only to discover that the fire was burning fiercely in another portion of the building. Particular care should be taken in multiple-story buildings to check the floors both below and above the fire floor, and also the basement and top floor. Don't forget the possibility of partition fires, fires in hidden areas, and the potential of fire spreading to other portions of the building through air-conditioning ducts or unprotected vertical openings. Keep in mind the thought that the fire could have been started by an arsonist who has also set fires in other sections of the building.

Determining Where and How Fast the Fire Will Travel.   The construction of the building is the best clue as to where the fire will spread. Thought should be given as to whether the building has interior walls and fire doors that will prevent or contribute to the spread of the fire. How about open stairways and other unprotected vertical openings? In addition to hindering or contributing to the spread of the fire, the construction features also play an important part in how fast the fire will travel. The walls and ceiling may act only as a pathway for the fire travel, or they themselves may burn and contribute to the rapid spread of the fire.

FIGURE 4.20 ◆ A good indication that the fire is in the attic. How far has it extended and where will it go? *(Courtesy of Rick McClure, LAFD)*

When evaluating where and how fast the fire will progress, thought should be given to the four methods by which heat travels from one body to another. Of the four (direct contact, conduction, radiation, and convection), direct contact, radiation, and convection play the biggest part in rapid travel of a fire. Fire will travel horizontally and vertically by any one of these methods, with vertical travel being influenced primarily by convection. Fire will move in every direction until stopped by a barrier and will be pulled toward any opening that provides an outside source of oxygen. Consequently, determining where the fire will go is generally one of locating open pathways for its travel and outside openings that will pull it in that direction. (See Figure 4.20.)

Making an estimate of where and how fire will travel is necessary in order to put into effect the second function for which an engine company is responsible—confining the fire. Consequently, the information should be used to determine at what locations stands will have to be made in order to stop the spread of the fire, and how many lines it will take to accomplish the task. Planning should include conceding some areas that are lost and some that are not yet burning in order to get ahead of the fire. It will therefore be necessary when determining where and how the fire will travel to estimate where the fire will be by the time lines are laid and moved into position. Thought should be given to how much water will be needed to hold or stop the fire and how much is readily available. The basic principle involved is always make plans to *stay ahead* of the fire. Don't get caught in the trap of chasing it.

Confining the Fire. From a practical firefighting standpoint, the procedures required to confine a fire will vary with the type of fire. For example, different procedures will be involved for confining a brush fire than for confining one in a truck containing LPG or for a fire in a structure. However, the basic principles involved are similar. Consequently, confining the fire in structures will be discussed in this chapter. The variances as applicable to individual types of fires will be covered in later chapters.

From a theoretical standpoint, confining the fire means to prevent its spread beyond the point of origin. From a practical point this is impossible unless, of course, the fire is discovered when ignition first takes place and an extinguishing agent can immediately be applied to it.

From a practical rather than a theoretical standpoint, two different objectives can be established for the tactics used to confine a fire in a structure.

1. Prevent the spread of the fire from the building of origin, or from the buildings involved at the time of arrival of the fire department. Achieving this objective is generally a matter of protecting the external exposures and patrolling downwind from the fire for flying brands. Protecting the external exposures is normally accomplished by keeping the temperature of the exposure below its ignition temperature. This is normally done by wetting down the exposure, setting up a protective curtain between the exposure and the heating source, reducing the amount of heat produced by the source, or a combination of the three.

2. Prevent the interior spread of the fire both horizontally and vertically beyond the point where the fire is met once hose lines have been laid and are in place.

In the development of tactics to achieve these two objectives, it should be kept in mind that at the majority of all fires *both objectives are achieved by a rapid, aggressive attack on the fire.* In essence, this means that at all fires, one of the most effective methods of protecting the exposures is to reduce the amount of heat produced by the heating source.

Preventing External Spread.   Preventing external spread is primarily a matter of protecting the external exposures. (See Figure 4.21.) Following is a typical example of the tactics used to achieve this objective.

The department arrives on the scene to find one or two rooms of a one-story, single-family dwelling well involved with fire. Fire is shooting out the windows and the radiant heat is threatening the house next door. To keep the fire loss to a minimum, it is necessary to prevent the exposed house from bursting into flames and the fire from extending to other portions of the involved building. Both objectives can normally be achieved through the use of 1½- or 1¾-inch lines. Consequently, the general tactic is for the first arriving engine company to lay a large supply line to the pumper and work the smaller lines off the pumper. One of the smaller lines should be

FIGURE 4.21 ◆ Engine companies are responsible for extinguishing the fire and protecting the exposures.

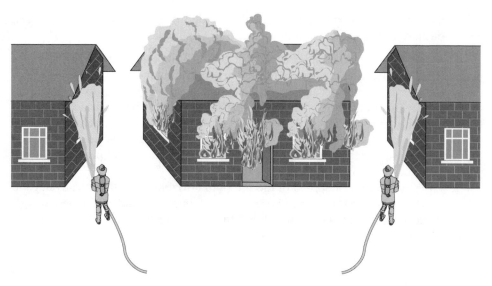

FIGURE 4.22 ◆ Use the first lines to protect the exposure.

advanced to a point where the firefighters can direct water onto the exposed dwelling while the second line is taken inside to prevent further spread and extinguish the fire.

Let's extend the example situation further. The first arriving company finds the dwelling well involved with fire. Fire is going through the roof and extending out windows on all sides. Houses on both sides of the burning building are in immediate danger of being ignited from the radiated heat. In this case, the survival profile indicates that there are probably no survivors and that very little in the burning building can be saved. Consequently, the objective is to prevent the adjacent buildings from becoming involved. Therefore, both of the small lines are used outside to protect the exposures. (See Figure 4.22.) One is taken in between the fire building and the exposure on one side with water being quickly projected onto the exposure. The other is taken to the opposite side and used in the same manner. The second engine company to arrive is given the responsibility of attacking the fire.

Many times this second type of operation requires that an explanation be given to the owner of the burning dwelling and perhaps also to the neighbors. The general public does not understand the principles involved and reacts only to what they have seen. They have probably been outside for a few minutes watching the fire get bigger, wondering why the fire department is taking so long to get there. The owners of the burning building are perhaps in tears, watching all they have worked for going up in flames. The fire department arrives on the scene, lays lines, and starts applying water on the buildings on both sides of the fire—and these buildings aren't even burning. The question immediately arises in the minds of the people watching: "What kind of fire department do we have?" It is much better to educate them now than to do so in the media after the unfavorable publicity has hit the press.

There is very little difference between the tactics used to protect the external exposures at a fire in a dwelling than those used at a fire in an industrial or commercial building. The big difference is in the size of lines required, the number of personnel needed, and the lack of experience of most firefighters with this type of operation. Let's pause for a minute on the last point. In some fire departments, or in some areas of larger cities with larger departments, it is possible for a firefighter to spend his or her entire career in the department and not handle a 2½-inch line or operate a heavy-stream appliance at a fire more than a half dozen times. This is not stated in a derogatory manner but more as a fact. Fortunately, the lack of experience is compensated for in most departments by training. Most of the initial training received by a recruit firefighter at the drill tower is devoted to preparing him or her to operate at a large fire. Furthermore, most of the monthly training in hose lays received at the company level is devoted to this objective. Consequently, although lacking in experience, operations at large fires do not come as a complete surprise to most firefighters.

Operations at a larger fire might progress in a manner similar to the following: The officer of the first arriving engine company finds a one-story industrial building partially involved with fire. Fire is shooting out the windows on the north side and threatening an adjacent one-story building. At this point the officer informs the dispatch center of the situation and requests the number of additional companies he or she thinks will be needed for the extinguishment and confinement of the fire. The officer may then lay a line and make an attack on the fire or use the line to protect the exposure, depending on his or her analysis of the seriousness of the threat to the exposure. If the decision is made to attack the fire, the officer will call for the next arriving company to protect the exposure. It may even be possible that two more companies will be needed for exposure protection.

When wetting down an exposure, it is generally best to direct the stream near the top of the wall and allow the water to run down the side. This procedure will keep the entire wall cooled. Of course, if any portion of the exposure is immediately threatened, it would be wetted down thoroughly prior to the above procedure being established.

The big difference in protecting this exposure and protecting the exposure at the dwelling is that 1½- or 1¾-inch lines could be used at the dwelling, whereas it is much more likely that 2½- or 3-inch lines or heavy-stream appliances will be needed at the industrial site. This means that a company that was capable of providing two lines at the dwelling fire will be restricted to handling one at the larger fire. (See Figure 4.23.)

In many cases the immediate threat to the exposed dwelling is more serious than the immediate threat of an industrial fire. The reason is that the exposure is closer to the source of heat at the dwelling fire and the exposure is generally made of material

FIGURE 4.23 ◆ Three firefighters are required on a 2½-inch line in order to maneuver it.
(Courtesy of Rick McClure, LAFD)

that will burn easier. However, at the industrial fire the exposed area is generally larger and the heat being given off by the source is more intense. In both cases the need and tactics are basically the same. Lay the line or lines required and keep the exposures wetted down until the threat of ignition has been eliminated.

Let's extend this fire even further. The first arriving officer finds the building well involved with fire and flames shooting out all sides and through the roof. There are exposures on all sides. The officer calls for the help he or she feels is needed and starts directing companies. (See Figure 4.24.) Protection of the exposures is achieved in this situation in the same manner as in the previous illustration. The difference is that it will take more lines and more personnel, and more decisions will have to be made as to the priority for the placement of lines.

The priority for placement of lines for exposure protection should be based on the order in which the exposures will become involved with fire if something is not done. Generally the fire on the leeward side of the fire will receive first priority; however, the distance between buildings, the construction of the exposures, and the intensity of the fire at various locations all will influence the priority arrangement. It should also be mentioned that with this type of fire, it is good practice to have an engine company patrolling downwind as a precautionary measure against flying brands. As a further precautionary measure, it is good practice to set up for the protection of buildings that are not exposures on arrival of the department but may become exposure problems before the fire can be successfully contained. (See Figure 4.25.)

The protection of exposures from fires in multiple-story buildings presents a somewhat different problem. Fortunately, as buildings rise in height, the distance between buildings generally increases and construction improves. The improvements, however, do not completely eliminate the exposure problem. Fire has demonstrated the ability to cross wide streets and ignite buildings on the other side. A prime example is the Burlington Building fire that occurred in Chicago a number of years ago. At this fire, radiated heat from a group of burning buildings carried across an 80-foot street. The result was the ignition of the upper 7 floors of a 15-story office building. Radiated heat was estimated to have reached a temperature of 1,800°F.

It is possible to protect the exposures to a multiple-story building fire by wetting them down if the exposures are within the reach of streams from water towers, ladder

Figure 4.25 ◆ Heavy streams can be used effectively to protect exposures. *(Courtesy of District Chief Chris E. Mickal, New Orleans Fire Department— Photo Unit)*

pipes, elevated platforms, or other heavy-stream appliances. The streams should be directed well above the heated area to allow the water to cascade down the wall and protect as much area as possible. It is normal for the exposure problem to exist over several floors with this type of fire.

Upper-story exposures can also be protected by working from within the exposed building. All exterior openings on the exposed side should be closed and, if needed, protective lines should be set up within the exposed building. It is best to lay several lines into the dry standpipe system of the exposed building whenever protective lines are to be set up from the interior of exposures over four stories in height. This will ensure an adequate supply of water at various floors for the use of the protective streams.

It is possible for an external exposure problem to exist above the reach of the streams from fire department equipment. When this situation exists, the protection of the exposures will be limited to the closing of external openings and the setting up of protective streams from the windows of the exposure. This will require more streams than if the protective streams from the exposed building are used in conjunction with lines wetting down the exterior of the exposure. In such situations it should be remembered that closed windows do not fully protect the interior of the building from radiated heat.

There is one last point to discuss on the protection of external exposures. Hose streams set up for the protection of exposures do not normally need to be directed continuously on the exposure. Although the lines should be maintained in position to provide the necessary protection to the exposure, the stream can periodically be directed on the main body of fire and redirected to the exposure as the need arises.

Preventing Internal Spread.   Preventing the interior spread of the fire is primarily a matter of stopping its progress both horizontally and vertically. The problem is normally somewhat simpler in a one-story building than it is in a multiple-story building because it is necessary only to stop its horizontal progression. However, regardless of the size or construction of the building, each fire is an entity within itself and no two are exactly alike. Consequently, it is not practical to try and establish hard and fast

FIGURE 4.26 ◆ Walls and doors prevent horizontal fire spread.

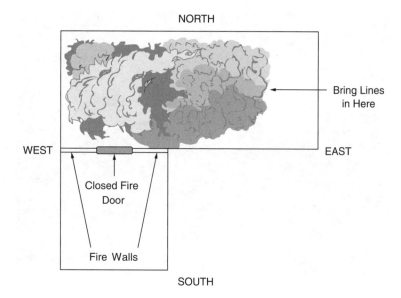

rules for the placement of hose lines used to cut off the progress of a fire; however, it is possible to examine some techniques that have proven to be valuable in confining an interior fire.

To stop the internal spread of the fire it is necessary to get cutoff lines into position at every location where the fire will travel if it is not stopped or extinguished. As a review, fire will travel horizontally in all directions and will travel vertically up any unprotected opening. However, this does not mean that it is necessary to bring lines in from all sides at every fire as the fire is normally prevented from spreading in one or more directions by walls, fire doors, and so on. For example, in Figure 4.26 the fire will spread horizontally only to the east, whereas in Figure 4.27 it will spread both

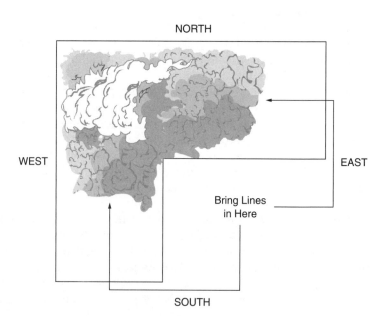

FIGURE 4.27 ◆ Lack of building construction aids increases the fire spread problem.

southerly and easterly. In the situation shown in Figure 4.26, it will be necessary to bring lines in from the east to stop the progression, but in Figure 4.27 it will be necessary to bring lines from both the east and the south.

When making plans to get lines into position to stop the spread, it is important to remember that it takes time to lay lines and move them into position. Consequently, it is essential to estimate how far the fire will travel while lines are being laid and moved into position. The fire will probably progress past the point where it was planned to stop it if the estimation were faulty. Additional lines will have to be laid and in some cases additional companies called to the fire if this happens.

If fire has an opportunity to travel both horizontally and vertically, its primary movement will be in a vertical direction. Not only will it move first in a vertical direction but it also will travel more rapidly in this direction due to its natural tendency to rise. Consequently, it is good practice to put the first hose line on the floor above the fire to cut off the vertical spread. Standpipes should be used for this purpose if they are available as this makes it possible to provide lines to other floors with little additional effort if they are needed. Extreme care should be taken when advancing lines above the fire. Firefighters should not be placed in a position where their retreat path could be cut off by the fire. (See Figure 4.28.)

Normally the most serious threat to vertical extension of the fire is open stairways. These vertical openings not only present an immediate danger from fire spread but also are a life hazard to anyone caught above the fire floor. It is important that the line taken above the fire be used to gain control of one of these openings. Sometimes

FIGURE 4.28 ◆ Lines should be taken to the floor above the fire to prevent the vertical spread by controlling the stairwell.
(Courtesy of Rick McClure, LAFD)

FIGURE 4.29 ◆ Gain control of the vertical openings.

two or more lines will be required for complete control in case more than one stairway extends from floor to floor. It is generally best to use a spray stream to stop the fire extension (see Figure 4.29); however, the line should not be opened until the heat developed on the lower floor becomes a threat to the upper floor.

If the fire is spreading horizontally at the same time in two or more directions, it generally will be necessary to make a decision as to where to place the first horizontal cutoff line. In making this decision, it is necessary to consider the speed of the fire spread, how far it can spread, the potential life hazard in its path, and what type of material will eventually become involved if the fire is not stopped. A general rule is that the first cutoff lines should be positioned to protect life. Second priority should be given to the protection of any hazardous materials in the path of the fire. Not only will the control problem be intensified if these materials become involved but their burning may present a serious threat to the health or lives of the firefighters.

If there is no life hazard and no hazardous materials in the path of the fire, then it is good practice to place the first cutoff line in position to save the greatest portion of the building. For example, in Figure 4.26 the first line should be positioned to protect the eastern portion of the building. There is, however, an exception to this concept. If there were extremely valuable contents in the smaller portion of a building, then the first cutoff line should be positioned to protect this area. As an example, if the fire loss would be much greater if the southern portion of the building in Figure 4.28 were destroyed than if the eastern portion were lost, then the first cutoff line should be positioned to protect the southern portion of the building.

Extinguishing the Fire. Most fires encountered by fire departments are of the Class A type. The primary extinguishing agent for Class A fires is water. Tactics used for extinguishment when water is used basically consist of determining the type of attack to be made, selecting the number and size of hose lines that will be required to extinguish the fire, and moving the lines into positions that will extinguish the fire most rapidly with the least amount of damage.

Water Application.   Water is generally projected onto the fire by the use of fire streams during firefighting operations. A **fire stream** may be defined as a stream of water from the time it leaves a nozzle until it reaches the point of intended use, or until it reaches the limit of its projection, whichever occurs first. As it leaves the nozzle, the stream is affected by such factors as nozzle pressure, nozzle design, and, with adjustable nozzles, the adjustment setting.

There are a number of different types of nozzles available for stream projection; however, all of them provide two basic types of streams: straight streams and fog streams. **Solid streams** may be produced both by smooth bore tips and by straight streams by fog nozzles adjusted for straight stream application. (See Figure 4.30.)

There is, however, one nozzle identified as a multipurpose nozzle that is capable of producing a straight stream and a spray stream at the same time. (See Figure 4.31.) Otherwise, **spray streams** are produced by fog nozzles. (See Figure 4.32.)

Smooth bore tips are available for use on both hand lines and master stream appliances. The standard nozzle pressure for those used on hand lines is 50 psi. The standard nozzle pressure for heavy stream appliances is 80 psi.

Following is the discharge from smooth bore tips used on hand lines at a nozzle pressure of 50 psi.

½"	53 GPM	1"	210 GPM
⅝"	82 GPM	1⅛"	266 GPM
¾"	118 GPM	1¼"	328 GPM
⅞"	161 GPM		

fire stream   A stream of water from the time it leaves a nozzle until it reaches the point of intended use, or until it reaches the limit of its projection, whichever occurs first.

solid stream   May be produced by both smooth bore tips and by straight stream/fog nozzles.

spray stream   One of the two basic types of streams produced by a nozzle.

FIGURE 4.30 ◆ Fire stream, straight stream.
(Courtesy of Akron Brass Company)

FIGURE 4.31 ◆ Combination nozzle.
(Courtesy of Akron Brass Company)

FIGURE 4.32 ◆ Fog nozzle.
(Courtesy of Akron Brass Company)

The discharge from smooth bore tips used on master stream appliances at a discharge of 80 psi is:

1⅜"	502 GPM	2"	1063 GPM
1½"	598 GPM	2¼"	1345 GPM
1⅝"	702 GPM	2½"	1660 GPM
1¾"	814 GPM	3"	2391 GPM
1⅞"	934 GPM		

FIGURE 4.33 ◆ Nozzles. *(Courtesy of Task Force Tips)*

Fog nozzles are available for use on both hand lines and master streams. The standard nozzle pressure for fog nozzles is 100 psi for both hand lines and master stream appliances. However, some fog nozzles designed for hand lines operate at a nozzle pressure of 50 psi or 75 psi. These nozzles are specially designed for use on high-rise fires or fires where it is desirable to restrict the nozzle pressure.

There are a number of different types of fog nozzles available.

Automatic nozzles are designed to maintain a constant nozzle pressure regardless of the flow. (See Figure 4.33.)

Fixed-gallonage or constant flow nozzles are designed to provide the same flow at a specific nozzle pressure regardless of the setting of the flow pattern.

Selectable-gallonage or **manually adjustable nozzles** provide a constant flow regardless of the stream pattern. These nozzles enable the nozzle operator to select a discharge rate that he or she deems best suitable for the task at hand. A series of flow settings is available that permits the selection of the desired flow at a predetermined nozzle pressure.

The selection of the size and number of lines to be used for the attack is a matter of determining the amount of water that will be required to remove heat from the fire faster than it is being generated. A number of departments have completely abandoned the use of smooth bore tips and rely entirely on fog nozzles for both hand lines and master stream appliances.

The amount of water discharged from a nozzle tip depends on the size of the tip and the discharge pressure. It is impossible on the fire ground to determine the exact amount of water being projected from various streams at any one moment; however, certain guidelines can be used to achieve the desired results. These guidelines were taken from the textbook *Fire Department Hydraulics* by Eugene Mahoney.

The discharge from 1-inch lines ranges from 20 GPM to 60 GPM with an average of approximately 40 GPM. These lines are effective for extinguishing fires in trash and grass fires.

The discharge from 1½-inch lines ranges from 40 GPM to 150 GPM with an average of approximately 100 GPM, whereas the discharge from 1¾-inch lines ranges from 95 GPM to 200 GPM with an average of about 150 GPM. Using two 1½- or 1¾-inch lines is normally sufficient to knock down a fire in one or two rooms of a dwelling or fires of similar size in commercial or industrial buildings.

automatic nozzles Automatic nozzles are designed to maintain a constant nozzle pressure regardless of the flow.

fixed-gallonage nozzles See "constant flow nozzles."

constant flow nozzles Nozzles designed to provide the same flow at a specific nozzle pressure regardless of the setting of the flow pattern. Also known as fixed-gallonage nozzles.

selectable-gallonage nozzles Nozzles that provide a constant flow regardless of the stream pattern. Also known as manually adjustable nozzles.

manually adjustable nozzles See "selectable-gallonage nozzles."

FIGURE 4.34 ◆ Each of the heavy streams shown here is throwing a minimum of 500 GPM into the fire area.
(Courtesy of District Chief Chris E. Mickal, New Orleans Fire Department—Photo Unit)

A hose stream from a 2½- or 3-inch line will produce an average of about 300 GPM. A 2½-inch line is the largest hand line used in the fire service. These lines should be applied to large fires in commercial and industrial occupancies and, in some cases, on dwellings that are well involved. They are particularly effective on well-involved dwellings when fog nozzles are employed.

Some departments prefer to use 2-inch lines rather than 2½-inch lines. The 2-inch lines produce large volumes of water when equipped with automatic nozzles and are more mobile than 2½-inch lines.

Standard tips used on heavy stream appliances discharge a minimum of 500 GPM. Both straight and fog tips are available for use on heavy-stream appliances. A 1⅜-inch straight tip will discharge approximately 500 GPM, a 1½-inch tip about 600 GPM, a 1¾-inch tip about 800 GPM, and a 2-inch tip about 1,000 GPM. The discharge from fog tips varies depending on their size, setting, and nozzle pressure. Some fog nozzles produce a constant volume of water. (See Figure 4.34.)

Good judgment must be used when selecting the size of line to be used. All the facts must be considered when weighing the value of a small line against that of a larger one. Small lines offer the advantage of mobility and ease of operations, whereas large lines offer volume, reach, and striking power at the cost of reduced mobility. One of the basic principles that should be considered when making the decision as to the size of line to use is *never underestimate the size of the fire*. It is better to overkill than have to back out because there is not sufficient water to do the job.

Attacking the Fire. There are two basic types of attacks that can be made on the fire for the purpose of extinguishment: direct attack and indirect attack.

FIGURE 4.35 ◆ The beginning of a direct attack. *(Courtesy of District Chief Chris E. Mickal, New Orleans Fire Department—Photo Unit)*

The Direct Attack.   A **direct attack** is one whereby water is applied directly to the fire. (See Figure 4.35.) The theory behind the direct attack is that the best method of extinguishing the fire is to eliminate it at its source, or what is referred to as the seat of the fire. The seat of the fire, as defined earlier, is the hottest part of the fire and is usually found at the point of origin. Because a large amount of heat is being generated at this point, water applied at this location will exert its most effective cooling action.

Both straight streams and fog streams are used to make a direct attack on the fire. Fog streams are normally used when 1½- or 1¾-inch handheld lines are used in a direct attack. A 1½-inch line can be physically handled by one firefighter; however, for safety purposes it is better that two firefighters be used to handle both 1½- and 1¾-inch lines. The stream pattern from a fog nozzle provides better protection for the firefighter and permits a closer attack on the seat of the fire than when straight streams are employed.

The direct attack method is almost universally used on large fires that have gained considerable headway, such as lumberyard fires, fires that have burned through the roof in industrial and commercial occupancies, fires that have burned through to the outside, or fires that are well involved. With these types of fires, both straight streams and spray streams are used on handheld 2- and 2½-inch lines. Both types of streams are also used when heavy-stream appliances are placed in operation.

direct attack Refers to fighting the fire itself with the use of water, throwing dirt on it, using beaters, or otherwise directly fighting the fire.

To be able to maneuver a 2½ inch handheld line effectively requires a minimum of three firefighters. (See Figure 4.36.)

The straight stream offers distinct advantages in this type of attack. It has the reach, penetration, and striking power necessary to bore into the fire through radiant heat that would boil off a spray stream before it reached the seat of the fire. A straight stream, however, forces the firefighters to operate from some distance back due to the heat produced by the fire. Large amounts of water are required on fires of this type because the water strikes the burning material and runs off too quickly to absorb a maximum quantity of heat. As a result, water damage may be considerable when attacking fires inside of buildings.

Many times a direct attack on the fire is made by the use of heavy-stream appliances. A heavy-stream appliance is a piece of equipment that discharges more water than can be projected from a handheld line. The minimum amount of water normally discharged from a heavy-stream appliance is about 500 GPM; however, nozzles producing a lesser amount of water are sometimes substituted for the standard tips. Streams from these appliances are referred to both as *heavy streams* and *master streams*. Heavy streams take time to set up, but there should be little hesitation to use them if the volume of the fire warrants or if life hazard is a factor. They are excellent for making a quick knockdown of the main body of fire, thus providing a means for gaining access to the fire by the use of hand lines. It is good practice to use heavy streams only as long as necessary, and *never* after firefighters with hand lines have entered the area. Firefighters inside must be removed first.

One of the basic tactical principles that should be used with the direct attack is to bring in hose lines from the uninvolved portion of the building and push the fire back into the involved portion. Although this principle seems simple, it is one of the most often violated. The result is that fire is frequently pushed unnecessarily into uninvolved areas of the building. The reason the principle is violated is that firefighters are basically aggressive and want to hit the fire where they see it. For example, a company pulls up in front of a dwelling that has one of the front rooms well involved with fire. Fire is shooting out the windows in front and on the side. Firefighters lay lines and fight their way to one of the windows, hitting the fire as

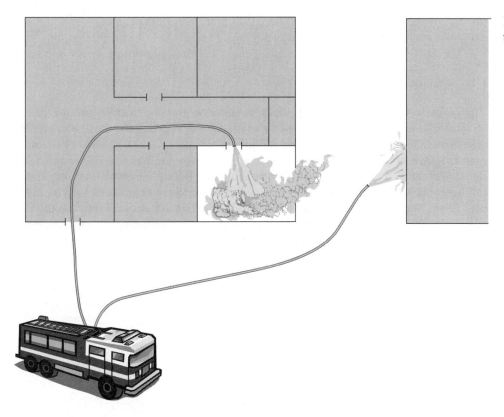

FIGURE 4.37 ◆ Push the fire out, not in.

they proceed. When they reach the window they direct the stream inside in an attempt to achieve a fast knockdown. The result is that they probably push the fire and its smoky by-products to other portions of the building. The proper tactic to use is illustrated in Figure 4.37.

One line is taken in through the front door and advanced down the hallway to the fire area. The fire is aggressively attacked and pushed out the window while being extinguished. The outside line is protecting the exposures, which includes nonburning portions of the involved building.

Another example of this tactical principle is illustrated in Figure 4.38. One end of an industrial building is well involved with fire. Lines are brought in from the opposite end and the fire is attacked. This procedure restricts the damage to the involved

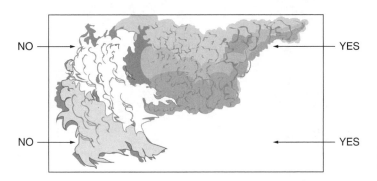

FIGURE 4.38 ◆ Bring lines in from the uninvolved portions of the building.

portion of the building. If lines had been brought in from the opposite end, the fire probably would have been pushed throughout the entire area and the building lost.

The principle should also be kept in mind for fires in multiple-story buildings. Figure 4.38 illustrates a room fire on the third floor of a five-story hotel. Two lines are being advanced simultaneously.

indirect attack With an indirect attack, the water is not directed at the seat of the fire as with the direct attack, but rather above the fire and into the heat that has built up at the ceiling level.

The Indirect Attack.   An **indirect attack** is made on the fire through the use of fog streams. The indirect method is based on the theory of securing the maximum cooling effect from the use of the water applied. With the indirect attack, water is applied under high pressure in the form of a wide-angle spray or in a fog pattern. The water is not directed at the seat of the fire as with the direct attack, but rather above the fire and into the heat that has built up at the ceiling level. The line is swirled about to project the water over as great a ceiling area as possible. Some departments prefer to use the "Z" pattern. This involves starting at the ceiling toward the left, sweeping across to the right, then down to the left, and across the lower level of the room to the right. The overall object is to cool the fire area to a temperature below the ignition temperature of the material burning. (See Figure 4.39.)

With the indirect method of attack, theoretically each drop of water soaks up the maximum amount of heat and is converted into steam. The extinguishing effect is somewhat twofold:

1. The temperature is lowered so rapidly that heat is absorbed faster than it is being generated.
2. Some degree of smothering effect is produced by the steam.

Some of the advantages claimed for the indirect method of attack are:

1. A minimum of water is used for extinguishment because the maximum cooling is being obtained from the water used.
2. Water damage is held to a minimum because nearly all the water used is converted to steam.
3. Firefighters are subjected to far less punishment because the attack is normally made from outside the fire area, such as from a doorway or window.
4. A faster knockdown is generally achieved.

Two conditions are essential for the most effective application of the indirect method. First, it must be a fire that is generating a considerable amount of heat.

FIGURE 4.39 ◆ The indirect method of attack.

Second, the fire must be burning in an area that is not vertically vented and preferably one that is not vented at all.

Compressed Air Foam Systems (CAFS) for Class A foam. The technology of **compressed air foam systems (CAFS)** has been available to the fire service for a number of years. Although a few departments have adopted the technology, there are still a large number who are doubtful of its claims. Proponents of the technology claim that the system is capable of extinguishing a fire using less water, provides a much greater knockdown capability, lightens the weight of hose lines used for extinguishment, reduces the stress placed on firefighters, and increases the overall efficiency of firefighting.

A CAFS consists of a water source, a pump, a foam proportioning system, an air compressor, and a well-designed system of controls that ties all the components together for effective pump operation.

The air compressor provides the energy that, gallon for gallon, propels compressed air foam farther than do aspirated systems or standard water nozzles. The reach of the fire stream is much greater than the reach of conventional fire streams. The small air bubbles produced are of uniform size and very durable.

There are five different foam consistencies available for producing Class A foam. These are classified as Very Dry, Dry, Medium, Medium-Wet, and Wet.

Type 1 foam is classified as Very Dry. The air-to-foam solution has a ratio of 44 to 1. This proportion rate produces a very dry, fluffy texture blanket that easily clings to vertical surfaces. The blanket drains off slowly and can be adversely affected by winds.

Type 2 foam is classified as Dry. This foam solution has a 22-to-1 ratio. The finished foam will have a consistency very close to that of shaving cream. It will drain faster than Type 1 foam but will not immediately run on vertical surfaces.

Type 3 foam is classified as Medium. This foam has a ratio of 15 to 1. It is considered the best foam for all-around use. It has a consistency similar to watery shaving cream and will not readily cling to vertical surfaces.

Type 4 foam is classified as Medium-Wet. The ratio of this foam is 11 to 1. It has a very fluid consistency and has an excellent ability to penetrate porous material.

Type 5 foam is classified as Wet. The ratio of this foam solution is 8 to 1. This type is generally considered the best to use for overhaul operations. It is very watery and readily runs off vertical surfaces.

There are a number of advantages claimed for the CAFS. One is that the hose lines carrying the foam are much lighter than those carrying other foams or plain water. It is estimated that the weight of the hose lines is reduced approximately one-half to two-thirds. This is an added advantage when lines are used on aerial devices to supply master streams. Additionally, the increased length of the stream allows an effective attack to be made on the fire from a much greater distance. These two factors result in less fatigue of firefighters and increase the safety to crews by allowing them to operate at a greater distance from the fire.

Some of the other advantages claimed for using a CAFS for attacking Class A fires, particularly structure fires, are:

1. Class A foam reduces the surface tension of plain water, which improves the water's wetting and penetrating capabilities. One estimate advanced is that the surface tension of plain water is reduced by two-thirds when a 3 percent Class A foam solution is used. This feature allows the water to spread out quicker, which provides a larger contact with the burning surface. The overall result is a quicker absorption of the heat.

compressed air foam system (CAFS) System for knocking down fires using less water.

2. A quicker knockdown is achieved. Some tests conducted by the Los Angeles County Fire Department indicated that the CAFS knocked down the fire in approximately one-fifth the time that plain water knocked it down.

3. Less water is required for knocking down the fire. The Los Angeles County tests also concluded that it took 4.7 times more plain water to extinguish a fire than when using a CAFS with Class A foam. The water savings was not achieved by using a lower fire flow but by knocking down the fire quicker. The overall result was that there was less damage to the building and contents, and there was less contaminated water runoff.

4. The Class A solution adheres to the burning Class A materials. This action produces a seal that insulates the burning material. The adherence feature is also valuable when using a hose line to protect exposures. Test results have shown that the application of Class A foam by a CAFS on exposed structures is up to twenty times as effective as using plain water.

5. The time required to overhaul is less, and the potential for rekindle is reduced.

There are, however, some disadvantages regarding the use of a CAFS.

1. The cost to add a CAFS to a pumper's specifications is considerable. This becomes a debatable factor when budgets are already extremely tight.

2. There is also a continuing cost for maintenance, for the replenishment of the foam concentrate, and for the training of personnel.

3. Class A foam concentrate is a very strong corrosive detergent. It can corrode metal tanks, pump parts, and similar items. If handled carelessly, it can damage paint and the finish on fire apparatus.

4. Regular exposure to the concentrate can cause drying and chapping of exposed skin on members. It can be very irritating to the eyes and respiratory tract. It is strongly recommended that members handling the material wear rubber boots, gloves, and eye protection.

5. The foam created is more slippery than plain water. Personnel working in areas where it has been discharged should be aware of this factor.

6. The additional mechanical components together with the human factor associated with the production of CAFS foam provide the opportunity for more failures by both the equipment and the personnel involved. The critical part is that any failure could compromise the attack stream for those handling the attack lines.

7. There have been instances in which the firefighters inadvertently mixed Class A and Class B foam concentrates into the same tank. An error of this type can severely damage the foam-proportioning equipment.

As the result of the tests they conducted, the Los Angeles County Fire Department published six tactical lessons it learned.

1. CAFS used for interior attacks should be applied at the same rate as plain water. CAFS saves water by knocking down the fire faster, not by using a lower flow rate.

2. The initial nozzle reaction of a charged CAFS line is substantial.

3. An interior CAFS attack can often be made through a door or window. Best results are achieved when the stream is directed toward the ceiling.

4. Large amounts of steam are generated when CAFS hits a fire.

5. Even though CAFS reduces interior temperatures faster than water, the upper portions of the rooms will remain quite hot.

6. Always overhaul. LACOFD prefers a wet CAFS that maximizes surface penetration.

Selecting the Proper Method of Attack. The choice of the type of fire stream, the size of the stream, and whether a direct or indirect attack is to be made on the fire will depend on the total knowledge gained during the size-up and prefire plan. The question that must be answered is: Which configuration will extinguish this particular fire with the least amount of loss to life and property? Basically, the number and size of lines required and whether a direct or indirect attack will be made will be determined by the amount of heat present, where the fire is, and where it is likely to go.

On interior fires, small lines of 1½ or 1¾ inches will usually suffice if the heat will not prohibit entering the building. These lines discharge a sufficient volume of water to handle the fire and will permit the degree of maneuverability desired. However, larger lines should be placed into operation without hesitation if there is any doubt whatsoever as to the ability of small lines to control the fire. It is easier and quicker to reduce a line than it is to lay a new one. When selecting the size of lines to use, consideration should always be given to the progress the fire will make during the time required for placing the hose lay into operation.

TRUCK COMPANY OPERATIONS

In some parts of the country truck companies are referred to as ladder companies. (See Figure 4.40.) Regardless of the terminology, the functions performed by these companies are the same. It might be noted that all fire departments do not have truck companies. In these instances, the truck company's functions might be carried out by the second or third arriving engine company or by the squad company. This might also be the case in some rural areas that do not receive a truck company on the first alarm response due to the extreme distances involved. Members of truck companies do not normally become involved in actual fire extinguishment operations. However, the functions performed by members of these companies are extremely important to the successful, rapid control and extinguishment of a fire, to the safety of the occupants of the fire building, to the safety of the firefighters at the emergency, and to keeping the overall fire loss to a minimum. If the truck company members carry out their tasks effectively in a timely and coordinated manner, the punishment taken by engine company members in a hostile environment is reduced or eliminated. However, if they fail to do their job as they should, firefighters on the scene will take a beating and the fire

FIGURE 4.40 ◆ A truck company member using a pike pole to hold open a sliding metal door while an engine company member directs water on the fire. *(Courtesy of Rick McClure, LAFD)*

loss will be increased beyond what it should be. It has often been said that those chief officers who know how to use their truck companies effectively do the best job on the fire ground.

Truck company members are different from engine company members. Training officers attempt to drive all initiative out of rookies during their initial training. The objective is to get them to work and think as a member of a team, as their initial assignments are generally to an engine company in which teamwork is essential.

When assigned to an engine company, a firefighter is normally under the constant supervision of a company officer at an emergency where he or she generally works as a team member in the laying and handling of hose lines. The scenario is different when a firefighter is assigned to a truck company. Truck company members are taught to use their initiative and be able to work and think as individuals. Many times a pair of truck company members is on their own while carrying out their duties at an emergency. A truck company officer will direct a couple of members of his or her company to open up a roof, or to go to the fourth floor and make sure everyone is out of their rooms, or maybe to shut off the utilities. When a pair of truck company members are ordered to open the roof, where and what size hole to cut is normally left to their initiative and judgment. The same condition exists when they are ordered to make sure everyone is out of a particular floor, or where and what tools to use to shut off the utilities. Consequently, it is the practice of most departments to assign the more experienced firefighters to the truck companies. Most officers agree that good truck company personnel are worth their weight in gold.

A truck company is assigned a number of varied tasks. The best way to remember these is by the word **LOUVER.** The word stands for the following tasks:

LOUVER This word stands for: ladder operations, overhaul, controlling the utilities, ventilation, forcible entry, and rescue.

1. *Ladder* operations
2. *Overhaul*
3. Controlling the *utilities*
4. *Ventilation*
5. Forcible *entry*
6. *Rescue*

The tasks of overhaul, ventilation, and rescue were thoroughly explored in Chapter Two. The functional tasks of laddering, controlling the utilities, and forcible entry will be examined in this chapter.

Laddering Operations. Laddering operations basically include raising or using ladders for the purpose of making physical rescue, gaining entrance into a building, providing a path for hose lines, gaining access to various portions of the building, and various other phases of firefighting and fire or water control. (See Figures 4.41 and 4.42.) Both **straight ladders** and extension ladders are carried on truck companies for performing needed operations. A large range of sizes and types of ladders are carried, varying from smaller ladders, which can be effectively handled by one person, to longer extension ladders, which require the coordinated efforts of six people.

straight ladders TK

It is not the intent of this section to illustrate the various methods of raising the different ladders carried on truck companies. Firefighters should learn this in their initial training as rookies. Rather, the intent is to indicate some basic principles for handling ladders that can be applied to various incidents on the fire ground.

FIGURE 4.41 ◆ A good illustration of a well-laddered incident.
(Courtesy of Rick McClure, LAFD)

FIGURE 4.42 ◆ A good example of a truck company raising ladders to both the fire building and to an exposure.
(Courtesy of District Chief Chris E. Mickal, New Orleans Fire Department—Photo Unit)

Types of Ladders. A truck company carries a variety of different types of ladders. Some of them are designed for general use whereas others are used for a specific purpose. The following are some of the types of ladders that can be expected to be carried on a truck company. Some of these types are also carried on other companies.

Single Ladders (Straight Ladders). **Single ladders (or straight ladders)** are nonadjustable in length and are constructed as a one-section ladder. These ladders come in lengths that generally vary from 12 to 24 feet. They are very stable and generally easy to raise into position for use. Straight ladders are ideal for climbing to the roof of one- and two-story structures.

Roof Ladders. **Roof ladders** are single ladders that are equipped at the tip with folding hooks. The folding hooks provide a means of anchoring the ladder over the roof ridge or other roof part. Roof ladders are designed to lie flat on the roof. This provides a safe working platform for a firefighter doing roof work. When

single ladders Single ladders (also called straight ladders) are nonadjustable in length and are constructed as a one-section ladder.

roof ladders Roof ladders are single ladders that are equipped at the tip with folding hooks.

FIGURE 4.43 ◆ A roof ladder being used as a straight ladder.
(Courtesy of Las Vegas Fire & Rescue PIO)

necessary, roof ladders can be used for the same purposes as straight ladders. (See Figure 4.43.)

extension ladders Extension ladders are adjustable ladders that can be raised to various heights.

Extension Ladders. **Extension ladders** are adjustable ladders that can be raised to various heights. These ladders have a base section and one or more fly sections. Their adjustable length removes some of the guesswork required when selecting straight ladders. These ladders are heavier and require more personnel for lifting. Extension ladders generally range in length from 12 to 35 feet.

Bangor ladders Bangor ladders are extension ladders that use stay poles for added leverage and stability.

Bangor Ladders. **Bangor ladders** are extension ladders that use stay poles for added leverage and stability. They are available with two to four sections. The maximum length of Bangor ladders is generally 50 feet. A 50-foot Bangor takes the coordination of six firefighters to raise it. NFPA 1931 requires that all extension ladders that are 40 feet or longer be equipped with stay poles. Few departments still use a 50-foot Bangor ladder.

combination ladders Combination ladders are designed so that they can be used as extension ladders or converted to stepladders.

Combination Ladders **Combination ladders** are designed so that they can be used as extension ladders or converted to stepladders. They range in length from 8 to 14 feet with the 10-foot ladder being the most popular. A 10-foot combination ladder is ideal as an inside ladder due to its versatility and length.

collapsible ladders Collapsible ladders are versatile ladders that are easy to carry and position by one firefighter.

Collapsible Ladders **Collapsible ladders** are versatile ladders that are easy to carry and position by one firefighter. When placed in position, they have the appearance of a straight ladder.

Raising Ladders. The best angle for climbing a ladder is 75 degrees. (See Figure 4.44.) This angle makes climbing easier and utilizes the maximum strength of the ladder. Experienced firefighters do a fairly good job of approximating this angle by making a quick evaluation of the situation. However, some guidelines have proven to be useful.

The distance that the ladder should be placed from the building in order to obtain a 75-degree angle can be approximated by dividing the length of the ladder by 4. (See Figure 4.44.) For example, a 20-foot ladder should be placed approximately 5 feet from the building (20 divided by 4), and a 30-foot ladder should be

FIGURE 4.44 ◆ The best climbing angle.

placed approximately 7½ feet from the building (30 divided by 4). Some departments use the system of dividing the length of the ladder by 5 and adding 2. This method provides a slightly different answer but is just as effective in obtaining fire ground results.

When ladders are raised to a windowsill for the purpose of making ingress, they should be placed to one side or the other of the window and extended approximately three feet above the windowsill (approximately three ladder rungs). Placing the ladder in this position will make it easier and safer for anyone leaving the ladder to enter the building, particularly if the person has a hose line on his or her shoulder. In fact, if lines are taken up a ladder and into the building through a window, it is generally better to place the ladder to the side of the window opposite of the shoulder on which the line is carried. As an example, if it is the practice of a department to carry the hose on the right shoulder, then the ladder should be placed to the left side of the window.

If the ladder is raised to a window for the purpose of making a rescue, it is generally best to place the tip of the ladder at or slightly below the windowsill. This position generally makes it easier to remove people from the building.

When ladders are raised to the roof, they should be extended approximately five feet above the roof (about five ladder rungs). This position makes it easier to leave the ladder to gain access to the roof, particularly if carrying a hose line. It also makes it easier to leave the roof and to see the ladder if departure must be hastened. It is good practice to paint the tips of all ladders white or fluorescent orange. Not only can the white or fluorescent orange tip be seen more clearly through smoke if it becomes necessary to make a rapid retreat from the roof, but it also makes it easier to see the end of the ladder when it is being lowered into a window or roof edge. (See Figure 4.45.)

For safety purposes, it is best that all ground ladders be footed when anyone is climbing or descending, unless the ladder is secured to the building. **Footing** means that a firefighter is stationed at the bottom of the ladder with his or her toes placed

footing Stationing a firefighter at the bottom of a ladder with his or her toes placed against the heel of the ladder or one foot placed on the bottom rung for the purpose of securing the ladder.

FIGURE 4.45 ◆ This ladder had been extended well above the roof parapet. Notice how safely a firefighter can step from the ladder to the parapet and then onto the roof. *(Courtesy of Rick McClure, LAFD)*

against the heel of the ladder or one foot placed on the bottom rung. The firefighter should grasp the beams to help steady the ladder. (See Figure 4.46.)

It is good practice to secure the ladder to the building at the top if it is to remain in one position for any length of time. This not only relieves a firefighter for other duties but also is a safety factor in the event someone may find it necessary to climb down the ladder when no one is available for footing.

It should be kept in mind that the 75 degree climbing angle and the distance that the ladder should be above the windowsill and roof are ideals; however, these ideals are seldom achieved on a regular basis on the fire ground. Generally, the more training and experience the members of a truck company have, the closer these ideals are approached at an emergency. It should also be remembered that safety takes precedence over the ideal angle. For example, better ladder security will generally be provided if the butt of the ladder can be placed in a crack in the cement rather than on a flat smooth surface. If such a spot is near the intended grounding position, then this should be taken even if there is a slight sacrifice in climbing angle.

A number of operational tactics should be used in order to provide safety for members operating ladders. Some of these are:

- ◆ Always wear protective gear, including gloves, when working with ladders.
- ◆ Make sure that ladders are not raised into electrical wires. This is extremely important when working with metal ladders.
- ◆ Make sure that the proper size ladder is selected for the job to be done.

FIGURE 4.46 ◆ Footing a ladder provides stability and safety for members climbing the ladder.
(Courtesy of Eugene Mahoney)

- Check to make sure that the ladder is adjusted for the proper climbing angle.
- Use the proper number of personnel for raising a ladder.
- Make sure that the pawls are properly secured over the rungs when an extension ladder is raised.
- Secure the ladder to the building if the ladder is to remain in one position for a period of time.
- Make sure the ladder is footed or secured to the building whenever a firefighter is climbing it.
- Do not overload a ladder by allowing too many firefighters to climb it at the same time.
- Be sure to lock in to a ladder with a leg lock or a ladder belt when working from the ladder.

Selecting a Ladder: A ladder should be selected that will achieve the ideal most closely; however, there are no set standards as to the location of windows, the distance between floors of buildings, or other factors that affect the proper selection of a ladder. Experience, training, and good judgment are required to compensate for the lack of standards.

As mentioned before, there is no hard and fast rule regarding the distance between floors of buildings, but certain guidelines might be useful in the selection of ladders. The average distance from floor to floor of residential buildings is approximately 9 feet, and the average distance from floor to floor in commercial buildings is 12 feet. Windowsills are generally about 3 feet above the floor in both types of occupancies. Parapets on roofs of commercial buildings are generally about 3 feet in height.

One of the reasons that the ideal is seldom achieved on the fire ground is that the exact-sized straight ladder required to reach the desired height is not carried on the apparatus. Another reason is that if it were carried, it might not be available at the time needed. Consequently, the best that truck personnel can generally do is to come as close as possible to achieving the ideal. The following, although not exact, will provide some guidelines for the selection of ladders.

In a residential building, a 16-foot ladder placed on the windowsill of the second floor will extend approximately 3 to 4 feet above the sill. A 24-foot ladder placed on the windowsill of the third floor will extend approximately 2 to 3 feet above the sill.

A 20-foot ladder placed on the second floor window sill of a commercial building will extend approximately 4 feet above the sill, whereas a 30-foot ladder placed on the third floor windowsill will extend approximately 2 to 3 feet above the sill.

There is a greater variance in roof heights in buildings than there is distance between floors. This makes it difficult to establish guidelines for ladders to be used to reach the roof. However, more than likely an extension ladder will be used for roof operations. Consequently, some guidelines can be established as to the approximate length of ladders required to reach roofs, and extension ladders can be adjusted accordingly. Keep in mind that the following are only approximations:

	Residential Roof	*Commercial Roof*
Two stories	24 feet	35 feet
Three stories	35 feet	47 feet

Truck companies carry both straight ladders and extension ladders. Straight ladders are used to best advantage for gaining access to roofs of one-, two-, and sometimes three-story structures and into second and third floor windows. Extension ladders can be used in almost all locations that straight ladders can be used. If there is a choice between using a straight ladder and an extension ladder, the extension ladder should generally be chosen because of the variable height factor. Extension ladders, however, are generally heavier than straight ladders, which may limit their use when the number of personnel is a critical factor.

One other thought should be kept in mind. After selecting and raising a ladder, if it does not reach its intended point, take it down. A ladder left in this position serves no useful purpose and presents a constant threat to firefighters.

Placement of Ladders.   It is good practice to ladder all exposures as well as the fire building. Ladders should be raised not only where they are needed at the time but also wherever it is anticipated that they might be needed. However, more ladders than are needed should not be raised at the expense of inside truck work.

Ladders should be raised as close as possible to the point of work. For example, they should be raised to the roof at such locations that firefighters will not have to travel long distances on slippery, steep roofs or over obstacles.

Ladders should not be placed over louvers, windows, or other openings that are likely to emit heat or fire. Always place ladders in as secure a position as possible. Constantly consider the possible risk to firefighters. When firefighters are working on the roof, ladders should be provided to ensure that there are at least two ways for firefighters to escape from the roof.

In addition to the placement of ladders, keep in mind that ladders can be used effectively for other purposes such as bridging from one building to another, and for breaking glass for the purpose of ventilation.

FORCIBLE ENTRY

Forcible entry is the process of using force to gain entrance into a building or secured area. (See Figure 4.47.) Using force at an emergency becomes necessary when the first arriving unit finds the building secured and entry must be made for the purpose of search and rescue, locating the fire, extinguishing the fire, or carrying out any of the functions required during firefighting or rescue operations. Forcible entry operations may also become necessary once entry is gained if doors or other areas inside the building are found locked or secured.

forcible entry The process of using force to gain entrance into a building or secured area.

In most cases, entry into buildings is made through doors or windows, but occasionally it may be necessary to use more drastic means to gain entrance. Also where, when, and how a building is entered will vary from one situation to another. Although there are no set rules that apply to every situation, several basic principles should be considered prior to using forcible entry techniques.

1. Before attempting to gain entrance into a building through a door, regardless of whether it is locked or unlocked, feel the door to make sure it is not hot to the hand. A hot door is

Figure 4.47 ◆ A firefighter using a haligan door opener to demonstrate the proper method of using it to force open a door.
(Courtesy of Eugene Mahoney)

FIGURE 4.48 ◆ Glass is less expensive than a door.

In Lieu of Forcing a Door, Break a Small Panel of Glass and Reach Inside to Open the Door.

an indication that a possible backdraft condition exists inside. Do not open the door under these circumstances until ventilation has commenced.

2. Make sure a door is locked before trying to force it open. This principle may seem basic, but it has been violated a number of times at fires. It is easy to assume that a door is locked when arrival is made at a commercial occupancy at 3:00 A.M. and it appears that no one is in the building.

3. Don't force a door if it is possible to break a small panel of glass and reach inside to open the door. The glass is much easier and cheaper to replace than the door, and this operation generally takes less time than forcing the door. (See Figure 4.48.)

4. Before using force on an entryway, make sure that other possible means of entering have been checked to see if they are open.

5. If time and conditions permit, before using force to gain entry to the fire floor of a multiple-story building, check to see whether it is possible to raise a ladder to an unlocked second floor window or entry point.

Principles, of course, are nothing but guidelines that must occasionally be violated. If the purpose of gaining entrance is search and rescue, then the primary considerations are time and safety. Entry must be made as quickly as possible, as even a few seconds may mean the difference between life and death to someone inside.

If entrance is to be made for the purpose of fire extinguishment, then a delay may be indicated. This is particularly true if there are any signs of a potential backdraft. Ideally, entry should be made almost simultaneously with the commencing of ventilation operations, but at actual emergencies this is seldom achieved.

There are many different types of entryways into buildings. There are also many different types of locks and security devices designed to make unauthorized entrance into buildings more difficult. Sometimes entrance is a simple matter of breaking a window, twisting off a padlock with a claw tool, or using an axe to force a single-hinged door. At other times, getting inside is slow and complicated. The extra care taken by occupants to secure their building against intruders has brought about the need for special tools and special knowledge to gain entrance into some buildings. Fire companies carry electric power tools, hydraulic tools, and gasoline-powered tools to assist in forcible entry operations. It is important that those firefighters responsible for forcible entry become familiar with the different problems that exist within their district, and learn to use the tools necessary to accomplish the objective.

A check for special problems should be made whenever a prefire planning inspection is being made of a building.

CONTROLLING THE UTILITIES

Control of the utilities should take place early during an emergency. This is not only for the potential saving of property but also as a safety factor to the firefighters working at the scene.

Utility shutoffs may be located almost anywhere inside or outside of a building. The ideal time to locate the various shutoff locations is, of course, during a prefire planning inspection. Not only should the location of the shutoffs be found and identified but the area of the occupancy that individual shutoffs control should be understood. Unfortunately, most fires occur in buildings that have not been subjected to a prefire planning inspection. In these cases it is generally necessary to cut off the utilities where they enter the building rather than taking the risks involved with delaying the shutoffs until interior control points can be found.

Water.   Broken water pipes within the building can cause unnecessary loss and could possibly cause a restricted water supply for firefighting purposes. It is best to shut off the water system to the building if there are signs of a significant leak or broken pipes in the system.

Water systems in buildings generally have a shutoff where the supply enters the building. In residences, this shutoff can usually be found at the outside faucet nearest the street. If it cannot be found quickly, then it is usually best to shut off the water at the street. There often is a cover in the sidewalk in front of the building that will identify this location. Generally, but not always, the water is on when the valve lines up with the pipe and off when the valve is perpendicular to the pipe. (See Figure 4.49.)

Gas.   The fuel supply to the building should be shut off if there is any possibility that there are broken pipes or a leak inside the building. Sometimes the broken fuel pipe or leak will be feeding the fire; at other times it will be setting up an explosive potential with the building. Gas entering the building can be shut off at the meter. (See Figure 4.50.) There is no standard location for gas meters. In single-family residents the gas meter is generally outside. Whether the gas is off or on usually can be determined by the position of the valve.

Some properties use LPG for heating and cooking. Storage tanks are almost always outside and are easy to identify. Shutting off the supply to the building can be made at the storage location. (See Figure 4.51.)

Electricity.   Burned or broken electrical wires are a potential life hazard to firefighters. If there is any possibility that such a condition exists within the building, then electricity to the fire portion of the building should be shut off. If the location of an inside control panel is known, and positive information is available as to the portion of the building

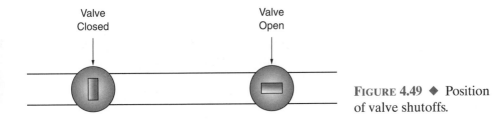

Valve
Closed

Valve
Open

FIGURE 4.49 ◆ Position of valve shutoffs.

FIGURE 4.50 ◆ The gas meter is generally the place to shut off the gas to an occupancy. *(Courtesy of Eugene Mahoney)*

FIGURE 4.51 ◆ Many homes in rural communities use LPG for heating and cooking. The shutoff valve is located under the cover. *(Courtesy of Eugene Mahoney)*

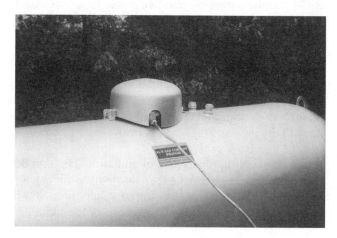

controlled by the various circuit breakers, then it is safe to shut off the electricity at that location. However, if there is any doubt as to the location of the control panel or what portion of the building it controls, then the electricity should be shut off as it enters the building. There might be a main circuit breaker at the control panel to accomplish this. Otherwise, to effect a positive shutoff, it is best to have the electric company cut the wires where they enter the building. The electric company can do this by cutting the wires at the loop where the power enters the building. In the event that there is no loop as the power enters the building, it may be necessary to have the electrical department cut the wires from the pole to the building.

◆ **SUMMARY**

The old saying "Sunday afternoon games are won on Monday mornings" is an important concept for firefighter to learn about. On Monday mornings, coaches begin the process of preparing for Sunday afternoon's game. So as it is for the fire service. We must preplan the building, complete building walk-throughs, or otherwise be ready for the "game" when the whistle blows.

In the fire service constant training is being completed each and every shift in order to better prepare for an incident before it even occurs. This planning process does aid tremendously when the time comes and prepares firefighters for what they may face in the field. "Prior planning prevents poor performance" is another statement firefighters use to describe the reason why they plan and train, to be ready for the game!

■ ■

Review Questions

1. What should be done prior to making a pre-incident planning inspection?
2. What information should be gathered about an occupancy during a pre-incident planning inspection?
3. How should the information gathered during the inspection be recorded?
4. What are the factors that should be considered during the response size-up?
5. Describe the objectives of a pre-alarm size-up.
6. What factors should be considered during the preliminary size-up on the fire ground?
7. What are the three tasks for which an engine company is responsible at the fire?
8. What is the procedure to use to determine the seat of a fire?
9. What are some of the methods that can be used to determine where and how fast the fire will travel?
10. What is the mistake most often made in extinguishing a fire?

11. What methods can be used to prevent the external and internal spread of a fire?
12. What is the extinguishing agent most often used on a fire?
13. What is meant by a direct attack on a fire?
14. How would the indirect attack on a fire be best explained?
15. What are some of the considerations that should be made regarding laddering operations?
16. What do the letters LOUVER refer to regarding a truck company's responsibilities?
17. What are some of the principles that should be considered regarding forcible entry?
18. Where will the shutoff valve for the water to an occupancy most likely be found?
19. Who should cut the electrical wires to the building in the event it is necessary to do so?

CHAPTER 5 Fire Ground Operations

Key Terms

CAD (computer-aided dispatch), p. 192
collapse zone, p. 207
command mode, p. 196
defensive mode, p. 206
defensive/offensive mode, p. 208

fast attack mode, p. 195
first-in district, p. 193
GIS (Geographical Information System), p. 193
investigation mode, p. 194

offensive/defensive mode, p. 208
offensive mode, p. 204
worker, p. 193

Objectives

The objective of this chapter is to examine the progress of a fire from the time the alarm sounds until the fire is extinguished. Upon completing this chapter, the reader should be able to:

- Explain what happens before the alarm sounds.
- Explain what happens when the alarm sounds.
- Describe what goes through a company officer's mind as the company responds to the fire.
- Describe the three conditions that can exist at a reported emergency when the first company arrives.
- Discuss the importance of taking command.
- Describe the methods the Incident Commander can use to estimate the number of personnel and the amount of water needed at the fire.
- Explain what happens when a change of command takes place.
- Explain the different fire attack modes.

Laddering the building, direct fire attack, collapse zones, and friction loss: These are terms that a firefighter not only hears about but also learns about. The importance of good fire ground operations cannot be taken lightly. The initial response and what you and your crew knows about a particular building before the event, can and does make a difference during the event. Fire ground operations and a sound, developed plan set the stage for a successful attack at a fire. Without good tactics and strategies, balanced with available resources and firefighters who understand their role and assignments, the fire ground can be chaotic. Learning about and applying good fire ground operations is essential to firefighting and ensuring the success of any fire operation.

◆ **BEFORE THE ALARM SOUNDS**

It is important that a firefighter have a basic understanding of the thought process the company officer goes through as that individual makes a decision as to the first action to take at a fire. Knowledge of the thought process provides the firefighter with a better understanding of the reason that certain decisions are made. This understanding makes it more palpable for a firefighter to carry out an order that otherwise might seem questionable.

Theoretically, some of the information on which a company officer makes a decision at a fire is gathered when the company makes a pre-incident planning inspection of the building in which the fire occurs. Unfortunately, although this is the ideal, a pre-incident planning inspection was not made in the buildings in which a great majority of the fires have occurred.

Pre-incident planning inspections are normally limited to a company's target hazards. A target hazard might be defined as an occupancy that constitutes a large collection of burnable valuables or an occupancy in which the life hazard is severe. (See Chapter Four.) These are the occupancies in which the maximum fire prevention effort of a department is concentrated, and the occupancies in which the strictest laws are generally applicable regarding built-in fire protection. As a result of the combination of strict laws and maximum enforcement, fires are kept to a minimum. A large majority of all fires that occur in most communities are in buildings other than target hazards.

◆ **WHEN THE ALARM SOUNDS**

The first information a company officer receives regarding a fire in a building is generally extremely limited. An alarm sounds throughout the engine house and a message similar to the following is received:

Engine 5, Engine 7, Truck 2, and Battalion 3—Respond to a structure fire at 347 W. Maple.

If the dispatcher has received additional pertinent information such as "The fire is reported on the third floor," or "It is reported that people are trapped in the building," he or she usually passes this information on to the companies.

GEOGRAPHICAL INFORMATION SYSTEM/GEOGRAPHICAL POSITIONING SYSTEM (GIS/GPS)

What happens at this point depends on whether the department has taken advantage of some of the technical advances available to the fire service. Some of the smaller departments continue to operate with a system that has not taken advantage of the major technology available. This may be due to financial restrictions or the department administration sees no particular advantage to upgrading. If this is the situation, prior to leaving quarters, the company officer of an engine company will generally check the map to review the location of the last hydrant available before arriving at the given location and the next available hydrant past the given location. (See Figure 5.1.) The company officer will also check to see whether there is an alley to the rear of the address and also behind the addresses across the street. The company officer knows by the address received whether the given location is on the north, south, east, or west side of the street; however, the company officer is also aware that many times a neighbor across the street will call in the alarm and give his or her own address.

CAD Letters refer to a computer-aided dispatch system.

If the dispatch office is equipped with a **CAD (computer-aided dispatch)** system, information on the dispatch may be sent to and received in the station on a printer. These printers are known as "tear and run" printers. This means that the information needed by the company officer is on the printer and the officer need only tear the printed sheet off the printer and respond to the alarm.

A more sophisticated CAD system goes beyond the "tear and run" principle and adds a mobile data terminal in vehicles for the officer's use. Some departments have installed permanently mounted screens on the apparatus and others provide both the companies and the command officers with laptop computers. Through a system

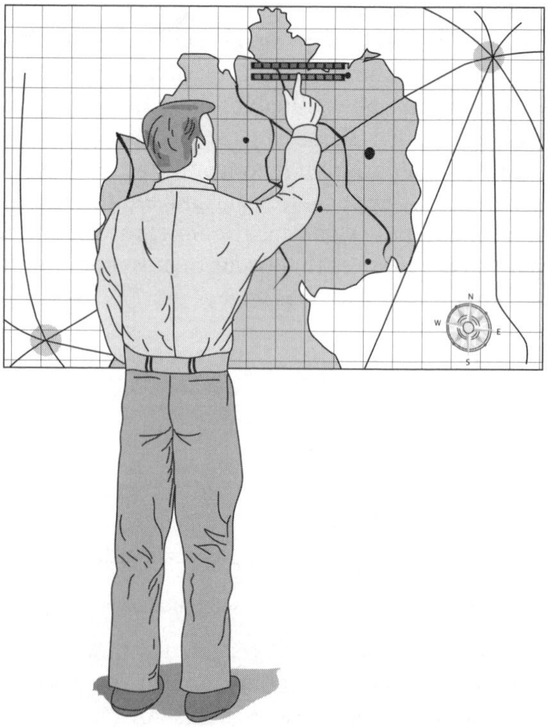

FIGURE 5.1 ◆ Check the map before leaving quarters.

known as GIS, the dispatch office is capable of transmitting a map of the area where the fire is located to the responding units. **GIS (Geographical Information System)** is both a database system with specific capabilities as well as a set of operations for working with the data. The maps generally include the hydrant locations and, in the more advanced systems, both the location of the emergency and the location and identification of the units responding. This provides the Incident Commander with the ability to assign units to various positions depending on the need.

GIS Refers to a Geographical Information System.

The CAD provides a great deal of flexibility to the dispatch office in providing information to on-scene units. The information the dispatcher is capable of sending to field units depends on what type of program has been placed in the database and what operating program has been placed in the dispatch office's computers. In some systems both the information on hazardous materials and the information on target hazards is in the dispatch database ready for transmission to units as needed.

◆ RESPONDING TO THE ALARM

A number of thoughts go through the company officer's mind as the apparatus pulls out of quarters. The officer has a general idea of whether the fire is in a single-family dwelling, a commercial occupancy, or whatever by mentally reviewing the type of area to which the company is responding. The mental review also takes into consideration what effect the time of the response and the existing weather will have on the life hazard, the traffic problems, and the fire spread. If the area to which the company is responding does not have a well-developed water supply system, the officer will start considering where the required water can be acquired after that on the apparatus is exhausted. Thoughts will flash through his or her mind such as: Have arrangements been made in the area for the development of auxiliary supplies? Are there locations near the reported address where drafting operations can be established? If the fire is extensive, might relay operations be required? Of course, if the department is responding a water tanker, this information would have been received at the time of the dispatch.

If the address is in the company's **first-in district,** the company officer knows that his or her company will probably be the first to arrive and that he or she will be in charge until a chief officer arrives. The officer's thoughts consider what other companies will be arriving and from what direction they will be approaching the fire location. He or she mentally calculates what companies will arrive on a second alarm, if the fire is of such a size to warrant one.

first-in district A district into which a company should be the first company to arrive at an emergency.

While responding, the company officer's eyes constantly are searching in the direction of the given location—watching for any sign that there may be a **"worker"** (a fire in progress). If smoke is sighted, the size and the color will give some clue as to what is burning.

worker A fire in progress.

◆ THREE POSSIBILITIES

One of three conditions (or a slight variation) will normally be observed by the company officer as the apparatus approaches the last available hydrant. The company officer will have to make an initial decision as to whether a line should be laid prior to passing that hydrant.

FIGURE 5.2 ◆ Condition number one—nothing showing.

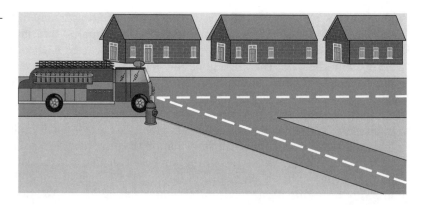

CONDITION NUMBER ONE

One of the three conditions is that nothing will be showing. (See Figure 5.2.) When this is the situation, the company officer should proceed directly to the reported address. A report should be made to the dispatch center while moving from the hydrant to the address. The call might go somewhat as follows:

> Dispatch from Engine 5. On the scene at 347 W. Maple. Nothing showing. Will investigate.

This report will alert the dispatch center that a company is on the scene and will provide additional information to incoming companies. On hearing this report, the company officer of the second engine company should proceed to the last hydrant available and wait for further instructions.

There normally will be someone to meet the apparatus as the company pulls up to the site. The company officer will listen to what the individual has to say and then will go inside to check the reported condition. It is good standard practice for a firefighter to take a water extinguisher and proceed inside with the company officer. The company officer should take in a portable radio that can be used to command the incident. In the majority of cases in which nothing is showing on arrival, if there is a small fire it can be extinguished by the use of the hand extinguisher or a small line.

It is possible that the company officer may go inside and find a small fire that will require the use of a booster line or a 1½- or 1¾-inch line for extinguishment. If this is the case, the officer will return to the apparatus and call for the needed line. He or she is in charge and will take command of the situation. If the officer feels that a supply line is needed, he or she will instruct the company waiting at the hydrant to lay one. It is good practice for the first-in officer always to have a supply line laid whenever a working line of any size is being used in the fire building. If the fire is of sufficient size, the officer may decide to have a company lay an attack line to the rear of the dwelling. In this case the officer would instruct the company officer waiting at the hydrant or perhaps the third-in engine company officer if three engine companies had been dispatched on the first alarm to do so. Existing conditions should be reported to the dispatch center as soon as practical.

Note: The Incident Command System refers to a situation in which nothing is showing on the arrival of the first unit as the **investigation mode.**

investigation mode A situation in which nothing is showing on the arrival of the first unit.

CONDITION NUMBER TWO

Another situation that the company officer may observe on approaching the reported address is a small amount of smoke or fire (or both) in the direction of where the fire

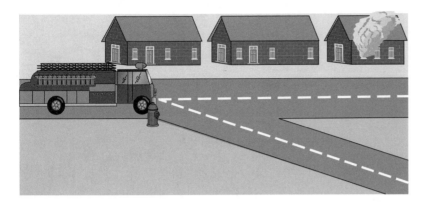

FIGURE 5.3 ◆ Condition number two—smoke and fire showing.

was reported. (See Figure 5.3.) Prior to passing the hydrant the officer will have to make a decision as to whether to lay an attack line or to proceed to the fire and take it with the tank. If this is the decision, the officer will most likely lay a supply line to the tank from the hydrant. There are no hard and fast rules on which the individual can base a decision. The decision will be left to one's subjective judgment and experience, using the basis of keeping the fire loss to a minimum. If the officer decides to handle the fire with the water in the tank, it is good practice that a supply line be laid as a backup. This can be laid by the first arriving company or by the second company. Regardless of the decision, the officer should report the condition to the dispatch center. The report may go something like this:

> Dispatch from Engine 5. At 347 W. Maple, we have a small amount of smoke and fire showing in a single-family dwelling. We will hold the first alarm assignment. I am taking the fire with my tank and laying a supply line. Engine 7, hold at the hydrant. Truck 2, move to the front of the building and wait for instructions.

Incidentally, if the fire happens to be at an address different from that originally received, the correct address should be reported to the dispatch center.

In this situation, as with all situations, the company officer of the first arriving company always should remember that he or she is in charge of the fire until relieved by a superior officer. Not only is it important to lay a supply line and have Engine 7 hold at the hydrant but also to have Truck 2 move to the front of the location. Having Engine 7 hold at the hydrant provides some flexibility in the event an additional line is required. It may be better that the line be brought into the back of the house or that one may be required to protect an exposure.

Note: The Incident Command System refers to this type of situation as the **fast attack mode.** In addition to using the fast attack mode for marginal offensive fire attacks, it is also used for incidents in which the safety of responding firefighters is a major concern or when there are rescues to be made immediately.

In a fast attack situation, the first-in officer should be capable of maintaining control of the emergency by the use of a portable radio. The fast attack mode normally lasts for only a few minutes. If the situation cannot be stabilized quickly, then the company officer should withdraw to the outside and establish a command post. However, in order to withdraw and leave the crew inside to handle the situation, there should be at least two crew members available for the inside work. If the situation is hazardous, the crew members should not be left inside unless they have radio communications capability and two additional members are outside as a backup team (RIT).

fast attack mode A term used for marginal offensive fire attacks or when the safety of responding firefighters is a major concern or when there are rescues to be made immediately.

FIGURE 5.4 ◆ Condition number three—fire showing.

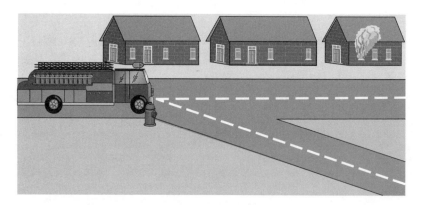

CONDITION NUMBER THREE

The third condition that the company officer may encounter is one in which there are definite signs of a fire as the reported address is approached. (See Figure 5.4.) The officer will then have to decide whether the fire can be controlled by the first alarm assignment or if help will be needed. In either case the company officer will lay a supply line going in and proceed to attack the fire or protect the exposures, depending on the exposure problems. If the individual decides that the fire can be controlled by the first alarm assignment, the report to the dispatch center will go something like this:

> Dispatch from Engine 5. At 347 W. Maple we have one room of a one-story single-family dwelling well-involved. We will hold the first alarm assignment. Engine 7, lay a line down the alley to the rear of the structure and protect the exposures. Truck 2, check to ensure that all people are out of the building and ventilate the roof.

command mode An operational mode whereby there is a need for an immediate strong, direct attack on the fire due to its size, complexity, or the potential for a rapid expansion.

Being in command, the officer should also instruct any other incoming company officers as to what they should do.

Because this incident is in a single-family dwelling, the ICS may refer to this incident as a fast attack mode or as a **command mode.** The general decision as to whether to go into the command mode is based primarily on the need for an immediate strong,

FIGURE 5.5 ◆ This situation showing on the arrival of the first company is definitely a condition number three situation. *(Courtesy of Las Vegas Fire & Rescue PIO)*

direct attack on the fire due to its size, complexity, or the potential for a rapid expansion. The situation shown in Figure 5.5 indicates that the incident should be handled in the command mode.

If a decision is made to go into the command mode, the company officer has several alternatives in regards to his or her own crew. A decision may be made to have the crew continue the attack on the fire as previously outlined. Or the officer may decide that the situation is too hazardous for the experience level of the remaining crew members and place them under the command of another company officer. A third option is to use his or her crew members to perform staff functions.

◆ MULTIPLE-ALARM FIRES

It may be apparent some distance from the reported location that the fire is of considerable size and probably beyond the control of the first alarm assignment or that the life hazard may be severe. A good example is a large amount of smoke showing in the direction of the alarm when the responding company is aware that the report is of a fire in a school during school hours. In such situations there is no harm to be done by ordering a second alarm assignment and extra companies to perform rescue operations. The actual call for the total number of companies desired, however, should be reserved until an adequate on the scene evaluation can be made. (See Figure 5.6.) The initial call to the dispatch center may go something like this:

> Dispatch from Engine 3. We have a considerable amount of smoke showing in the direction of the reported alarm. As a precautionary measure, give me a second alarm assignment and three additional companies to perform rescue operations. Let me know as soon as possible what companies I'll be receiving. I'll give you an update report on arrival at the scene.

The term "second alarm" assignment will differ from one city to the next. In some cities it might refer to the same number of engine companies that were dispatched on the first alarm, but one fewer truck company and perhaps lacking a squad company if

FIGURE 5.6 ◆ This fire developed into a fifth alarm requiring 12 engine companies, three ladder companies, three squad companies, six chief officers, and numerous support units.
(Courtesy of District Chief Chris E. Mickal, New Orleans Fire Department—Photo Unit)

one were dispatched on the first alarm. In other cities a second alarm might duplicate both the number of engine companies and the number of truck companies that were dispatched on the first alarm. The first arriving officer will be familiar with the meaning of the term as it is used in his or her community.

TAKING COMMAND

In most cases the company officer of the first-in company will arrive on the scene prior to additional help being requested. It is usually an engine company that arrives first but it could be a truck company, a squad company, or another type of company if one happened to be out in the area when the alarm was received. The company officer, being the first to arrive, is in charge until relieved by a superior officer.

If the fire is definitely one in which the initial operation should be conducted in the command mode, one of the first steps that should be taken by this officer is to establish a command post and inform the dispatch office of both who is the Incident Commander and where the command post has been established. (See Figure 5.7.) An example of the report to the dispatch office might go something like this.

> Dispatch from Engine 6. The command post and the staging area for the fire at 567 So. Main have been established at the corner of Main and Broadway. Engine 6 will be the Incident Commander.

If the first arriving company is an engine company or truck company, the company commander's superior officer will be a chief officer. Once the first company arrives, the company officer will become the Incident Commander. This individual should carefully evaluate the situation at a large fire, turn his or her company over to an acting officer providing the company is adequately staffed and experience qualified, and take command of the fire. It is poor practice to wait for the chief officer to arrive to make the necessary decisions. It is possible that the chief officer could have an accident on the response to the fire and never arrive, or that his or her vehicle might get held up in a traffic jam. This will leave the company officer in charge of the fire for an extended period.

FIGURE 5.7 ◆ Sometimes the hood of a vehicle serves as the operating position for the command post.
(Courtesy of Rick McClure, LAFD)

While there is a general rule that the first arriving officer is responsible for taking command of the fire, at times it is logical to make an exception to this rule. An example is a situation involving a serious life problem or one in which there is not sufficient personnel on the first arriving company to provide the requirements established for the two-in/two-out rule. This takes a staffing of five: two in, two out, and the apparatus operator. This kind of staffing is rare on some departments today.

If the second company has arrived, then it may be better for the first arriving officer to turn over control of the fire to the second arriving officer and proceed with the operation of his or her company. However, it must be kept in mind that control of the fire cannot be turned over to an officer who is on the way but has not yet arrived at the fire.

One of the first acts that should be conducted is to request the additional help that the Incident Commander deems necessary, keeping in mind that the progress of the fire will probably advance before additional companies arrive on the scene. It is not good practice merely to call for a second or third alarm. When this is done, the Incident Commander is basically saying, "I know that help is needed, but I'm really not sure how much." The Incident Commander should keep in mind that there is every excuse for overestimating the amount of help needed, but little excuse for underestimating. It is much more professional to return companies to quarters because they are not needed than to have to call for additional help later because an insufficient amount was not originally requested.

Estimating the Number of Personnel Needed.   It is not the number of companies at a fire that is important but the number of personnel on the scene that can be used to complete the tasks that are required. All officers should have some guidelines they can use to estimate the number of personnel needed at the fire scene. There is no single system that can be considered as best under all conditions. Different officers use different systems with similar results. The important thing is to have a system that is practical and workable.

It is important to remember that a system based on the determination of the number of engine companies required might be flawed. Some engine companies may arrive at an emergency and not have a sufficient number of personnel to handle a loaded 2½-inch line. Others may be able to accomplish this with no trouble. Fortunately, in larger cities the first arriving officer is knowledgeable regarding the number of personnel that are used to staff the various types of companies on the department. However, if the call for additional help involves the response of mutual-aid companies, the staffing of responding companies is not always known. Consequently, it is better for the officer who is making the determination to estimate quickly where and what size lines are required at various locations. Using the following will provide him or her with an estimation of the number of personnel required at the scene to make an aggressive attack on the fire and provide the hose streams required for exposure protection.

- Small outside lines can generally be handled by one firefighter. (See Figure 5.8.)
- Both a 1½-inch line and a 1¾-inch line inside lines require two firefighters.
- While it may be possible to maneuver a 2-inch line with two firefighters, for safety's sake it is better to use three firefighters.
- A 2½-inch line requires three firefighters to maneuver it. (See Figure 5.9.)
- Once lines have been laid, a heavy-stream appliance can be operated by one or two firefighters; and on some types of fires, such as tank fires, it can be secured effectively and left to operate on its own. (See Figures 5.10 and 5.11.)

FIGURE 5.8 ◆ Smaller lines can be handled by one firefighter. *(Courtesy of District Chief Chris E. Mickal, New Orleans Fire Department—Photo Unit)*

FIGURE 5.9 ◆ 2½-inch hose lines require three firefighters to move them. *(Courtesy of District Chief Chris E. Mickal, New Orleans Fire Department—Photo Unit)*

FIGURE 5.10 ◆ Stationary large lines can often be handled by one firefighter. *(Courtesy of District Chief Chris E. Mickal, New Orleans Fire Department—Photo Unit)*

FIGURE 5.11 ◆ Two engine company members using a portable monitor to provide a heavy stream. *(Courtesy of Rick McClure, LAFD)*

Using this information as a guideline, an officer can estimate quickly the number of personnel required at the fire for fire extinguishment and control. For example, if an officer estimates that two 1½- or 1¾-inch lines and two 2½-inch lines will be needed for extinguishment, and that two 2½-inch lines and a ladder pipe will be needed to protect the exposures, then the officer knows that he or she will need a minimum of 17 firefighters for handling lines. Additional personnel will also be needed for the RIT, staffing, and other staffing positions. It is good practice to order more than is first estimated, and if any wind it blowing, it is also good practice to add an engine company that can be used downwind as a precautionary measure against flying brands whenever wood shingles are involved at the fire. All of the personnel do not need to be supplied by engine companies. Available personnel from other companies such as truck companies or squad companies can be used.

The National Fire Academy advocates a method for estimating the amount of water required to extinguish and control the fire. The system being taught at the academy is based on the size of the fire building. The basic fire flow is determined by dividing the square feet in the building by 3 and then reducing this figure according to the percentage of the building involved in fire. In formula form, the system is expressed as:

Fire flow (in gallons per minute) = length × width ÷ 3

For example, the fire flow for a 30' by 50' building would be:

$$30 \times 50 = 1500$$
$$1500 \div 3 = 500 \text{ GPM}$$

This is the amount of water estimated as being needed to control a fire if the building is well involved with fire. If the building is 50 percent involved, the amount required would be 50 percent (½) of 500 GPM, or 250 GPM. If 25 percent of the building is involved, the amount required would be 25 percent (¼) of 500 GPM, or 125 GPM. Some officers will say that this system underestimates the amount of water required.

The amount of water required for exposure protection is figured separately. It is estimated that 25 percent, or ¼, of the water required inside the building will be needed for each side of the building that requires exposure protection. For example, if the 1,500 square foot building in the preceding example is well involved with fire, which means that 500 GPM is required for interior use, and two sides of the building require exposure protection, then ¼ of 500, or 125 GPM, would be required for each of the two sides that require exposure protection. This works out to be that a total of 250 GPM of water is required for the exposures and that the total amount required for the fire would be:

$$Interior = 500 \text{ GPM}$$
$$Exposures = 250 \text{ GPM}$$
$$Total \text{ flow requirement} = 750 \text{ GPM}$$

It should be noted that the preceding guidelines are normally not satisfactory for use at high-rise fires, basement fires, or other fires that will normally extend over a long period of time. These fires generally require that the firefighters at the end of the line be relieved periodically because of the limited time available on their breathing apparatus and the total physical excertion required for initially placing hose lines into operation. These fires generally require three crews for each line placed in operation inside the building. One of the crews is working on the fire, one is waiting to relieve those working on the fire, and one is changing bottles on their breathing apparatus. If a 2½-inch line is being worked on the fire, then the Incident Commander should estimate that nine firefighters (3 × 3) should be available at the fire for each of the lines placed in operation. (See Figure 5.12.)

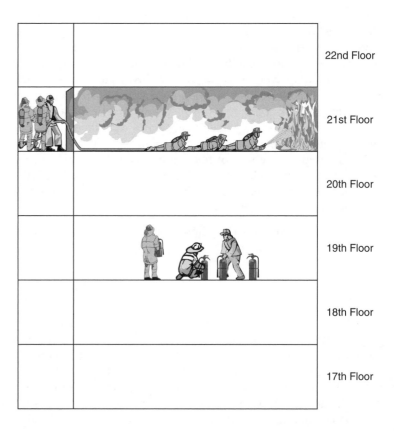

FIGURE 5.12 ◆ The three-for-one situation.

Estimating the Number of Other Companies Needed.   Consideration should also be given to the number of personnel required to perform truck company operations, salvage operations, and emergency medical care operations. It is usually more difficult to estimate the number of personnel required for these operations than it is to determine the number required for handling hose lines. Regardless, all officers should have some guidelines they can use at a fire to estimate the number of personnel that are needed to perform the other functions required at the fire. The following is offered as a guideline for assistance in the development of a workable system.

At structure fires, a company staffed with four firefighters to perform truck operations is normally needed for every three companies at the fire performing the operations of an engine company. If there is a known life hazard, one four-member firefighter company conducting truck operations should be estimated as necessary for every two engine companies handling hose lines at the fire; and in cases in which the life hazard is extreme, a one-for-one match would not be excessive. There should be a minimum of one four-member company performing truck operations at every structure fire of any size. Two members are needed for the inside work and two for the outside work.

At least one company doing salvage work is needed whenever lines are being used above the first floor. At a fire of any size, one company doing salvage work is necessary for every three companies handling hose lines inside the building. (See Figure 5.13.) Additional companies for salvage operations will probably be required if the use of water is excessive or an extensive water problem exists within the building. At some fires, such as those in sprinklered buildings, the number of companies required for salvage operations may far exceed the number required for handling hose lines.

Emergency medical service capability should be available at every working fire. A company should be on the scene that has no other responsibilities other than that of performing emergency medical care. Ambulances should be ordered if there is any possibility they may be needed.

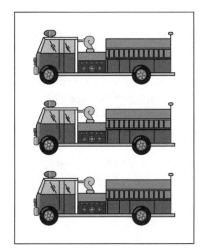

ENGINE COMPANIES

TRUCK COMPANY

FIGURE 5.13 ◆ The three-to-one ratio.

It should be part of the standard operating procedures (SOPs) of every department to implement the ICS at every fire of any size. Early thought should be given at greater alarm emergencies to the need for staff assistance. As the minimum, there will be a need for a staging area, a safety officer, a liaison officer, an information officer, and a company to perform the duties of the RIT. These needs should be requested at the same time that additional engine and truck companies are requested.

CHANGING COMMAND

When the chief officer arrives at the fire, he or she will report to the command post. The Incident Commander will provide the chief with all the information he or she has on the fire and explain what tactics have been put into place. The chief may want to change these tactics or continue to operate using those developed by the Incident Commander. Once the chief officer is assured that he or she has all the information needed to assume command, this officer will tell the Incident Commander that he or she is taking command. It is good practice for the officer assuming command to utilize the Incident Commander he or she replaced in a staff position. Once he or she has decided to assume command, the chief officer should notify the dispatch office of the change. The message might go something like this.

> Dispatch from Battalion 3. Battalion 3 has assumed command of the fire at 567 So. Main.

◆ FIRE ATTACK MODES

The decision as to what attack mode to use on a fire is the responsibility of the first officer to arrive at the fire. This officer must weigh the risks to firefighters versus the gains that can be achieved by the use of the different modes. Although there are more risks involved in an offensive mode, normally quicker and safer results are achieved. There are four modes from which he or she can choose. The mode most often used is the **offensive mode.**

offensive mode A fire situation in which a direct attack is made on the fire. It requires firefighters to go inside and put out the fire.

OFFENSIVE MODE

In simple language, this mode involves making a direct attack on the fire. (See Figure 5.14.) It requires going inside and putting out the fire. Although saving lives is the primary goal of a fire department, normally more lives can be saved by an aggressive attack on the fire than by any other means. An aggressive attack also reduces the amount of time required to remove all occupants out of the building. An aggressive attack, however, does involve some risks. The Incident Commander should carefully weigh the risks against the potential goals prior to committing companies for an attack. (See Figure 5.15.)

When people are still in a building, or exiting a building on fire, the first lines should be placed between those in the building and the fire. Lines should also be used to maintain control of all exit facilities, particularly the stairways. Gaining control of the stairways means gaining control of the building.

The rear of the building should not be overlooked. This area of the building has the potential for a rapid fire spread and also might require the immediate rescue of a number of occupants. It is good practice to assign at least one engine company and one company doing truck work to this area.

FIGURE 5.14 ◆ This fire is being attacked using the offensive mode.
(Courtesy of District Chief Chris E. Mickal, New Orleans Fire Department—Photo Unit)

FIGURE 5.15 ◆ A direct attack on the fire places this incident in the offensive mode.
(Courtesy of Rick McClure, LAFD)

FIGURE 5.16 ◆ This fire is being fought in the defensive mode.
(Courtesy of District Chief Chris E. Mickal, New Orleans Fire Department— Photo Unit)

DEFENSIVE MODE

defensive mode A fire situation in which the fire is fought from outside of the building.

There is less risk to firefighters when the **defensive mode** is used. This mode refers to fighting the fire from the outside of the building. (See Figure 5.16.) It should be considered whenever:

- The risks to firefighters outweigh any gains that can be achieved.
- There are insufficient companies or resources on the scene to achieve the desired results.
- A sufficient amount of water is not available to make a positive attack.
- The fire is beyond the control of handheld lines, therefore making the use of master streams mandatory.
- There is no life hazard to building occupants.
- The structural conditions of the building are questionable.

On Scene Scenario

A fire was reported in a vacant vacation lodge on a Sunday evening in 2003 in Lanagan, Missouri. By the time the first unit of volunteer firefighters arrived on the scene, the building was in flames. The chief ordered the firefighters to keep a safe distance away and to concentrate on protecting the exposures on the north side of the resort. According to the *Joplin Globe,* the chief said, "When a fire is that hot and a building is that far gone, there is no sense getting anyone hurt or any equipment damaged trying to save what's already lost."

The building was eventually completely destroyed.

OFFENSIVE/DEFENSIVE MODE

offensive/defensive mode A transition mode whereby operations are changed from an offensive mode to a defensive mode.

Both the offensive/defensive mode and the defensive/offensive mode are considered as transition modes. The **offensive/defensive mode** refers to the initial attack being made using the offensive mode with a later withdrawal from the building and the setting up of a defensive mode. An example is the use of an aggressive attack in order to defend a portion of the building that must be protected in order to save lives. The attack is made with the full knowledge that lines will have to be withdrawn after the objective is achieved.

FigURE 1-1 ◆ A wildfire is a good example of a fire on the rampage.
(Courtesy of FEMA)

FigURE 1.10 ◆ Radiant heat is being projected in all directions from this fire.
(Courtesy of District Chief Chris E, Mickal, New Orleans Fire Department—Photo Unit)

FigURE 1-17 ◆ This attic fire shows all signs of a potential backdraft.
(Courtesy of District Chief Chris E. Mickal, New Orleans Fire Department—Photo Unit)

FIGURE 2-20 ◆ A fire in the attic. A truck company member preparing to ventilate the roof using a chain saw.
(Courtesy of Rick McClure, LAFD)

FIGURE 2-17 ◆ Firefighters conducting overhaul operations.
(Courtesy of Rick McClure, LAFD)

FIGURE 3-29 ◆ September 27, 2001. The remaining section of the World Trade Center is surrounded by a mountain of rubble following the September 11 terrorist attack.
(Courtesy of FEMA)

FIGURE 4-6 ◆ Sometimes it is necessary to breach a fence to lay an attack line to the fire.
(Courtesy of Las Vegas Fire & Rescue PIO)

FIGURE 4-41 ◆ A good illustration of a well-laddered incident.
(Courtesy of Rick McClure, LAFD)

FIGURE 4-40 ◆ A truck company member using a pike pole to hold open a sliding metal door while an engine company member directs water on the fire.
(Courtesy of Rick McClure, LAFD)

FIGURE 4-42 ◆ A good example of a truck company raising ladders to both the fire building and to an exposure.
(Courtesy of District Chief Chris E. Mickal, New Orleans Fire Department—Photo Unit)

FIGURE 4-45 ◆ This ladder had been extended well above the roof parapet. Notice how safely a firefighter can step from the ladder to the parapet and then onto the roof.
(Courtesy of Rick McClure, LAFD)

FIGURE 5-11 ◆ Two engine company members using a portable monitor to provide a heavy stream.
(Courtesy of Rick McClure, LAFD)

FIGURE 5-15 ◆ A direct attack on the fire places this incident in the offensive mode.
(Courtesy of Rick McClure, LAFD)

FIGURE 5-10 ◆ Stationary large lines can often be handled by one firefighter.
(Courtesy of District Chief Chris E. Mickal, New Orleans Fire Department—Photo Unit)

FIGURE 6-44 ◆ Quick action is required on this fire to protect the exposures.
(Courtesy of Las Vegas Fire & Rescue PIO)

FIGURE 7-1 ◆ Car fires. Small lines generally are used to extinguish car fires.
(Courtesy of Las Vegas Fire & Rescue PIO)

FIGURE 6-45 ◆ High-rise fires produce many problems for firefighters that are not common in other types of structure fires.
(Courtesy of Rick McClure, LAFD)

FIGURE 7-2 ◆ Aircraft crashes. Aircraft with a flammable liquid spill can occur in any community at any time. Foam has been used on this incident.
(Courtesy of Rick McClure, LAFD)

FIGURE 8-4 ◆ Brush fires. A fast-moving brush fire.
(Courtesy of Rick McClure, LAFD)

FIGURE 8-36 ◆ West Glenwood, Colorado, June 11, 2002—The Flathead Hotshot crew has set a backfire to limit the consumption of more forest land.
(Photo by Andrea Booher/FEMA News Photo, Courtesy of FEMA)

FIGURE 8-37 ◆ A fixed-wing aircraft making a water drop.
(Courtesy of Rick McClure, LAFD)

FIGURE 8-39 ◆ Los Alamos, New Mexico, May 4, 2002—Mopping up a wildfire is a dirty and exhausting job.
(Photo by Andrea Booher/FEMA News Photo, Courtesy of FEMA)

FIGURE 8-61 ◆ A helicopter making a water d
(Courtesy of Rick McClure, LAFD)

FIGURE 8-44 ◆ A small portion of the Simi/Val Verde fire.
(Courtesy of Rick McClure, LAFD)

Several things need to be considered when changing from an offensive mode to a defensive mode. Some of the factors that should be given thought by the Incident Commander are:

1. Make sure that the ladder pipes, elevated platforms, and deluge guns that will be needed are set up and ready before they are actually required. These heavy-stream appliances should not be set up in the collapse zone.
2. Make an announcement to all those working at the fire that the operational plans will be changed. It is good practice for a department to set up a standard signal that indicates that all personnel should leave the building. One suggestion is to sound the SOS of three short blasts, followed by three long blasts, and followed by three short blasts using air horns on the apparatus.
3. Remove all working lines and personnel from inside the building. The lines should be shut down and drained. They should then be moved to an area outside of the building and outside of the collapse zone. The nozzles should remain shut off but the lines not disconnected. Pump operators should be advised not to load any of these lines without the permission of the Incident Commander.
4. After their lines have been properly placed, have all members that were handling attack lines, and those assigned to other duties, report to the staging area.
5. Do not permit any freelancing.
6. If not already assigned, assign an engine company to patrol downwind as protection against flying brands.
7. Make sure all units are following the operational plans established by the Incident Commander.
8. As a last step, make sure everyone is out of the building and in a safe area prior to opening up the master streams.
9. Make sure a personnel accountability report has been taken.

COLLAPSE ZONE

The **collapse zone** should be equal to one and a half times the height of the building. No personnel or apparatus should be allowed to operate in the collapse zone except

collapse zone A safety zone in which no operations should be performed or anyone allowed within it.

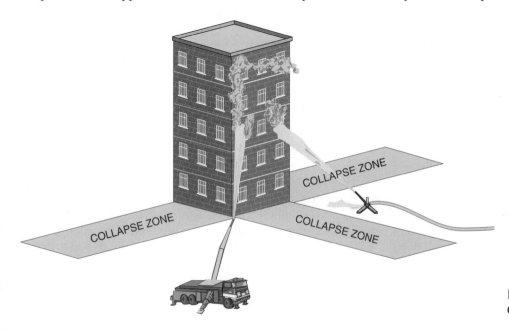

FIGURE 5.17 ◆
Collapse zone.

to place unmanned master stream devices. As shown in Figure 5.17, heavy streams can be worked in the area between collapse zones. The safest locations for operating streams is adjacent to the corners of the building. The corners of the building are considered the strongest portion of the structure.

Collapse zones should be roped off once they have been established. Fire zone tape can be used for this purpose. No personnel should be permitted within the zoned area.

On Scene Scenario

The following incident of an offensive/defensive operation was reported in the February issue of the Los Angeles Firemen's Relief Association publication *The Grapevine.*

On Tuesday, December 17, 2002, at 0014 hrs (12:14 A.M.), twenty-four companies of Los Angeles Firefighters, four Battalion Command Teams, and two LAFD Rescue Ambulances, under the direction of Assistant Chief Terry Manning, responded to a Major Emergency Structure Fire at 956 N. Seward Street in the Hollywood area. First units on the scene reported a two-story structure with heavy fire on the second floor and attic. Firefighters using handlines immediately mounted an aggressive interior attack on the fire. Due to the large volume of fire in the "Hollywood Centers Studios" on the second floor and in the attic, it was determined that a portion of the 75' × 400' structure had been compromised and the order to withdraw from the structure was given. In a defensive position on the exterior, Firefighters used heavy streams and handlines to control and confine the fire in one hour and twenty-one minutes. No injuries were reported and the cause of the fire is listed as under investigation. Fire damage is estimated at $2,000,000+.

DEFENSIVE/OFFENSIVE MODE

defensive/offensive mode A transition mode whereby operations are changed from a defensive mode to an offensive mode. It is generally used where the initial size-up of the fire indicates that the fire is beyond the capability of hand lines.

The **defensive/offensive mode** is used when the initial size-up of the fire indicates that the fire is beyond the capability of hand lines. Heavy streams are set up and a blitz attack is made on the fire. After the knockdown, the building can be inspected to ensure that it is structurally safe. If it is found to be, then an offensive attack can be made.

The mode is also used when the Incident Commander initially decides that there is an insufficient number of personnel at the scene to use attack lines inside the building. The fire will be fought from the outside until the Incident Commander is assured that a sufficient number of people have arrived to commence an offensive attack.

◆ SUMMARY

The basics of fire ground operations begin with an understanding of what conditions exist. A good initial report of what the fire officer observes on scene is important. The use of new technologies such as GIS or GPS as well as computer-aided dispatching allows firefighters to receive and interpret information more rapidly in order to make good, sound decisions. Understanding basic fire ground terminology such as "friction loss" and "smoke showing" will aid the firefighter to relate the required tasks better to the situation on hand. Good fire ground operations are a critical part of a successful operation on any fire emergency situation.

Review Questions

1. What normally happens before the alarm for a fire sounds?
2. What normally happens when the alarm sounds?
3. What goes through a company officer's mind when responding to an alarm?
4. What are the three different types of conditions that can exist on arrival at the scene of a reported fire?
5. What type of radio call should the first officer to arrive at the scene make to the dispatch officer?
6. What method can the Incident Commander use to estimate the number of personnel needed at a fire?
7. What method can the Incident Commander use to estimate the amount of water need for fire extinguishment?
8. How can an Incident Commander estimate how many companies other than engine companies are needed at a fire?
9. What happens during a change of command at a fire?
10. How would you define an offensive fire attack mode?
11. How would you define a defensive fire attack mode?
12. How would you define an offensive/defensive fire attack mode?
13. How would you define a defensive/offensive fire attack mode?

CHAPTER 6 Fires in Buildings

Key Terms

base, p. 264

basement, p. 216

cellar, p. 216

center-core design, p. 258

dwelling fires, p. 226

enunciator, p. 262

fire curtain, p. 245

flame, p. 267

high-rise, p. 000

high-rise fire attack
team, p. 261

high-rise staging area, p. 262

interior (wet) standpipe
system, p. 215

lobby control officer, p. 263

partition fire, p. 000

smoke, p. 267

staging area, p. 263

strip mall, p. 249

taxpayer building, p. 246

Objectives

The objective of this chapter is to introduce the reader to firefighting tactics as they apply to building fires. Attic fires, basement fires, chimney fires, partition fires, dwelling fires, garage fires, mercantile fires, industrial fires, fires in places of public assembly, taxpayer fires, strip mall fires, church fires, fires in buildings under construction, and high-rise fires are explored. Upon completing this chapter, the reader should be able to:

- Explain the procedure for fighting attic fires.
- Describe the procedure for fighting basement fires.
- Explain the procedure for extinguishing chimney fires.
- Explain the procedure for finding and extinguishing partition fires.
- Explain the procedure for extinguishing single-family dwelling fires.
- Explain the procedure for extinguishing multiple-story dwelling fires.
- Explain the extinguishing of fires in attached garages of two-story dwellings.
- Explain the procedure for extinguishing mercantile fires.
- Explain the procedure for extinguishing industrial fires.
- Explain the procedure for extinguishing fires in places of public assembly.
- Explain the procedure for extinguishing fires in taxpayer structures.
- Explain the procedure for extinguishing fires in strip malls.
- Explain the procedure for extinguishing church fires.
- Explain the procedure for extinguishing fires in buildings under construction.

◆ Describe the construction features of high-rise structures.
◆ Explain the firefighting problems of fighting high-rise fires.
◆ Discuss the duties of the first-in company officer at a high-rise fire.
◆ Explain the duties of the second-in company officer at a high-rise fire.
◆ Describe the rescue operations at a high-rise fire.
◆ Explain how occupants should be relocated in a high-rise fire.
◆ Describe the movement of personnel and equipment at a high-rise fire.

Fires in Buildings

FIGURE 6.1 ◆ High-rise fires present a multitude of problems.
(Courtesy of Rich McClure, LAFD)

FIGURE 6.2 ◆ A well-developed strip mall fire generally will require several additional alarm assignments. *(Courtesy of Rick McClure, LAFD)*

(Continued)

Fires in Buildings (*Continued*)

FIGURE 6.3 ◆ Some attic fires are deceiving. This one developed into a third alarm fire.
(Courtesy of District Chief Chris E. Mickal, New Orleans Fire Department— Photo Unit)

FIGURE 6.4 ◆ Industrial fires can be extensive and can require a large number of units and personnel.
(Courtesy of District Chief Chris E. Mickal, New Orleans Fire Department—Photo Unit)

FIGURE 6.5 ◆ Fires in buildings under construction can quickly develop into a huge bonfire.
(Courtesy of Las Vegas Fire & Rescue PIO)

FIGURE 6.6 ◆ Dwelling fires can occur in single-family or multifamily units.
(Courtesy of Rick McClure, LAFD)

FIGURE 6.7 ◆ A well-involved two-car attached garage with a car inside. There was no need to open the garage door on this fire.
(Courtesy of Rick McClure, LAFD)

(Continued)

Fires in Buildings (*Continued*)

FIGURE 6.8 ◆ A well-involved fire in a commercial occupancy with fire extending through the roof. *(Courtesy of Rick McClure, LAFD)*

◆ INTRODUCTION

As mentioned in previous chapters, no two fires are exactly alike; however, all fires are somewhat alike. Their similarity is the cornerstone for the establishment of the basic tactics of locating, confining, and extinguishing. Their differences are the foundation for the requirement of having to know more than the basics in order to become effective in fighting fires.

The location of the fire and the type of occupancy in which it is found are two of the many factors that contribute to the differences between fires. This chapter will attempt to highlight some of the variables found in various types of buildings and occupancies that influence the fire problem.

Prior to discussing fires in different locations and different types of occupancies, it is well to review several important points that apply to all types of fires in buildings. First, it is worth keeping in mind that most fires in buildings cannot be put out from the sidewalk until the building has collapsed and has become one big bonfire. It is necessary to go inside and put water on the seat of the fire. The quicker this is done, the sooner the fire will go out. This type of action requires that the Incident Commander operate in the offensive mode. Of course, the fire may have to be fought from the outside if the building is in danger of collapse or some other factor exists that requires it. A decision to do this means that the operations will be conducted in the defensive mode.

The Incident Commander is constantly alert to the fact that there is no building worth the life of a firefighter. He or she is also always aware that conditions can change that will influence his or her decisions and permit a switch to be made to the offensive/defensive mode or the defensive/offensive mode.

If it is possible to go inside to work on the fire, it probably will require that personnel move into locations where it is hot and smoky, sometimes so smoky that the firefighters cannot see their hands before their faces. OSHA requirements ensure that fully protective gear including breathing apparatus and PASS is worn under these conditions. This permits personnel to operate with much greater safety than was done in past years.

A majority of fires in buildings normally can be extinguished by the water from a single 1¾-inch or smaller line. Small lines are easy to maneuver and are normally effective for making a quick knockdown. It is important to remember, however, that a supply line always should be laid from the hydrant or other water source to ensure that an ample amount of water is available whenever a small line is used for the initial attack on a building. This is required just in case the small line cannot complete the job.

Another good practice at building fires is to take advantage of whatever built-in protection is provided. **Interior (wet) standpipe systems** contain water under pressure at all times and have racks attached containing linen hose. These lines can be used effectively to hold a fire in check while fire department hose lines are being laid. Lines laid into the exterior (dry) standpipe system will make water available at every floor and eliminate the need to lay lines up fire escapes or weave them up interior stairways. Supply lines laid into a sprinkler system will ensure an adequate supply of water on the fire or will hold the fire in check until department lines can be brought into play. Lines laid into the sprinkler system will be particularly effective at basement fires.

interior (wet) standpipe system   A built-in fire protection system that contains water under pressure at all times and has racks attached containing linen hose.

◆ ATTIC FIRES

Attic fires in dwellings can be caused by defective chimneys, by faulty wiring, from an extension of a fire below, or occasionally from a flying brand landing on a wood shingle roof. The first objective of firefighters on arrival at the structure is to make sure that all occupants are out of the building.

Attic fires should be attacked as quickly as possible with the use of 1½- or 1¾-inch lines equipped with fog or spray nozzles. The lines should be taken inside the building and advanced to the attic through the scuttle hole, attic stairway, or other means of ingress. (See Figure 6.9.) If none of these openings is available or can be found readily, then the ceiling should be opened from below and a ladder extended through the opening provided.

FIGURE 6.9 ◆ Attack attic fires through the scuttle hole.

As little water as possible should be used to control the fire. If possible, salvage operations should be started on the floor below the fire simultaneously with the line advancement into the attic. There should be ample time available to spread covers before the water comes through the ceiling if water is used sparsely.

Attic fires can also be attacked without initially advancing a line into the attic by using a piercing nozzle in combination with a CAFS using Class A foam or plain water. This type of attack often can affect a quick knockdown with the use of a minimum amount of water.

It may not be necessary to open the roof for ventilation if the fire is small. The roof normally should be opened for larger fires to help confine the fire and enable the firefighters handling the hose lines to gain access into the attic. Lines should not be directed through the hole cut in the roof except in extreme cases. To do so would drive the fire back into the faces of the firefighters advancing the line. However, a line should be taken onto the roof to protect the exposures or knock down any fire that spreads to the roof covering.

Gaining access to the attic from below is sometimes difficult in large finished attics with gable or arch roofs. Effective work can be done in these cases by opening the roof near the peak for ventilation and then making an opening near the eaves for the purpose of advancing a line.

Any loose shingles or roof covering should be removed during the overhaul process. A thorough check should be made for any downward extension of the fire. Particular care should be taken if hanging ceilings are involved. The hole in the roof should be covered during the overhaul process if there is any chance that the weather may cause further damage.

Strategic Goals

- Remove occupants from building.
- Advance a line into the attic.
- Use as little water as possible.
- Ventilate the roof if required.
- Cover furniture under the fire.
- Overhaul and leave the property in good condition.

◆ BASEMENT FIRES

basement The lowest story of a building or the one just below the main floor, usually wholly or partially lower than the surface of the ground.

cellar A room or group of rooms below ground level and usually under a building, often used for storing fuel, provisions, and so on.

According to Webster, there is very little difference between a basement and a cellar. Webster defines a **basement** as "the lowest story of a building or the one just below the main floor, usually wholly or partially lower than the surface of the ground." A **cellar** is defined as "a room or group of rooms below ground level and usually under a building, often used for storing fuel, provisions, etc." In this book, both of these subgrade locations will be referred to as basement fires. However, it should be kept in mind that a subcellar is usually farther below grade than a basement and is the most difficult below-grade-level fire extinguishment problem found in the fire service. A subcellar presents the problem of having only one entrance and no windows. This combination makes ventilation extremely difficult.

Fires in basements present some of the most difficult problems firefighters encounter. These fires take place in confined areas with very little access locations and present challenges to ventilation operations. Taking a line down into a well-involved basement fire is like climbing down a chimney with a good fire going in the fireplace.

The difficulty of fighting basement fires is partly due to the complexity of the problem. All types of material can be found stored in basements. Fires starting in basements generally burn slowly and many times will smolder over a long period of time before being discovered. Because of the limited supply of oxygen in the basement area, these fires will give off large quantities of heat and smoke that not only will contribute to the spread of the fire but also will make access to the fire difficult. It is possible for the fire to spread very rapidly out of the basement through various vertical openings such as stairways, air-conditioning systems, laundry chutes, dumbwaiters, and wall partitions. Although these vertical openings are not commonly known to firefighters, many times they are readily apparent on a visual inspection.

One of the first things that should be done on arrival at a basement fire is to make a survey to see whether heat or smoke is spreading to the upper floors or to the attic through any of the vertical openings. The building should always be evacuated as a precautionary measure against this possible spread.

The first lines should be brought into the ground floor and used to prevent any upward spread through the vertical openings. Fog nozzles should be directed into the vertical openings and a check should be made to see whether any fire is running through the partitions. Plans should be made as to the best method of attacking the fire while operations are being conducted to control the upward spread of the fire.

The electricity, gas, and water should be shut off prior to advancing lines into the basement. It is not unusual to find high-voltage electrical equipment or exposed gas lines in the basement. These may become extreme hazards if exposed to water or fire.

Another precautionary measure that should be taken prior to advancing a line into the basement is to have a ladder sitting off to one side in close proximity to the stairwell opening. There have been instances in which the stairwell collapsed as firefighters tried to advance into the basement area. Having a ladder close by provides a method of quickly removing the firefighters to a place of safety in the event this happens.

The method of attack will depend on the number and types of access openings into the basement. The task is easiest when there are openings of sufficient size to the outside of the building at opposite ends of the basement. Once lines are in position for the attack, the access on the leeward side of the building should be opened, followed by opening the access on the windward side. (See Figure 6.10.) Lines are advanced down into the basement from the windward access and the fire and heated gases pushed out the other side. Several lines should be worked abreast if the fire is of any size.

Two charged lines with two firefighters on each line should be positioned at the entrance to the basement as backups. A RIT (or RIC) should also be stationed in the immediate vicinity. Lines should also be stationed on the leeward side where the fire and smoke are being pushed out to protect the exposures. Never bring lines in through both openings in basements that have two openings as just described. This would result in each of the advancing crews fighting one another and would make the entire area untenable because the heat and gases would have no place to escape.

The problem is much more complex when there is only one access to the basement area. Sometimes there are glass blocks in the sidewalk that can be broken to provide a second opening; however, the breaking of deadlights is very time-consuming

FIGURE 6.10 ◆ The proper method for attacking a basement fire.

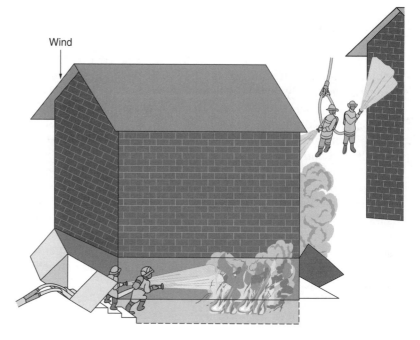

and will require the use of a heavy battering tool. It may be necessary to make an opening in the floor above the basement. If this must happen, select a location at the opposite end of the basement from the access. The opening should be near a window leading to the outside. The first floor should be cross ventilated and loaded lines placed into position at the location where the hole will be made. The lines will be used to push the heated gases and smoke out the window as the hole is cut. A fan placed at the entrance to the basement and one on the first floor adjacent to the hole cut in the floor may prove to be beneficial in the ventilation process. Lines can be advanced into the basement through the access area once the hole is cut and smoke starts moving out. (See Figure 6.11.)

It may not be possible to advance lines into the basement area if the area is well involved with fire. In this case it will be necessary to effect extinguishment by the use of cellar pipes and circulating nozzles.

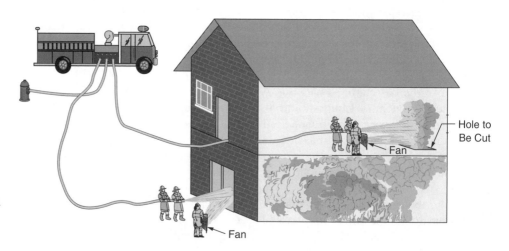

FIGURE 6.11 ◆ Using a fan during a basement fire.

FIGURE 6.12 ◆ A 2½-inch circulating cellar nozzle. *(Courtesy of Elkhart Brass Manufacturing Company)*

Figure 6.12 is a 2½-inch circulating cellar nozzle with four fog heads. Two of the heads produce horizontal streams. One of the heads produces an upward stream while the other one produces a downward stream. The total flow from the four heads is 260 GPM. The flow covers a 55-foot diameter.

Figure 6.13 shows a 2½-inch cellar applicator with a cellar nozzle attached. The applicator has folding support arms for unattended use.

Several holes will have to be cut in the floor above the fire. Where possible, holes should be cut over the hottest part of the fire. Many times these locations can be determined by feeling the floor with the hand. Care should be taken to have protective lines in place as the holes are cut as heat and gases can be expected to be discharged from the cut holes. Every effort should be made to keep the heat and gases in the basement area. (See Figure 6.14.)

FIGURE 6.13 ◆ A 2½-inch cellar applicator with four fog heads. *(Courtesy of Elkhart Brass Manufacturing Company)*

If the basement is well involved with fire, there is a good possibility that the floor above the basement can become so weakened by the fire that collapse is possible. Therefore, it is good practice to remove all personnel from the building once cellar devices are discharging water into the fire area. Of course, it goes without saying that all personnel should be removed from the building any time the floor gives any indication that it is weakening to the point that it might collapse, regardless of whether or

FIGURE 6.14 ◆ Use of cellar nozzles.

not water is being directed onto the fire. In these cases, it will be necessary to do whatever is possible to stop the fire from outside of the building.

When cellar appliances are used to extinguish the fire, their use should be discontinued once access to the basement can be made with hand lines.

Sometimes fires occur in basements that are protected by automatic sprinkler systems. In such cases the fire usually does not make much headway unless it is traveling in concealed spaces that are inaccessible to the flow of water from the sprinkler system. The sprinkler system should be supplemented by department pumpers whenever fires occur in sprinklered buildings. The system should be kept in operation until hand lines are in place and there is assurance that the fire will not spread once the sprinkler system is shut down.

Although all departments do not have the capability of providing it, high-expansion foam has proven to be beneficial in the control of some basement fires. It expands at a rate of 1,000 (or more) to 1 and will completely fill a basement in a short period of time. It has the capacity to smother the fire, is nontoxic, and leaves no residue. It is good practice to use high-expansion foam if it is available, particularly in those cases in which access to the basement area is limited. It is particularly effective in small basements.

An effective attack may be possible using Class A foam rather than water to extinguish the fire if an apparatus equipped with a CAFS is on the scene. It is possible that the fire can be knocked down quicker and with the use of less water.

As a last resort, the basement can be completely flooded. At some basement fires it may be the only method of gaining complete control.

A couple of last thoughts on basement fires are worth considering. Basement fires are death traps. Conditions are almost certain to be ripe for a backdraft if a fire has been smoldering any length of time in a basement. It is possible for firefighters working in this type of atmosphere to be cremated if the area suddenly gets an unexpected supply of oxygen. The backdraft potential should always be eliminated before personnel are sent into the area.

Another hazard to personnel at basement fires is the utilities. It is not unusual to find high-voltage electrical equipment or exposed gas lines in the basement. These may become extremely hazardous if exposed to water or fire. It is therefore imperative to shut off all gas and electrical lines to the building prior to attacking fires of any size in the basement area.

Strategic Goals

- Remove all occupants from building.
- Advance a line into first floor to stop any extension of the fire.
- Check for possible extension of the fire.
- Ensure that the floor above basement is safe to use.
- Check basement entry to determine whether a potential backdraft exists.
- Cut a hole in the first floor and attack the fire with the use of cellar pipes and circulating nozzles if fire is well advanced or backdraft conditions exist. Shut off lines when it is apparent that the fire can be attacked with hand lines.
- If it appears that the fire can be attacked, provide a ventilation opening on the leeward side of building.
- Advance two lines abreast into basement and push the fire out the ventilation opening. Use Class A foam if available.
- If available, consider flooding the basement with high-expansion foam.
- As a last resort, consider using water to flood the basement completely.

Incident Command System Considerations

- Incident Commander
- Safety officer
- Rapid intervention teams
- Staging officer
- Search and rescue group
- Ventilation group
- Interior division
- Exterior division
- Medical unit

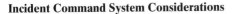

On Scene Scenario

On February 17, 2001, a 29-year-old male volunteer lieutenant and a 32-year-old male volunteer firefighter died while fighting a basement fire. Both victims were part of a crew searching for fire extension when they were suddenly surrounded by intense heat and fire. Upon the crew's exit of the structure, it was discovered that the two victims did not exit. The Incident Commander ordered additional firefighters to enter the basement and search for the victims. After extensive rescue efforts, both victims were removed from the structure and transported to a nearby hospital where they were pronounced dead.

For further information on this incident, refer to the NIOSH Fatality Assessment and Control Evaluation Investigative Report #F2001-08.

◆ CHIMNEY FIRES

Combustion is never complete when wood burns in a fireplace. There are two products that remain. One is the ash that has to be taken outside and carefully disposed of. The other is a residue called creosote. Creosote is formed by the smoke traveling up the chimney.

Creosote will condense on the surfaces of chimney flues at temperatures below 250°F. The creosote will be thick and sticky and appear the same as tar when the temperature falls below 150°F. The creosote traps carbon from the smoke that dries and bakes inside the flue. The result is a flaky, flammable substance.

Over time, the creosote will build up to a considerable thickness. The amount of creosote forming will be determined by the amount of smoke that moves up the chimney. Greater amounts of smoke produce greater amounts of creosote. Greater amounts of creosote mean a chimney fire is waiting to happen.

Fires in chimneys and flues usually do not present too much of a problem, but they should not be taken lightly. The problem usually is not in the extinguishment of the fire but in making sure that the fire has not extended into hidden portions of the building.

One of the first things that should be tried on a chimney fire is the use of dry chemicals. Approximately a pound of dry chemical from an extinguisher should be projected up the chimney and then the fresh air to the chimney should be eliminated by closing the damper and glass doors, if the fireplace is so equipped. Hopefully, this will extinguish the fire. Leave all of the air inlets to the chimney closed and check the building for extension of the fire. (See Figure 6.15.)

FIGURE 6.15 ◆ Cleaning out the chimney after the fire has been extinguished. *(Courtesy of Las Vegas Fire & Rescue PIO)*

If any fire remains within the chimney it is best to allow it to burn itself out. If it is necessary to use water, it should be used very sparsely so as not to crack the masonry. If available, a small spray from a garden hose nozzle is usually sufficient. If the chimney is stopped up, it will be necessary to use a ball and chain or some other weighted object to create a clearance. The ball or weighted object should be tied to a rope and then worked up and down in the chimney until the obstruction has been cleared. Any burning fuel in the fireplace should be removed and a salvage cover or something similar should be used to cover the fireplace opening prior to commencing this operation. This will prevent soot or pieces of the obstructing material from getting into the room. (See Figure 6.16.)

A check should always be made to ensure that the fire has not extended into other parts of the building. If available, a heat sensor or thermal camera can be used to check for hidden hot spots. It will be necessary to examine all intervening floors if the chimney extends from the first floor to the top of the building. The attic should also be examined to ensure that the fire has not extended into this area by conduction or

FIGURE 6.16 ◆ A good job has been done of protecting the furniture prior to cleaning out the soot from a chimney fire. *(Courtesy of Las Vegas Fire & Rescue PIO)*

through cracks in the bricks. Smoke in the attic or in other portions of the building from a chimney fire is a pretty good indication that there are cracks in the chimney lining.

Fires in masonry chimneys can crack the tile or mortar and provide a pathway for flames to reach the combustible wood frame of the house. Fires in prefabricated metal fireplaces with factory-built metal chimneys can severally damage the chimney. The damage is usually in the form of buckled or warped seams and joints in the inner liner of the metal chimney. The chimney no longer should be used and must be replaced when this occurs.

Strategic Goals

- Remove occupants from the building.
- Use dry powder to attempt extinguishment.
- Check for extension of fire.
- Clean out the chimney of any remaining fire.
- Protect furniture by covering the fireplace opening when cleaning out the chimney.
- If water is used, use sparingly to avoid cracking the chimney lining.

Incident Command System Considerations

- Incident Commander
- Rapid intervention team

◆ PARTITION FIRES

A **partition fire** is a concealed fire in a wall. These fires can be overlooked and become the source for a rekindle in a building fire. They are dangerous from the standpoint that they normally do not display themselves until considerable damage has been done. The possibility of partition fires is more prevalent in older buildings in which wood lath has been used in the wall construction and there is an absence of fire stopping. The lath becomes dry and the roughness can hold small sparks of fire over a long period of time. The problem also exists in buildings that have been renovated. Always be suspicious of a possible partition fire when renovations are noticed during the overhaul.

A good method of detecting a partition fire is with the use of a thermal camera. (See Figure 6.17.) However, many departments do not have this luxury. In the absence

partition fire A concealed fire in a wall.

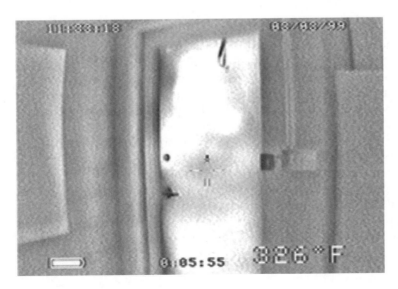

FIGURE 6.17 ◆ The white in the picture taken with a thermal camera shows heat on the other side of the door. *(Courtesy of International Safety Instruments)*

of a camera, one of the best visual indications of a partition fire is the discoloration of the wallpaper or paint; however, this sign is seldom evident. Another visible sign is smoke issuing from the baseboard, from the window moldings, and from around electrical outlets and switches. The one method of detection that is always available is to feel the walls carefully with the hand. If the wall is too hot to touch there is probably a fire inside. The walls on the fire floor, the floor below the fire, and the floor above the fire should be checked for partition fires. It is particularly important to check the floor above if fire originated in the basement.

It is necessary to open up the wall whenever indications of a possible partition fire have been found. The wall should be opened first between the studs near the baseboard. A small spray of water should then be directed into the opening if any signs of fire are found. The stream should be directed upward to stop any extension of the fire and the spray allowed to trickle down the inside of the wall to extinguish any fire found there. The wall should then be further opened between the studs until all of the burned area has been exposed.

Some departments are reluctant to open up walls to check for partition fires. The excuse is that certain types of plaster walls hold the heat for a considerable period of time and can give indications of a possible partition fire. This is true. It is also true that many walls have been opened and no evidence of fire found inside. However, the expense of the repair is minor compared with what could happen if a fire in a partition were allowed to go undetected.

Strategic Goals

- Check all areas for hidden fires.
- Extinguish any fire found.
- Open the wall until all charred areas are exposed.

Incident Command System Considerations

- Partition fires are a part of a larger operation.

◆ DWELLING FIRES

Dwelling occupancies vary from small one-story, single-family residences to multistory tenements, apartment houses, and hotels. (See Figures 6.18 and 6.19.)

Construction may be wood, stucco, brick, or various types of masonry. Some are considered as fast burners; others are classified as fire resistive. Although older structures are generally thought to be more of a fire problem, some of the newer, larger dwellings with their high ceilings and open spaces provide the setting for a rapid spread of the fire and a greater potential for a larger loss. For example, the estimated fire loss for the buildings and contents of a single-family dwelling that occurred in Malibu, California, in 1989 was $7.5 million. This is nearly six times more than the entire fire loss for the conflagration that swept through Chelsea, Massachusetts, in 1973 in which 300 buildings were involved.

In the opinion of the author, the trend toward larger and more expensive homes has increased the fire problem in dwellings. This is primarily true of those that are two stories or more in height and built in the hazardous brush areas of some

FIGURE 6.18 ◆ A dwelling fire with the fire through the roof. Firefighters are about to enter the structure to remove the occupants and attack the fire. *(Courtesy of Rick McClure, LAFD)*

FIGURE 6.19 ◆ The damage to this multifamily dwelling occupancy is extensive but a good stop has been made. *(Courtesy of Rick McClure, LAFD)*

FIGURE 6.20 ◆ Homes like this are being built all over America in areas removed from adequate water supply. *(Courtesy of Engene Mahoney)*

communities, or those constructed a long way from a well-developed water system. (See Figure 6.20.)

dwelling fires The term dwelling fires refers to all types of structures in which people live.

Dwelling fires require the employment of almost all the basic firefighting tactics discussed in previous chapters. Of course, the primary problem is the life hazard. This is apparent when one considers the fact that a large majority of all those who die in fires die where they live. For example, the 2002 annual report on fire loss by the NFPA states that home fires were responsible for 79 percent of the total civilian fire deaths in the United States. According to NFPA records, in 1999, home fire deaths peaked between 4:00 A.M. and 5:00 A.M. During the 1994 to 1998 period, they peaked between 2:00 A.M. and 3:00 A.M. This is a strong indication that most home fire deaths occur during the sleeping hours, despite the fact that the peak hour for home fires is between 6:00 P.M. and 7:00 P.M.

The life hazard is always present and should be given first priority at every fire. An immediate search should be made of all smoke-filled areas, with the search beginning as close to the fire area as possible. Ladders may be required for evacuation of upper floors of larger occupancies, and it always will be necessary to search every room. Although the life hazard takes first priority, it should be kept in mind that in a majority of cases the most effective means of saving life is an aggressive attack on the fire.

SINGLE-STORY, SINGLE-FAMILY DWELLING FIRES

Fires in one-story, single-family residences vary from those that can be put out with a fire extinguisher to those in which the entire building is well involved with fire. (See Figure 6.21.) Single-room fires, however, are much more common than well-involved dwellings. Single-room fires generally can be extinguished with a 1½- or 1¾-inch line, backed up by a larger supply line. The line should be brought in from the uninvolved portion of the dwelling with the objective of confining the fire to the single room. If possible, it is good practice to lay a floor runner from the dwelling entrance to the room fire. A company performing the duties of a truck company should execute this operation and also coordinate the ventilation efforts with the extinguishment efforts.

Well-involved, one-story dwellings of average size generally can be attacked with the use of 1½- or 1¾-inch lines; however, it may be necessary to use larger lines.

FIGURE 6.21 ◆ A well-involved fire in a single-story motel room.
(Courtesy of Rick McClure, LAFD)

With this type of fire, it is normal for a department to operate in the defensive mode with the primary objective of keeping the exposures from becoming involved. Dwellings of this size are built fairly close together in some parts of the country.

The exposure problem can be severe if the siding is wood. The normal procedure for this type of fire is for the first arriving engine company to make an attack with two 1½- or 1¾-inch lines. The lines are used to protect the exposures on either side of the fire with the first line going to the leeward side. The actual attack on the fire is made by later arriving companies. (See Figure 6.22.) The heat given off by larger one-story dwellings can be severe due to the tremendous amount of combustibles involved. Often, there is a greater distance between dwellings of this size, which reduces the exposure problem. However, if the fire building is equipped with wood shingle roofs, consideration always should be given to the possibility of flying brands starting roof fires downwind. In this case it is advisable to have an engine company patrol downwind as a safety factor against this threat.

FIGURE 6.22 ◆ The primary objective in dwelling fires of this magnitude is to protect the exposures.
(Courtesy of District Chief Chris E. Mickal, New Orleans Fire Department— Photo Unit)

If the fire is extensive, it is good practice to commence the attack using 2- or 2½-inch lines equipped with fog nozzles. It may be necessary for the firefighters to operate the lines in a straight stream pattern until the fire is knocked down to the point that a closer approach can be made. These lines require three firefighters on each line due to the difficulty of maneuvering the line. The 1½- and 1¾-inch lines are more maneuverable and can be handled by one or two firefighters. The use of smaller lines releases personnel that can be used elsewhere on the fire.

The attack can be made with Class A foam if the department has an apparatus equipped with a CAFS. This type of attack not only has the ability of knocking down the fire quicker with the use of less water but also provides the firefighters with lighter, more maneuverable lines. The operation is also less fatiguing to firefighters.

If the building is not entirely involved, fires in attics or basements of dwellings should be handled as discussed in previous portions of the book. Attic fires require a coordinated salvage/fire extinguishment effort in order to keep the loss to a minimum.

Strategic Goals for the Offensive Mode

- Conduct search and rescue to ensure all occupants are removed from the structure.
- Advance a 1½-inch or 1¾-inch line in from the uninvolved area to extinguish the fire.
- Lay a floor runner from the entry door to the fire area.
- Lay lines to protect exposures.
- Ventilate the fire area.

Incident Command System Considerations for Offensive Mode

- Incident Commander
- Search and rescue group
- Rapid intervention team
- Ventilation group
- Staging officer

Tactical Goals for Defensive Mode

- Lay 2½-inch lines to protect the exposures.
- Use heavy-stream appliances if needed to protect exposures.
- Extinguish the fire.

Incident Command System Considerations for Defensive Mode

- Incident Commander
- Safety officer
- Staging officer

MULTIPLE-STORY DWELLING FIRES

Firefighters are particularly concerned about fires of any size in multiple-story dwellings. (See Figures 6.23 and 6.24.) This is particularly true when the fires occur at night, primarily during sleeping hours. There is always the life hazard and the constant threat to firefighters in their attempt to make rescues and carry out their assigned functions.

The trend throughout the United States is for larger homes. Because of the reduced cost per square foot, a large portion of the newer homes are two or two and a half stories in height. Although the floor plans for these homes differ, there are some similarities whether the home is located on the West Coast, in the Midwest, or along the eastern seaboard.

FIGURE 6.23 ◆ An attic fire has self-ventilated in a multistory dwelling. The truck crew has laddered the exposure to the right.
(Courtesy of Rick McClure, LAFD)

Most of these homes have a front and back door on the first floor or perhaps a front and side door. The living room, kitchen, dining room, and family room generally are found on the first floor. The second floor is reserved for sleeping quarters. If the structure has a basement, it is normally accessible from stairs in the kitchen or family room or from a side door located on the exterior of the home. The major weakness from a firefighting standpoint is the open stairway.

Fires may vary from those in two-story, single-family dwellings to larger hotels housing hundreds of transients. The most threatening fires are often found in the older, multistory buildings with their wooden construction, open stairwells, balloon construction, and other vertical openings. These offer the constant threat of a rapid fire spread.

Buildings of complete wooden construction are generally limited to four stories in height, but so-called tenement buildings having wooden doors, floors, beams, and stairways with brick walls extend upward to seven stories.

FIGURE 6.24 ◆ Lines have been taken into a third floor apartment. The fire is through the roof.
(Courtesy of Rick McClure, LAFD)

FIGURE 6.25 ◆ Fires in multiple-story buildings may require both outside heavy streams and inside hose lines.
(Courtesy of Chicago Fire Department)

Rescue, salvage, and ventilation should be carried out at these fires in accordance with the procedures earlier established. As with all fires, rescue should be given first priority. These functions should be coordinated with a rapid and aggressive attack on the fire. (See Figure 6.25.)

The first lines should be advanced simultaneously to the fire floor (or floors) and the floor above the fire. A backup line should also be extended to back up those firefighters actively attacking the fire. This line is essential in order to protect the firefighters attacking the fire in the event there is a disruption of water to their line. Maximum protection is provided when the attack line and the backup line are supplied from different pumpers. Where possible, the attack line should be advanced onto the fire from the noninvolved portion of the building to keep the fire from spreading horizontally.

It is best to use the interior stairway for advancing lines to the fire floor if the fire is located on the second or third floor. This means of advancing the line is generally quicker and safer. The stairway also should be used for the backup line.

Lines can be taken to the floor above the fire by way of the fire escape or ladders. The inside stairway can also be used if the fire and other lines are not blocking this pathway; however, a second entry point provided by a ladder is much safer as it is best to keep the stairway reasonably clear for the rescue team.

Extreme care must be taken to ensure that an escape pathway is maintained for the firefighters staffing the line taken above the fire. This line should be used to gain control of the vertical openings and to hold the fire to the fire floor.

The top floor should be checked as soon as possible to ensure that the fire or smoke has not spread to this area or the cockloft (attic) before lines were brought into position to stop the vertical spread. If possible, a line should be advanced to this floor as a safety factor.

If large volumes of fire are visible on arrival, it might be wise to use heavy streams to knock down as much fire as possible while hand lines are advanced into position. These lines should be shut down, however, once hand lines have been taken into the building.

It is difficult to suggest one standard operating procedure that would apply to all fires in multistory dwellings. The fire may be of such a size that it can be handled effectively by the first alarm assignment with the use of just small lines and may have a limited need for truck operations. On the other hand, it could be of such size and magnitude that it would require a second or third alarm to control the fire with a multitude of companies doing truck work to provide the need for rescue operations. It might be a fire that initially can be attacked aggressively using an offensive mode or it might be one that initially will require a defensive mode until a sufficient number of personnel and equipment arrive on the scene to change from a defensive mode to an offensive mode. Therefore, the only attempt that will be made to establish the strategy and tactics for fighting fires in multistory dwelling operations is to list some of the goals that need to be considered for both an offensive attack and a defensive attack and the wide range of considerations that should be used for the incident management of the fire.

Strategic Goals for an Offensive Attack

- Conduct search and rescue operations for the entire building.
- Gain early control of the stairways.
- Provide both horizontal and vertical ventilation.
- Provide the proper size and number of lines that will be required both for firefighting and the protection of the exposures.
- Provide backup lines for firefighters attacking the fire.
- Stop the spread of the fire.
- Protect the exposures.
- Secure the number of personnel and the amount and type of equipment that will be needed.

Incident Command System Considerations for Offensive Mode

- Incident Commander
- Safety officer
- Rapid intervention team or teams
- Search and rescue group or groups
- Interior divisions
- Exterior divisions
- Ventilation groups
- Salvage groups
- Staging officer
- Logistics

Strategic Goals for a Defensive Attack

- Set up heavy-stream appliances both to extinguish the fire and protect the exposures.
- Lay 2½-inch lines for fire extinguishment and to protect the exposures.
- Set up a collapse zone.
- Ensure that all firefighters are out of the building.
- Provide a company downwind to protect the exposures from flying brands.
- Ensure that a personnel accounting report has been made and is accurate.

Incident Command System Considerations for Defensive Mode

- Incident Commander
- Safety officer
- Staging officer
- Exterior divisions
- Logistics

A Common Dwelling Situation

FIGURE 6.26 ◆ Many dwelling fires first show in a single window. This one has extended to the roof and it appears that the attic may also be involved. *(Courtesy of Las Vegas Fire & Rescue PIO)*

Figure 6.26 shows a situation that commonly occurs throughout the United States. It might happen in a single-family residence or in a multiple-family residence. However, it generally occurs in a bedroom. For lack of a better name, it is referred to here as a well-involved fire in one room.

Several things should be taken into account when sizing up a fire similar to the one shown in the photo. Notice that the fire is extending out of the room and impinging on the structure overhead. Also note the heavy black smoke above the house and somewhere near the rear of the dwelling. The black smoke is a clear indication that the fire has extended out of the room to other portions of the house or into the attic. It is more likely from what is seen that both conditions exist. As previously stated, there is one thing that should be avoided and several goals that need immediate attention.

First, DO NOT hit the fire directly from the outside. This would have the effect of pushing the fire farther into the house. Some of the immediate goals are:

1. As a precautionary measure, request assistance. The amount required will depend on the first alarm assignment.
2. Remove all people from the structure.
3. Advance a line outside to protect the exposure that is the upper portion of the structure itself.
4. Advance two lines inside: one to make an aggressive attack on the fire and the other to stop the extension.
5. Check the attic to determine whether the fire has extended into this undivided space. If so, get a line into the attic to extinguish the fire.

◆ GARAGE FIRES

Extinguishing a fire in a garage may appear to be a simple process when discussing tactics. There are, however, a couple of problems that are worthy of consideration. (See Figure 6.27.)

Garages in single-family dwellings are increasing in size. One-car garages were once considered as standard, but today three- and four-car garages are no longer

FIGURE 6.27 ◆ An example of a well-involved garage fire. *(Courtesy of Gavin Kaufman, LAFD)*

unusual. Their locations have also changed. Attached garages have almost become a standard, whereas they once were considered quite rare. The increase in the size of the garage and the change in the location have drastically affected both the attack and exposure problems.

The exposure problem from a well-involved detached garage fire is generally solved by advancing a 1½- or 1¾-inch line between the garage and the house and keeping the house wetted down. The exposure problem from an attached garage might involve taking a line both inside the house and working another outside the house. If all or part of the house extends over the garage, a third line may be required on an upper floor of the house.

The fire-resistive strength of the door leading from the house into the garage is established in the building code for most communities. It may be only a 20-minute door. It provides some protection from the possible extension of the fire from the garage to the inside of the house if it is closed. However, it is possible that this door could have been open or partly open when the fire started, or the fire could have burned through it. Even a small amount of smoke showing inside the house is an indication that one of these conditions might exist. It is very important that the first attack line be taken into the house and kept at this door until the fire is under control. The actual attack on the fire should not be made until this line is in position, as such an attack could have the result of pushing the fire and smoke into the house.

The Progress of a Garage Fire

FIGURE 6.28 ◆ A well-involved fire in one side of a double garage. *(Courtesy of Las Vegas Fire & Rescue PIO)*

FIGURE 6.29 ◆ The fire has extended outside of the garage and has ignited a car. Flames are impinging on other portions of the structure. *(Courtesy of Las Vegas Fire & Rescue PIO)*

Tactics on a garage fire differ considerably from day and night. During sleeping hours it is extremely important that the inside protective line be in place as soon as possible and a complete search of the house be made to ensure that everyone is out before the attack is started on the fire or simultaneously with the attack. All of the sleeping quarters are generally on the second floor in a two-story home, but it is still possible that someone may be sleeping on the couch in the living room or at some other location on the lower floor. Some two-story homes have the master bedroom on the first floor. This is particularly true when they are built for older citizens. Regardless, the entire house must be searched. The Incident Commander should be informed as soon as the search party is assured that everyone is out.

The attack on the fire itself is often made by advancing a 1½- or 1¾-inch line to the front or side door of the garage. (See Figures 6.28, 6.29, 6.30, and 6.31.) If the garage is equipped with a side door, it is normally easier to commence the attack

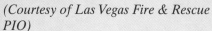

FIGURE 6.30 ◆ Firefighters have arrived on the scene and are advancing lines to commence an aggressive attack on the fire.
(Courtesy of Las Vegas Fire & Rescue PIO)

FIGURE 6.31 ◆ Total damage is extensive, but the firefighters have made a good stop on the fire.
(Courtesy of Las Vegas Fire & Rescue PIO)

through this door as this entry is easier and presents less hazards. However, some departments make it standard practice to force up the overhead garage door and make a direct attack on the fire from the front direction.

The garage door is normally held open by springs; however, these springs should not be depended on during firefighting operations. The door will come down in a hurry if the springs break. This could possibly injure firefighters or trap them inside the garage with the fire. Consequently, standard operations should require that the door be propped open with the use of an 8-foot pike pole or similar object. The line that will be used for the attack should protect the firefighters opening the door until it is secured in place.

Once the door is lifted and a pike pole inserted under the door to keep it open, it is a good idea to place a firefighter by the pike pole to ensure that it is not knocked down accidentally. If this does happen, it could result in firefighters being trapped inside the garage with the fire.

FIGURE 6.32 ◆ Prevent the fire from spreading into the dwelling.

Caution should be exercised when the door is first opened. There is a possibility that conditions inside the garage could be ripe for a backdraft. If one occurs, the fire-fighters opening the door would be caught right in the middle of the explosion. (See Figure 6.32.)

The preceding procedure assumes that the garage door can be opened easily. However, this is not always the case. If the door is equipped with an electric door opener or locked on either the outside or the inside, then forcible entry methods will have to be employed. If the garage has a side door, this may be a better way to gain entrance than trying to force open the larger door. Another alternative is to cut a small hole in either the side door or main door and direct a stream into the interior of the garage using the indirect method of attacking the fire.

Attached garages may be attached directly to one-story residences and to or under two-story residences. Those next to a one-story dwelling will require that a line be taken inside the house to prevent extension of the fire into this area. A 1½- or 1¾-inch line generally will be sufficient for this purpose. A two-story residence with the garage attached at the first floor and under the second floor will require that two lines be taken inside, and 1½- or 1¾-inch lines generally will suffice. If the stairway from the garage leads directly to the second floor (which is unusual), a line should be taken inside the house and moved to the top of the stairway to prevent extension of the fire in that direction. If a door opens directly from the garage into the house on the first floor, a line should be taken inside and positioned at the door opening into the garage. The door should not be opened until the fire is under control.

Extreme care should be used when making the attack on the main body of the fire. Full protective gear including breathing apparatus always must be worn and a backup line always should be provided for the main attack line. This is necessary because it is impossible to know in advance what type and amounts of material will be involved in the fire. In addition to ordinary combustible materials, garages can be expected to act as a storage location for gasoline, kerosene, pesticides, fungicides, propane tanks, spray cans, and other materials that could become an extreme life and health hazard to firefighters. Every effort should be made to keep any containers containing gasoline, kerosene, propane, or similar material cooled down during the entire firefighting process. If a hazardous material team is available, it is a good idea to have its members evaluate the seriousness of the material encountered.

It also can be expected that magnesium may become involved. It is not unusual to find wheels on lawn mowers and motorcycles to be made of this material. Its involvement would be indicated by a brilliant white light in the area of its burning. Care should be taken when directing water into this area, as small explosions can be expected when the water comes into contact with the burning magnesium.

At times some of the combustibles stored in the garage will be found on overhead rafters or in a loft storage area. This could be a severe hazard to firefighters working in the garage in the event a fire could burn through the rafters or loft storage and dump the stored material onto the firefighters working the line.

One other precaution should be observed on garage fires. On some occasions the heating and ventilation system and the hot water heater are located in the garage. If this is the case, the fire can expect to have extended throughout the house through the ductwork and/or vents. A thorough check must be made for extension into these hidden spaces.

A good strategy to adopt when fighting fires in garages is always to expect the unexpected.

Strategic Goals

- Remove all occupants from the building.
- Advance line inside to prevent extension of the fire.
- Check for possible backdraft conditions.
- Advance a line through the side door if the garage is so equipped.
- Prop open the garage door, and have attack lines in place.
- Check the overhead for storage.
- Make attack using two 1½- or 1¾-inch lines.
- Make attack with Class A foam, if available.
- Lay 1½- or 1¾-inch lines to protect exposures.
- Check for the extension of the fire if the heating and ventilation system is located in the garage.
- Expect the unexpected.

Incident Command System Considerations

Unless the dwelling becomes fully involved, this fire will be fought in the offensive mode.

- Incident Commander
- Search and rescue group
- Rapid intervention team
- Safety officer
- Ventilation group
- Interior division
- Exterior division
- Staging officer

FIGURE 6.33 ◆ Fires in mercantile occupancies are common in many cities. *(Courtesy of District Chief Chris E. Mickal, New Orleans Fire Department—Photo Unit)*

◆ MERCANTILE FIRES

Mercantile establishments vary from those that are relatively small to some that are extremely large. The typical store has a large undivided sales area in the front and a small service area in the rear. If the building has a basement, the entrance generally will be somewhere near the rear entrance. (See Figure 6.33.)

Most fires start in the service area. If the fire has not progressed into the sales area by the time the department arrives, it usually can be confined to the rear of the building by bringing lines in through the front and making an aggressive attack on the fire. Opening doors and breaking windows in the rear of the building will assist in keeping the smoke and fire out of the sales area.

Life safety can be a problem if the fire occurs during the hours the business is open. Most shoppers will attempt to leave the store through the entrance they came in. This is usually in the front of the building, the same place where the fire department will be bringing in hose lines. Large show windows are usually found in the front of the occupancy. Although it is better to ventilate in the rear of the building or by opening the roof over the fire, if necessary, the large front windows can be removed for access and ventilation. (See Figure 6.34.)

The potential for a fast-moving fire generally exists in mercantile establishments due to the large quantities of flammable material found in the sales area. There is also a potential for a large loss due to the value of the merchandise on display. Every effort should be made to commence salvage operations immediately. Care should be taken during the overhaul process to collect burned stock in such a

FIGURE 6.34 ◆ The color of the smoke indicates that water is reaching part of the fire. *(Courtesy of Rick McClure, LAFD)*

manner that as accurate an inventory as possible can be made of the damaged material.

Smoke can cause a considerable loss to clothes and similar items. Ventilation should be started early and made in such a manner as to keep smoke out of the sales area, if possible. Roof ventilation is usually preferred. If conditions permit, positive pressure ventilation should be employed. (See Figure 6.35.)

FIGURE 6.35 ◆ Firefighters have opened the roof of this commercial building in two different locations. *(Courtesy of Rick McClure, LAFD)*

The problem is magnified in large commercial occupancies. However, the tactics remain the same—locate, confine, and extinguish. If the location of the fire is apparent when the department arrives, the Incident Commander must make a decision quickly as to the best method of attacking the fire. Bringing a line in through the wrong door can result in the fire being pushed into the uninvolved areas, which will compound the extinguishment process, the safety of shoppers, and the fire loss.

The number of lines required for extinguishment and control depends on the size and magnitude of the fire. If the fire is small, it may be possible to extinguish it with a few lines. If the fire has a good start and travels fast, it is possible that the entire occupancy can become involved.

While it is difficult to imagine a typical commercial fire, it is easy to visualize a well-involved fire with the fire going through the roof on the arrival of the fire department. Such a condition is most likely to occur late at night in a building that is not sprinklered.

Whether the initial action will be conducted in an offensive mode or a defensive mode depends on such factors as the size of the fire and the time of the fire. If a fire occurs in a commercial occupancy during the day, there is the possibility of a serious life hazard. The life hazard has been pretty well eliminated if the fire occurs at night when the store is closed and locked. The tactics involved change considerably in the two scenarios.

Strategic Goals for an Offensive Attack

- Conduct search and rescue operations for the entire building.
- Advance the number of hand lines that will be required for both firefighting and protection of the exposures.
- Set up heavy-stream appliances both for extinguishment and to protect the exposures.
- Stop the spread of the fire.
- Request the number of companies that will be required to conduct salvage operations.
- Request the number of personnel and the amount of equipment that might be required.
- Provide both horizontal and vertical ventilation.
- Use the CAFS for fire extinguishment, if available.
- Provide for water removal.

Incident Command System Considerations for Offensive Mode

- Incident Commander
- Safety officer
- Rapid intervention teams
- Search and rescue group
- Ventilation group
- Salvage group
- Staging officer
- Interior divisions
- Exterior divisions
- Medical unit
- Logistics

Strategic Goals for Defensive Attack

- Set up heavy-stream appliances both for fire extinguishment and to protect the exposures.
- Lay 2½-inch lines for firefighting and to protect exposures.
- Set up a collapse zone.

- Set up a personnel accounting report.
- Set up a patrol downwind to protect exposures from flying brands.

Incident Command System Considerations for Defensive Mode

- Incident Commander
- Safety officer
- Staging officer
- Exterior divisions
- Information officer
- Medical unit
- Logistics

◆ INDUSTRIAL FIRES

Serious fires in industrial plants are less frequent than in mercantile establishments due to a fire-conscious management and better built-in protection. The potential loss, however, is severe due to the high values involved. Industrial plants generally contain large open areas that allow rapid extension of the fire once it gains any headway whatsoever. Stopping the spread and extinguishing the fire present some serious problems. (See Figure 6.36.) Many times it is necessary to make the initial attack using heavy-stream appliances until hand lines can be brought into position. If the fire has gained considerable headway it may be necessary to continue the use of heavy streams due to the limited reach of handheld streams.

Firefighting tactics in large industrial fires consist of protecting the exposures and extinguishing the fire using as many lines and firefighters as necessary. There are, however, some problems peculiar to industrial plants that should be considered when smaller fires are encountered or in the early stages of larger fires. See Figure 6.37.

Large quantities of flammable liquids in containers may be found in some occupancies. Every effort should be made to keep the containers cool to prevent their

FIGURE 6.36 ◆ Some industrial fires require the use of a large number of heavy streams.
(Courtesy of Rick McClure, LAFD)

FIGURE 6.37 ◆ This industrial building was well-involved on the arrival of the fire department requiring that it be attacked in the defensive mode.
(Courtesy of District Chief Chris E. Mickal, New Orleans Fire Department— Photo Unit)

rupture or explosion and the subsequent rapid spread of the fire. Spray streams are particularly useful for this purpose. (See Figure 6.38.)

Every effort should be made to keep acids or other corrosive materials from coming into contact with combustible material. Although federal law requires that all firefighters be trained in hazardous materials emergency response, it is possible that hazardous materials may be found on the premises that are not familiar to anyone at the scene. A hazardous material team always should be requested, if available, whenever materials of this nature are encountered.

Fire in areas where molten metals are found should be extinguished using high-pressure fog streams. Solid streams played on molten metals could result in violent explosions and splattering of the metal.

Fires in large sawdust or dust-collecting bins should be knocked down initially using spray streams. After the fire appears to be extinguished, the contents of the

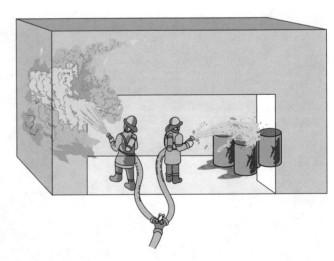

FIGURE 6.38 ◆ Protect exposed flammable liquid containers.

FIGURE 6.39 ◆ Dump the sawdust through a fog or spray stream.

bin should be dumped through spray streams to prevent a possible dust explosion. (See Figure 6.39.)

Vats of hot salt baths present a similar problem to that of molten metals. If possible, the vats should be covered prior to extinguishing any fire around them. The fire should be knocked down by using high-pressure fog nozzles if the vats cannot be covered.

Fires in dip tanks or large vats of combustible paints can be extinguished by the use of water but care must be taken not to use so much water that it will overflow the container. This can generally be prevented by using high-pressure fog nozzles for extinguishment.

Fires in lumberyards or in the lumber storage area of industrial plants frequently require heavy streams for extinguishment. This is due not only to the size of the fire but also to the tremendous heat that is being given off. The heat may be of sufficient intensity that hand lines cannot be advanced close enough to spray water on the fire. There is almost always the hazard of flying brands from this type of fire, which requires that the area downwind from the fire be patrolled.

Strategic Goals for an Offensive Attack

- Conduct search and rescue operations for the entire building.
- Advance the number of lines that will be required for both firefighting and protection of the exposures.
- Stop the spread of the fire.
- Request the number of companies and additional equipment that might be need for fighting the fire and protecting the exposures.
- Request the number of companies required for conducting salvage operations.
- Provide both horizontal and vertical ventilation.
- Anticipate the special firefighting tactics that will be required for the special processes used in the building.
- Consider the use of CAFS for fire extinguishment, if available.

Incident Commander System Considerations for Offensive Mode

- Incident Commander
- Safety officer
- Rapid Intervention team or teams
- Search and rescue group or groups
- Staging officer
- Ventilation groups
- Salvage groups
- Interior divisions
- Exterior divisions
- Information officer
- Medical unit
- Logistics

Strategic Goals for a Defensive Attack

- Set up as many heavy-stream appliances as will be needed to protect the exposures and extinguish the fire.
- Advance 2½-inch hose lines for fire extinguishment and to protect the exposures.
- Set up a personnel accounting report.

Incident Command System Considerations for Defensive Mode

- Incident Commander
- Safety officer
- Staging officer
- Exterior divisions
- Medical unit

◆ PLACES OF PUBLIC ASSEMBLY

Some of the most disastrous fires from the standpoint of loss of life have occurred in places of public assembly. The five deadliest public assembly fires in U.S. history were:

1. The Iroquois Theater, Chicago, Illinois, December 30, 1903.
   602 killed.
2. Cocoanut Grove Night Club, Boston, Massachusetts, November 28, 1942.
   492 killed.
3. Conway's Theater, Brooklyn, New York, December 5, 1876.
   285 killed.
4. Rhythm Club dance hall, Natchez, Mississippi, April 23, 1940.
   207 killed.
5. Rhodes Opera House, Boyertown, Pennsylvania, January 12, 1908.
   170 killed.

Panic was a big contributor to the loss of life in each of these fires. Thus, it follows that one of the primary objectives of the fire department at public assembly fires is to prevent or eliminate panic.

Panic is normally the result of fear. A single shout of "fire!" can convert a calm group of people into a raging mob. Anything a fire company can do to reduce panic will have an overall effect on the saving of life. Of course, if there are signs of an active fire when the first company arrives, it can be expected that panic has already become a part of the exit process and the department will have to act accordingly.

One of the first actions that should be taken on arrival is to see that all exits are thrown open immediately. Lines must be advanced to protect the exits, and ladders should be raised to every fire escape and every window where people are showing. This action will assist in emptying the building as quickly as possible. Ventilation should be started to pull the smoke and heat from the exit pathways and to help relieve the panic-contributing atmosphere that exists inside. Ambulances should be requested as they no doubt will be needed.

The first lines should be used to protect the exits until everyone is safely out of the building. It may be that the best method of protecting the exits is by a rapid attack on the fire. This would certainly be the case if the fire was threatening to move in a direction that would cut off exit pathways.

A thorough search of the building should be started as soon as possible after arrival to ensure that no one has been trapped or overcome by smoke. It should never be assumed that the rescue problem is over when people quit moving out of the exits.

Thought should be given early in the emergency to the possibility that the fire will involve the electrical equipment and throw the building into darkness. This not only would increase the problem of rescue but also would incite the crowd into a panic situation if such condition did not already exist. Consequently, portable lighting equipment should be taken into the building and plans made for its use, even if it appears that lighting will not be a problem.

Theaters are better prepared to cope with a fire situation than other places of public assembly such as nightclubs, dance halls, and hotel ballrooms. The most likely place for a fire to start in a theater is in the backstage. The stage area is designed with a vent that should open automatically in case of fire. A fire of any size should also cause the fusible link on the fire curtain to part, which will cause the curtain to drop. The **fire curtain** is designed to prevent fire, heat, and smoke from entering the auditorium area.

fire curtain   A fire curtain in a theater is designed to prevent fire, heat, and smoke from entering the auditorium area.

When responding to a backstage fire, the Incident Commander should ensure that the curtain has dropped and the vent is opened. Lines must be brought into the auditorium to prevent extension of the fire into this area. Additional lines should be brought into the stage area from both sides if possible. Although theaters are generally made of fire-resistive material, there is normally sufficient combustible material backstage to generate a good-sized fire.

Strategic Goals for an Offensive Attack

- Remove all occupants from the structure.
- Conduct a thorough search and rescue effort to ensure that no one was overlooked.
- Do whatever is possible to calm down patrons.
- Make an aggressive attack on the fire using as many hand lines as possible.
- Stop the extension of the fire.
- Provide horizontal and vertical ventilation.
- Request additional companies and ambulances.
- If occupancy has a stage, make sure the protective curtain has been dropped; then advance hand lines backstage from both sides.

Incident Command System Considerations for Offensive Mode

- Incident Commander
- Safety officer

- Search and rescue groups
- Rapid intervention teams
- Staging officer
- Interior divisions
- Exterior divisions
- Ventilation groups
- Salvage group
- Medical units
- Information officer

Tactical Goals for Defensive Attack

- Set up heavy-stream appliances both for fire extinguishment and to protect the exposures.
- Lay 2½-inch lines for fire extinguishment and to protect exposures.
- Set up a collapse zone if the building is two or more stories in height.
- Set up a personnel accounting report.

Incident Command System Considerations for Defensive Mode

- Incident Commander
- Safety officer
- Rapid intervention team
- Search and rescue groups
- Staging officer
- Documentation unit
- Information officer
- Medical units

On Scene Scenario

On February 20, 2003, a band playing at The Station nightclub in West Warwick, Rhode Island, used a pyrotechnics display as a portion of its act. The pyrotechnics ignited nearby combustibles. The club was engulfed by flames within minutes of the fire breaking out. Approximately 100 people were killed in the fire. Some were killed by the fire and smoke; others were reported to have been killed as they rushed for the exits.

The West Warwick fire chief said that the entire club was consumed by flames within three minutes. The chief also said that in their rush to escape, people had neglected to use three fire exits. "As is human nature," he said, "they tried to get back out the way they came in."

◆ TAXPAYER FIRES

taxpayer building A cheaply constructed building that was constructed in the early 1900s for the purpose of saving money.

The term **taxpayer building** came about in the early 1900s to identify a building that was constructed for the purpose of saving money. An investor would buy a piece of vacant land that he or she anticipated would increase sharply in value within a few years. The individual would then construct a building as cheaply as possible to provide sufficient income to pay the taxes during the interim required for the land to increase in value.

The term has been expanded to incorporate a type of building with common problems. The building is normally one or two stories in height and contains a number of small businesses under a common roof. If it is two stories or higher in height, the ground floor will be operated as commercial entities while the second and higher floors are utilized for such purposes as offices and/or apartments. The primary feature is an undivided attic or an occasional dividing wall of weak construction. If the building has a basement, it normally will also be undivided with wire or flimsy walls separating the storage area of one business from that of another.

The type of construction and the contents of the various businesses provide the potential for a rapid spread of fire; however, the primary problem is the undivided attic and possibly undivided basement. The key to fire control is to get ahead of the fire and stay ahead of it. Of course this is a basic principle of all fires, but with this type of fire, failure to attack rapidly and aggressively will probably result in the loss of the entire building.

On arrival it is sometimes difficult to determine exactly which store is involved. Smoke will fill the attic space and will be pushing out from all parts of the building. Visible flames help in locating the fire. Several actions should be taken once the location of the fire in a single-story structure has been identified. A large hole should be cut in the roof directly over the fire. It is good practice to make this hole at least 10' by 10' in size. Simultaneously, a hole needs to be cut in the ceiling of the second store on either side of the involved area. Ladders must be placed into the openings and lines advanced into the attic to stop the spread. Lines of 1½ or 1¾ inches normally will be sufficient for this purpose. (See Figure 6.40.)

The problem will be increased if the building is two stories in height. The same general attack procedure will prevail if the fire is in a store on the second floor. However, the salvage problem has been intensified. Merchandise in the store below the involved area should be covered first. Salvage operations need to continue in the adjacent stores as soon as possible.

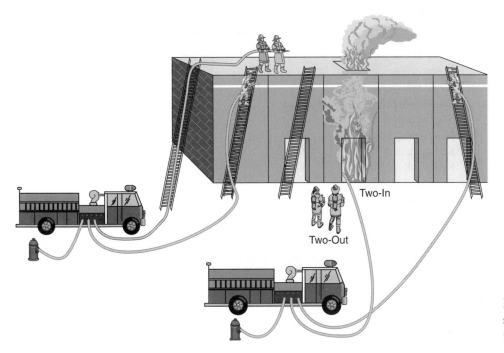

Two-In

Two-Out

FIGURE 6.40 ◆ Stop the spread and extinguish the fire.

If the fire is of any size in the basement area, it can be anticipated that the entire basement area will probably be lost. Standard procedures for extinguishing fires in basements should be employed with the hope that the fire can be confined to the basement area.

Depending on the size and extent of the fire, the Incident Commander may decide to operate in either offensive mode or defensive mode.

Strategic Goals for an Offensive Attack

- Regardless of the time of the fire, if the building has dwellings on the second or higher floors, an immediate search and rescue operation should be conducted on all floors above the first floor. This should be given first priority with the exception that first priority should be given to search and rescue of those on the first floor if there are indications that anyone located there is in immediate danger.
- If the fire occurs during business hours, it will be necessary immediately to perform search and rescue operations on the first floor to ensure that all occupants and shoppers are removed from the building.
- Call for additional companies and/or resources if needed.
- If the fire is in one store of a one-story taxpayer building, open the ceilings in the occupancies two stores away from the fire on both sides. Then raise ladders to the openings and advance lines into the attic to stop the spread of the fire in the event it has entered the common attic.
- If the fire has entered the attic area, ventilate the roof above the fire if the roof is safe for firefighters to perform the ventilation. If an elevated platform is available, secure a safety rope to any firefighter working to ventilate the roof.
- Advance a charged line to the roof for the protection of the firefighters and to extinguish any fires that might occur as a result of cutting a hole in the roof.
- Make sure there are two ways for firefighters to escape from the roof.
- Provide for salvage operations if the fire is on the second floor or above.
- Provide for horizontal ventilation on all floors.
- Anticipate the possibility of an early failure of the roof if the building is one-story in height.

Incident Command System Considerations for Offensive Mode

- Incident Commander
- Safety officer
- Rapid intervention team
- Search and rescue group or groups
- Staging officer
- Ventilation group
- Salvage group
- Interior divisions
- Exterior divisions
- Medical unit

Strategic Goals for a Defensive Attack

- Set up a collapse zone and ensure that it is clear.
- Set up both hand lines and master streams to attack the fire and protect the exposures.
- Set up a personnel accountability report.

Incident Command System Considerations for Defensive Mode

- Incident Commander
- Safety officer

- Rapid intervention team
- Staging officer
- Exterior divisions
- Medical unit

◆ STRIP MALL FIRES

Strip malls are a part of the changing face of America. (See Figure 6.41.) While residential properties have moved from the central city to the suburbs, the shopping areas have moved from the downtown area to large malls and strip malls.

Strip malls are generally located along the principal streets of a community. Many of these streets have been rezoned from residential to commercial. A strip mall contains a row of small businesses housed under a common roof. There might be six to eight businesses with no semblance of compatibility. One section of the building might be occupied by a medical unit while two doors down might be a shoe store or a small cafeteria. However, they are all constructed basically the same and are all connected by a common attic. Most strip malls are one-story in height but some two-story malls exist. In front of the mall is a common parking lot.

The walls between occupancies are normally constructed of 1½-inch drywall. Older strip malls may have a truss roof covering the entire assembly; however, the roof construction in newer strip malls is generally flat or slightly sloping and constructed of lightweight components. There normally is either a supported or an unsupported fascia.

These fascias may extend only a few feet from the main building's front wall or as much as 12 or 14 feet. The wider fascias may or may not be supported. In almost all cases, a fascia is an undivided area that extends from one end of the building to the other, permitting a rapid spread of the fire in the event it enters this area. A general principle regarding facias is that it is necessary for firefighters to move

strip mall A row of small businesses housed under a common roof.

FIGURE 6.41 ◆ Strip malls, with all their firefighting problems, are part of the changing face of America. *(Courtesy of Eugene Mahoney)*

FIGURE 6.42 ◆ Fire has invaded the fascia and is spreading through the length of the strip mall.
(Courtesy of Rick McClure, LAFD)

underneath them in entering or leaving the building. However, it is best that firefighters not work under them due to the possibility of an unanticipated collapse. (See Figure 6.42.)

Rapid construction failure and lateral spread of a fire in a strip mall is predictable. The most likely contributor to a fast fire spread is the common attic. In certain instances a heavy snowstorm has caused the roof of a strip mall to collapse. Heavy air conditioners or refrigeration units on the roof can become a hazard to firefighters during fire operations.

Generally, a newer strip mall is built on a concrete slab; however, the possibility of the mall having a common basement separated by flimsy partitions should not be ignored. Overall, the potential firefighting problems of a strip mall can pose a challenge. They are very similar and have the same problems as fires in taxpayer buildings discussed in the preceding section. In fact, in some parts of the country, strip malls are referred to as taxpayers.

Strategic Goals for an Offensive Attack

- If the fire occurs during business hours, it will be necessary to perform search and rescue operations immediately to ensure that all occupants and shoppers are removed from the building. This should be given first priority even if the initial fire is confined to one store on arrival. An aggressive attack on the fire may be the most effective method of protecting life.
- Call for additional companies and/or resources if needed.

- Open the ceiling in the occupancies two stores away from the fire on both sides, raise ladders to the openings, and advance lines into the attic to stop the spread of the fire in the event it has entered the common attic.
- If the fire has entered the attic area, ventilate the roof above the fire if the roof is safe for firefighters to perform the ventilation. If an elevated platform is available, secure a safety rope to any firefighter working to ventilate the roof. (See Figure 2.25.)
- Advance a loaded line to the roof for the protection of the firefighters and to extinguish any fires that might occur as a result of cutting a hole in the roof.
- Make sure there are two ways for firefighters to escape from the roof.
- Anticipate the possibility of an early failure of the roof.

Incident Command System Considerations for Offensive Mode

- Incident Commander
- Safety officer
- Rapid intervention team
- Staging officer
- Search and rescue group
- Salvage group
- Ventilation group
- Interior divisions
- Exterior divisions
- Medical unit

Strategic Goals for Defensive Attack

- Set up a collapse zone and ensure that it is kept clear.
- Set up master streams to both attack the fire and protect the exposures.
- Ensure that a personnel accountability report is taken.

Incident Commander System Considerations for Defensive Mode

- Incident Commander
- Safety officer
- Rapid intervention team
- Staging officer
- Exterior divisions
- Medical unit

On Scene Scenario

A fire broke out in a strip mall on a Friday evening in Kenai Peninsula, Alaska. The Kenai Fire Department responded immediately. The total number of units used at the fire including mutual-aid companies was four engine companies, one truck company, three medic units, and three command vehicles. The strip mall was occupied by six businesses.

Heavy smoke and flames were projecting from the west end of the building when the first unit arrived on the scene. The firefighters cut a trench at the east end of the building to stop the fire spread. A water curtain was set up in the trench. However, within one hour all firefighters were removed from the building and the Incident Commander resorted to a defensive mode due to hazardous conditions.

Two firefighters were injured during the fire as a result of falling from ladders. Both were treated on the site and did not require further care.

◆ CHURCH FIRES

Church fires are different from almost any other type of fire to which the fire department may respond—different from the standpoint that it can almost always be anticipated that if the fire is of any size at all, the entire building will probably be lost. There is more of the probability that the fire department will end up operating in the defensive mode at a church fire than there is in any other type of fire. The primary reason for this is the construction of the church itself. (See Figure 6.43.)

The interiors of most churches, and in particular the older ones, are made of wood with wood also used in the pews and benches. Wooden choir lofts and wooden balconies are not uncommon. There is usually a high-pitched roof with a large-vaulted ceiling and a steeple. The ceiling is normally too high to be opened if fire invades that area. The wide-open spaces within the church create drafts that have the potential of whipping fire in all directions.

Concealed spaces run rampant throughout a church. In addition to those in the vaulted ceiling and organ loft, many will be found in the walls. Additional construction is added to hide heating pipes. These pipes run through the main auditorium as well as through rooms to the rear of the church and in the organ gallery area. Walls, recesses, and pipe channels are many times furred out to give a smooth appearance. This provides concealed spaces that run a foot or more in depth. Unfortunately, most of the furred areas terminate in the vaulted ceiling or the attic.

Most churches have basements. Statistics indicate that a large number of church fires originate in this area. This might be due to the fact that the heating unit normally is located in the basement. A fire originating in this location has a good chance of spreading rapidly throughout the entire church. However, it may be possible to hold the fire to the basement area by a rapid attack on the fire if the fire department receives an early call. Concealed spaces in walls should be opened and lines used to prevent the upward spread. It is likely that the fire will pass the auditorium area and extend directly into the huge hanging ceiling area or the concealed spaces of the enclosed attic if the fire is not immediately controlled. Once this happens, the building is lost.

Fortunately, most fires in churches start when the church is unoccupied. This reduces the life hazard but does not eliminate it entirely. Many of the churches have an

FIGURE 6.43 ◆ Church fires present unique problems to firefighters. *(Courtesy of Chicago Fire Department)*

unprotected passageway that extends from the church to a rectory where it is common to find people at all hours. This passageway will permit fire to travel in either direction. Lines should be advanced into this passageway from the uninvolved area if a fire of any size is apparent in either the church or the rectory when the department arrives on the scene.

Fires that have gained any headway in a church should be attacked with the use of heavy lines. If the fire has entered the hanging ceiling or enclosed attic area, there is little chance of extinguishing it. The ceiling is too high to open and it is too hazardous to place firefighters on the steep roof to ventilate. Elevated platforms or ladder pipes should be set up to attack the fire once it burns through the roof; however, care should be taken as to where these apparatus are placed. If the fire has extended or is likely to extend into the steeple, it is almost a certainty that this portion of the church will eventually collapse. The steeple acts like a chimney flue and will pull fire quickly through its entire interior. Plans should be made for this collapse with thought given to the hazard created by the heavy bells that are located in the steeple area.

It is normally dangerous to operate inside the church if the fire is of any size whatsoever. The possibility of heavy lighting fixtures falling and the potential failure of balconies and organ loft structures create too high a risk for firefighters.

If the fire is of any size whatsoever, the Incident Commander will have to make an early decision as to whether to commence operations in the offensive mode or the defensive mode. If the fire is in the steeple, or an inside fire has entered the hanging ceiling or attic, serious thought should be given to commencing operations in the defensive mode. However, if there are indications that the fire is confined to the basement, an offensive mode might be appropriate.

Strategic Goals for an Offensive Attack

- Rescue operations should be started immediately if there is any possibility that life might be threatened.
- Attempt to control and extinguish the fire using hand lines.
- If there are interconnecting passageways between the church and the rectory or between the church and other buildings on the property, lines should be advanced into these locations to stop any spread of the fire to these potential exposures.
- All potential vertical pathways through which fire could spread to the attic area should be checked and lines advanced to stop any such spread.
- Attempt to establish ventilation. Evaluate the possibility of using forced ventilation. Do not hesitate to break glass, including stained glass widows, if deemed necessary.
- Call for additional companies and resources if needed.
- Anticipate the possibility of light fixtures or other objects dropping from above within the church area.
- Anticipate the early collapse of the steeple.
- If possible, remove valuable artifacts, relics, and other irreplaceable items from the church to a place of safety.
- Keep well-intentioned people, including the pastor, out of the church and harm's way.

Incident Command System Considerations for Offensive Mode

- Incident Commander
- Safety officer
- Rapid intervention team
- Staging officer
- Search and rescue group
- Interior divisions

- Exterior divisions
- Salvage group
- Medical unit

Strategic Goals for a Defensive Attack

- Set up a collapse zone and ensure that it is kept clear. Consider the fact that the steeple could fall in any direction.
- Set up master streams to both attack the fire and protect the exposures. Make sure that no master-stream appliance that requires the constant supervision of a firefighter is set up within the potential collapse zone.
- Remove all firefighters from the interior of the building, even if it appears safe to operate inside.
- Set up a patrol downwind to check for flying brands.
- Ensure that a personnel accountability report is taken.

Incident Command System Considerations for Defensive Mode

- Incident Commander
- Safety officer
- Rapid intervention team
- Staging officer
- Exterior divisions
- Medical unit

◆ FIRES IN BUILDINGS UNDER CONSTRUCTION

A fire in any frame building under construction will present unusual problems to firefighters; however, the most troublesome appear to be those large complexes of two-story apartments, condominiums, or townhouses. Some of these developments extend over an area of several blocks. It is not unusual to find these complexes being built in the midst of a well-developed area of dwellings with wood shingle roofs. They are particularly vulnerable to rapid fire spread when they are in the framing stage prior to drywall or other coverings being installed. A fire starting at this stage of construction can quickly become a two-story bonfire with all the potential of developing into a major conflagration. This is particularly true during periods of low humidity and high winds.

Not only do fires in these complexes present a constant threat of rapid fire spread but firefighters are normally hampered by an inadequate water supply and an extremely difficult access problem. It is not uncommon to find these complexes in the framing stage prior to water mains being installed. Additionally, deep trenches, piles of dirt, and stacks of construction material may block apparatus access and hamper the advancement of hose lines. The problem can become acute right after a rain that fills trenches with water and makes the entire area one big puddle of mud.

In reality, the only effective method of fighting a fire in these types of complexes is to start before the first nail is driven. This means that a city must pass adequate ordinances and ensure proper enforcement. No construction should be started until the water mains have been installed, the system tested, and the streets properly paved. On-site hydrants need to be provided if the distance from street hydrants is excessive. Inspections should be made during the entire construction process to ensure that access pathways are kept clear and that unusual problems do not develop that would hamper firefighting operations.

FIGURE 6.44 ◆ Quick action is required on this fire to protect the exposures. *(Courtesy of Las Vegas Fire & Rescue PIO)*

This is the ideal; unfortunately, it is not always achieved. Not only are some cities lax in passing needed ordinances but also many of these complexes are built in rural areas that do not have adequate building requirements. Consequently, fire officers have to attack the fire using the resources they have and under the conditions they encounter. Additional help beyond the first alarm assignment will undoubtedly be needed if the fire has gained any headway whatsoever. The first arriving officer should not hesitate requesting assistance. In fact, no harm can be done when a second alarm is requested if any size of working fire is observed in the vicinity of the complex while apparatus is responding.

The primary problem will be the rapid fire spread with the resultant threat to exposures. If the complex is isolated, the threat may be confined to that of flying brands. However, in most cases the threat will include that of flying brands along with the problem of protecting the exposures from the tremendous radiant heat that will develop. The exposure problem from radiant heat may be confined to units in the complex or may include the threat to other dwellings surrounding the complex. The primary problems will often exist downwind. (See Figure 6.44.)

Exposures can be protected by using basic firefighting tactics. This type of fire will call for large lines and the employment of heavy streams. A sufficient number of companies should be requested to provide the fire streams required to protect the exposures and make an aggressive attack on the fire. The first streams should be directed on the exposures and should be heavy streams. Several companies should be patrolling downwind to protect exposures from flying brands. It may be necessary to relay water from some distance if an adequate supply is not available in or close to the complex. Consideration should be given for the additional companies that may be needed for relay purposes.

Strategic Goals

If a fire occurs in a complex under construction that is in the later stages of framing and no dry wall or other coverings have been installed, there is little chance that the fire will be fought in offensive mode. Consequently, the strategic goals are for a fire using the defensive mode.

- Call for help immediately if the fire has gained any headway whatsoever.
- If possible, secure a good water supply.

- Consider the possible need for a relay operation.
- Anticipate early collapse of the structure.
- Set up a collapse zone and ensure that it is kept clear.
- Set up master streams to both attack the fire and protect the exposures.
- Assign companies downwind to protect exposures from flying brands.

Incident Command System Considerations

- Incident Commander
- Safety officer
- Exterior divisions

◆ HIGH-RISE FIRES

high-rise Any multiple-story building that requires the use of high-rise firefighting tactics for effective extinguishment.

High-rise is a term coined by the media to refer to those buildings that are reaching toward the heavens. A number of definitions have appeared since the initial use of the term. Some fire departments have defined it to mean any building over 75 feet in height, primarily for the purpose of controlling by ordinance the built-in fire protection during the construction of such buildings. Others have defined it to mean any building beyond the reach of the fire department's aerial equipment or, to put it another way, those buildings too high to rescue all occupants by the use of the department's aerial equipment. (See Figure 6.45.) This definition included buildings of less height than the aerial equipment if the buildings were set back a considerable distance from the street, which

FIGURE 6.45 ◆ High-rise fires produce many problems for firefighters that are not common in other types of structure fires.
(Courtesy of Rick McClure, LAFD)

restricted the use of the aerial equipment to its maximum height. Regardless of the number of definitions, it appears safe to say that a high-rise building is any multiple-story building that requires the use of high-rise firefighting tactics for effective extinguishment. Present high-rise tactics have been designed primarily to cope with fires in the newer, higher buildings. These tactics merge the Incident Command System, the experience gained over the years in fighting fires in multiple-story buildings, and the lessons learned from the many disastrous high-rise fires that have occurred around the world.

The tallest building in the United States is the Sears Towers in Chicago. Built in 1974, this building contains 110 stories and is 1,450 feet in height. On December 19, 2003, the design of the Freedom Tower that is to be built at the site of the devastated World Trade Center in New York was unveiled. It is planned that it will be 1,776 feet in height as a symbol of America's freedom and the signing of the Declaration of Independence. A broadcast antenna attached to the building's tower will raise the structure to a height above 2,000 feet.

At the present time, the tallest building in the world is the Taipei 101 Tower completed in 2004 in Taipei, Taiwan. The building is 1,667 feet in height and contains 101 stories. Originally scheduled for completion in 2004 is the Shanghai World Financial Center in China, which is planned to have 95+ stories and reach a height of 1,509+ feet. Another interesting high-rise building planned for the future is the Center of India Tower to be built in India with 224 stories and a proposed height of 2,222 feet.

CONSTRUCTION FEATURES

According to almost any of the accepted definitions, high-rise buildings include many older multiple-story buildings having unprotected vertical openings in addition to the newer structures that are reaching 100 stories and more. Although the newer, taller buildings present an excessive number of problems, most have the advantage of fire-resistive construction and many built-in protection aids for both the occupants and the fire department. For example, the high-rise ordinances of some cities require that the building be equipped with the following:

1. A sprinkler system throughout the building.
2. An on-site water supply system with pumps capable of delivering water to the top-most floor through a combination standpipe system. The outlets on the system are 2½ inch. This provides the capability for the fire department to take lines aloft and work directly off the system. The system includes fire department inlets, which make it possible for the fire department to help boost the pressure if necessary.
3. A local fire warning system that is actuated either manually by smoke detectors or by water-flow switches. Some ordinances require that the building be thrown into an alarm mode when the system is actuated. The alarm mode will sound a localized alarm and then shut down the heating, ventilation, and air-conditioning systems on the fire floor and in some cases on adjacent floors. All stairwell doors will be automatically unlocked and the stairwells pressurized. If the smoke detector in the elevator vestibule is actuated, the elevators on the affected bank will be recalled to the ground floor.
4. A one-way communication system that permits fire department officials to communicate with occupants in all or any single public area of the building.
5. A building control station specifically designed for fire operations. It houses a fire alarm enunciator panel, the building communication system controls, an elevator status panel, and the smoke-handling controls.
6. At least two means of egress from each floor other than the elevators.
7. A fire department lockbox that contains keys to operate the automatic elevators and the fire alarm system, and to gain access to floors from the various stairwells.

8. An emergency power system.
9. An emergency helicopter landing pad on the roof.

Because of the type of construction found in the high-rise buildings, with the exception of the Twin Towers collapse on 9/11/2001 in New York, there is no fire on record involving the newer high-rise buildings that resulted in any type of building collapse. Although there is an advantage in firefighting in knowing that the ceiling is not going to fall in or the floor will not collapse, the collapse of the Twin Towers has raised a number of doubts.

It should be noted, however, that the superior qualities of fire-resistant construction found in high-rise buildings also cause a severe problem to firefighters. When an entire floor is involved, the contained heat creates an oven-like atmosphere that could result in a weakening of the building if any structural steel were to become exposed. The heat also subjects firefighters to severe punishment when making an attack.

Two basic types of high-rise buildings have been designed for human occupancy: residential and commercial. Hotels, apartment houses, condominiums, hospitals, and physical care facilities are included in the residential type. The commercial type is occupied by offices and many types of businesses.

Most residential high-rise buildings are characterized by center corridors and a large number of interior compartments; however, some are designed around the center-core concept. These buildings are normally occupied on a 24-hour basis.

Commercial high-rise buildings are characterized by a center-core construction with circuit corridors around the core. Many contain a large number of open spaces.

The **center-core design** concept features office areas or living areas surrounding a "core" containing stairwells, elevators, and utilities. This design presents an unusual problem to firefighters. Firefighters advancing a line onto a fire can suddenly find themselves trapped by the fire behind them that has pushed its way around the core. Protection against such a happening can be provided by advancing a line in the opposite direction of the original. However, this line should not be used to attack the fire, as it would result in the two streams fighting one another. (See Figure 6.46.)

center-core design The center-core design concept features office areas or living areas surrounding a "core" containing stairwells, elevators, and utilities.

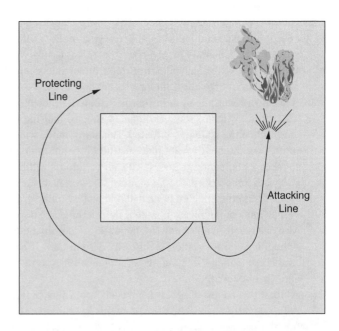

FIGURE 6.46 ◆ Provide a safety line as protection against a wraparound.

The population density is usually much greater in commercial buildings than in residential buildings, with the greatest concentration occurring during business hours. Some buildings contain 5,000 or more people during this period. It takes a considerable period of time to remove 5,000 people from a building via the stairways, even under ideal conditions.

One construction feature that can hamper firefighting operations is the manner in which the ceilings are designed. The ceiling serves as the return air plenum for the heating, ventilation, and air-conditioning systems. It provides a hidden space where the fire can spread rapidly and move undetected. This area should be checked constantly when fighting a fire on any floor. (See Figure 6.47.) A pike pole or similar tool can be used to remove one of the panels. A quick burst of a hose stream will normally do the job if such a tool is not available.

FIREFIGHTING PROBLEMS

Fires in high-rise buildings present some unique problems to the fire service. One of the primary problems is the time required to get to the fire. If the fire is high in the building it may take a considerable time to reach it. This gives the fire additional time to grow in intensity and spread if avenues are available. It also means that planning ahead is a must. An example is the First Interstate Bank fire that occurred on the twenty-first floor in Los Angeles in 1988. Although the fire was well fought, 36 minutes elapsed from the time of receipt of the alarm until water was first applied to the fire. The time frame for getting water on the fire was not considered excessive for a fire in a high-rise building.

Another problem is the number of personnel needed to fight even a moderate fire in a high-rise structure. The minimum number of companies required to set up the basic high-rise incident command structure of fire attack, staging, lobby control, and base is four. Using the Iowa State University method for calculating water flow rates for fighting a fire, a high-rise with a single floor having 10,000 square feet involved with fire would require 1,000 GPM of water. Using hose streams flowing 200 GPM would mean that 45 firefighters would be required to staff and maintain five hose lines. To this must be added numerous companies to support those fighting the fire. A rule of thumb is that three support companies are required for every fire attack company.

Having sufficient personnel to fight the fire is an acute problem. In fact, the total strength of most departments is not sufficient to cope adequately with a high-rise fire.

It takes many times more people to fight a fire high in a building than it would if the fire were on the ground floor of a one-story building. It has been estimated that at least 50 firefighters are required to extinguish a high-rise fire of any size.

Unfortunately, only a few of the larger cities have sufficient personnel and equipment necessary to do an effective job. For example, in 1988 the Los Angeles Fire Department used 64 fire companies and 383 firefighters on the First Interstate Bank fire. Most departments will be hard pressed if a fire of any magnitude is encountered in their city. By necessity, smaller cities having high-rise buildings will have to depend on mutual aid for assistance in coping with these fires. It is extremely important that plans be made and training sessions held as success depends on everyone at the fire being familiar with the tactics involved. Failure to do so will subject firefighters and occupants to an undue risk and most likely result in a predictable disaster.

Extreme heat is an additional factor. The newer high-rise structures are of fire-resistive construction and are sealed buildings designed for interior climate control. This combination will retain tremendous amounts of heat. When taking into consideration the need for forced ventilation, the large open spaces, and the flammable contents, the effective work time for a firefighting team is reduced to a mere 15 minutes.

The rescue problem is considerably different at a high-rise fire. At most fires rescue consists of achieving a complete evacuation of the building. This is almost an impossibility in some high-rise buildings. Some of these buildings house the population of a small city. The time factor, combined with the limited number of people available to perform rescue operations, may not permit complete evacuation. Consequently, other measures may have to be taken.

Water supply can be a problem. The problem is somewhat limited if the building is equipped with its own water supply system and an adequate pumping arrangement. However, many of the taller buildings were constructed prior to the requirement for an adequate pumping system. This means that the responsibility for getting water to upper floors is completely that of the fire department.

Communication is always a problem at fires, but at high-rise fires the problem is compounded. Because of the type of construction it is almost impossible to get messages out of some buildings by the use of portable equipment. Yet it is imperative that the Incident Commander be constantly informed of situations as they develop throughout the building. This means that the building's interior communication system such as stationary telephones, public address systems, and elevator intercoms may have to be used. Cellular phones or regular telephones may have to take priority over radios.

INITIAL RESPONSE RESPONSIBILITIES

The following is offered as a guideline for fighting fires in high-rise buildings. This guideline will undoubtedly have to be modified to fit the personnel and equipment capabilities and individual problems common to any particular city. The guidelines have been developed around the concept that any fire in a high-rise building could develop into a major problem.

The First Alarm Assignment. Many cities do not differentiate between a reported structure fire in an ordinary building and one in a high-rise building. This is a mistake. The initial response to a reported high-rise fire is very critical to the overall success of the operations. The initial response not only should be sufficient to provide for a prompt investigation of the reported fire and the ability to make an initial attack on the fire but also should consider the personnel required for support operations such as base, lobby control, staging, and stairwell support. A suggested minimum assignment for the first alarm is a total of six companies with a combined complement of at least 24 firefighters.

Table 7.2.1 in the nineteenth edition of the *NFPA Fire Protection Handbook* suggests a typical initial attack response capability assuming interior attack and operations command capability for high-hazard occupancies (schools, hospitals, nursing homes, explosive plants, refineries, high-rise buildings, and other high life hazard or large fire potential occupancies) to be:

> At least 4 pumpers, 2 ladder trucks (or combination apparatus with equivalent capabilities), 2 chief officers, and other specialized apparatus as may be needed to cope with the combustible involved; not fewer than 24 fire fighters and 2 chief officers.
>
> Extra staffing of units first due to high-hazard occupancies is advised. One or more safety officers and a rapid intervention team(s) are also necessary.

This dispatch is sufficient to establish the basic incident command structure required for a high-rise fire. Without the basic incident command structure being in place, firefighter accountability is jeopardized and the establishing of the incident command structure from the second alarm is difficult.

Duties of the First-in Officer. The first officer to arrive at the scene will initially be the Incident Commander. The officer's first duty is to size up the situation from outside the building and give a report to the dispatch office and to incoming companies. The report should include information on the size of the building, the type of occupancy, the conditions observed on arrival, any safety concerns, and the action the company will take. If any smoke or fire is showing from the outside of the building, this officer immediately should request additional resources to be dispatched. As a minimum, the request should include four additional companies.

After making the initial report to the dispatch office, the officer should lead his or her company into the building to locate and attack the fire. This company will be known as the **high-rise fire attack team.** It is essential that the fire attack team have a minimum of four members in order to comply with the two-in/two-out concept. The primary responsibility of the attack team is to locate and identify the emergency and determine its scope.

The company officer should determine that the crew has secured the proper equipment prior to entering the building. The ascent to the fire floor may be extensive and fatiguing; therefore, it is important that the crew is not overloaded. The tools that should be taken are SCBA, high-rise hose packs with nozzles, forcible entry tools, portable lights, radios, and, if available, an infrared-imaging camera.

Entrance to the building should be made through the lobby. Once inside the lobby, the officer should obtain information from the security personnel or building management as to the location and nature of the fire. The individual who knows the most about the building is the building engineer. This makes locating him or her an initial concern. The officer should determine whether the building is equipped with an enunciator panel or a control room and obtain the phone number in the lobby. This number should be written rather than trying to commit it to memory. It may be very useful in the event other methods of communication fail.

The control room is normally located off the lobby area and contains a fire enunciator panel, HVAC controls, and the building's communication systems. If an

high-rise fire attack team
The first company to arrive at a high-rise fire that has the primary responsibility to locate and identify the emergency and determine its scope.

enunciator A piece of equipment that indicates on what floor or floors a problem exists and whether it was tripped by a manual pull station, a smoke detector, a heat detector, or a water flow.

enunciator panel is available in the control room, it should be checked. The **enunciator** should indicate on what floor or floors the problem exists and whether the enunciator was tripped by a manual pull station, a smoke detector, a heat detector, or a water flow. Tripping of more than one device usually indicates that a problem exists. Elevators that have returned to the ground level because of a fire alarm activation are also a clue that a problem exists. Once information has been gathered as to the fire floor, the officer should bring dispatch and incoming companies up to date on the conditions and prepare to advance to the fire floor.

If there is a fire department lockbox in the lobby, the officer should take out one set of keys and a copy of the building inventory sheet, if available. He or she should leave the remaining contents for the lobby control officer. The officer and crew are then ready to determine the best and safest means of getting to the fire. The elevators should not be used unless it has been factually established that the entire elevator shaft is not threatened by fire.

Finding the stairways is not always an easy task. The entrance to some stairways blends in with the walls. This effectively camouflages the stairway entrance. The doors to the stairways normally can be distinguished from other types of doors by their double-door width and key cylinder lock.

It should be remembered that all stairways do not serve every floor and extend through the roof. For example, scissor-type stairways may serve only every other floor. Some stairways extend only partway up the building, making it necessary to exit and find another stairway to continue the ascent.

Some stairways are equipped with a numbering system inside the stairway that indicates which floors will be available and if the stairway extends to the roof. This information is invaluable when determining which stairway to use.

The company is ready to start climbing once the determination has been made as to the best stairway to use. Before starting the climb, the company officer should relay the stairway identification to incoming companies. If it is impossible to contact incoming units via the hand radio, the officer should send the apparatus operator outside and have him or her make the contact.

If possible, the company officer should keep incoming units informed of what is found during his or her ascent. It is good practice to make an appraisal of conditions every four or five floors by opening the stairwell doors and checking inside. The ascent should be paced so that company members will not be completely exhausted on reaching the fire floor. If so, they will be in no condition to attack the fire.

high-rise staging area The collection point for the equipment and personnel that will be used on the fire.

The second floor below the reported fire floor should be inspected to determine whether it will be suitable for the **high-rise staging area.** The inspection should also include obtaining the knowledge of the entire floor layout. This will give the officer valuable information regarding the floor layout on the fire floor. The company officer should relay the survey results to dispatch or to the Incident Commander in the event a change of command has been made. A change of command should have taken place by this time.

On reaching the fire floor, the company officer should determine what is burning and the potential for vertical and horizontal extension. The officer should assess whether any occupants are endangered and the best route for resources to move from the staging area to the fire floor. This information should be relayed to the Incident Commander. The relayed information should include the floor number, the type of occupancy used on the floor, what percent of the floor area is involved with fire, any rescue problems, what action his or her crew is taking, and the request for additional crews if needed and how they will be used. An example of the message might be:

> Incident Commander from Engine 7. We have arrived on the 14th floor. This floor is occupied by offices and is approximately 60 percent involved with fire. There are no apparent rescue problems. Engine 7 has attached to the standpipe and is attacking the fire. Send me two more engine companies and a truck company. One of the engine companies will assist Engine 7 on this floor and the other two companies should be sent to the 15th floor to stop the extension of the fire and provide any needed rescue assistance.

The high-rise fire attack team may start setting up and connecting hose lines to attack the fire. While this is being accomplished, the officer should proceed to the floor above to see whether the fire has extended and then return to the team before it leaves the stairwell. The two-in/two-out system should be established. The officer and one member should remain in the stairwell and feed the line and relay radio messages to incoming companies.

The company officer then assumes the duty as the division commander. A division in a high-rise is usually one floor and identified by the floor level. As an example, the division commander on the 14th floor will be identified as Division 14.

On assuming these duties, the division commander will attack the fire. If the fire cannot be contained with the available resources, the company should make every effort to protect the vertical openings and contain the fire until help arrives.

Elevators are the most effective method for delivering personnel and equipment to the fire. However, as previously stated, the elevators should not be used until they have been determined to be safe. Because of this, it is important for the fire attack team to determine whether the elevator vestibules on the fire floor have been impinged with fire, thus making the elevators inoperable. The information gathered should be relayed to incoming companies.

Duties of the Second-in Company Officer. On arrival, the second-in company officer will enter the lobby and assume the duties of the **lobby control officer.** This officer is responsible for controlling all vertical access routes (including the elevators), controlling the HVAC, and coordinating the movement of personnel and equipment between the base and the staging area. He or she should use the building engineer, if one is available, for assistance in controlling the HVAC.

Records should be maintained adequately in regards to all personnel entering or exiting the building. The lobby control officer should designate which stairways will be used for particular purposes and post personnel at each location to control entry and to direct occupants exiting the building. He or she should also call all elevators to the lobby by the use of the emergency service control. The elevators should be secured there until it is determined that they are safe to use.

When the elevators have been determined to be safe for use, the lobby control officer should designate which ones are to be used by firefighting personnel. A firefighter should be assigned to operate each of the elevators so designated. Any elevators not designated for fire department use should be placed out of service.

lobby control officer At a high-rise fire, this officer is responsible for controlling all vertical access routes (including the elevators), controlling the HVAC, and coordinating the movement of personnel and equipment between the base and the staging area.

Duties of Other First Alarm Officers

Staging. One of the other first alarm companies should be used to set up the staging area. After the fire attack team identifies the floor for staging, the third arriving company usually sets it up. The **staging area** is the assembly point where a reserve of personnel and

staging area At a high-rise fire, the staging area is the collection point for the equipment and personnel that will be used on the fire. It normally is located two floors below the fire floor.

equipment is awaiting assignment within the building. The minimum amount of reserve personnel and equipment to be kept at the staging area will be established by the Incident Commander. It is the responsibility of the staging officer to request additional resources whenever those in the staging area fall below the minimum.

The staging area normally is established two floors below the fire floor. There is an advantage for locating it at this point. It is close enough to the fire that personnel can get to the fire in a timely manner and still have energy to mount an aggressive attack. Water and salvage operations may affect the personnel and equipment if it is too close to the fire. The third floor below the fire generally will serve as an alternative if the second floor is not suitable because of a large collection of machinery, stock, or other material.

The medical unit should establish a medical treatment station in staging to provide medical treatment and rehab care for incident personnel. The staging area is under the control of a staging area manager, who reports to the Incident Command or to the operations chief, if this position is established.

Resources are dispatched from staging at the direction of the Incident Commander or operations chief. Additional resources are requested by the staging area manager through the Incident Commander or operations chief to the logistics chief any time reserves fall below the specified level.

Rapid Intervention Team. A rapid intervention team (RIT or RIC) should be available in the staging area. Having a RIT (or RIC) in staging will ensure that a team is ready and available to make an immediate and rapid search and evacuation of any firefighter injured or lost. The RIT (or RIC) in staging should consist of sufficient numbers to be able quickly to staff and support two entry teams. The RIT (or RIC) is led by the rescue group supervisor.

The RIT (or RIC) should be stationed near the stairwell used for ascent to the fire floor. In some cases it may be necessary to create additional RICs under a single rescue group supervisor and place them at different locations to ensure rapid deployment.

Each RIT (or RIC) should be equipped with a 45- or 60-minute SCBA, infrared-imaging cameras, ropes to mark the routes, and an additional SCBA for downed firefighter use. The RIT (or RIC) needs to have sufficient radios to monitor all necessary frequencies including the tactical channel, the emergency channel, and the rescue channel.

If a RIT (or RIC) is deployed, the rescue group supervisor must obtain information continually as to the company's or members' operating area. This will assist in narrowing the search area and increase the chances for success. In addition, the rescue group supervisor must request additional companies to create relief teams, backup teams, and additional rescue teams.

Base. The base should also be established by one of the first alarm companies. Normally the fourth arriving company officer assumes this position. The officer assigned to the base is known as the base officer.

base The base is the reporting point for incoming companies and serves as the collection point for personnel and equipment pending transfer to the staging area.

The **base** is the reporting point for incoming companies and serves as the collection point for personnel and equipment pending transfer to the staging area. It should be located a minimum of 200 feet from the fire building and be in an area sufficiently large enough to handle all the apparatus. Two hundred feet is generally a sufficient distance to avoid the congestion around the building and is a safety factor against falling glass.

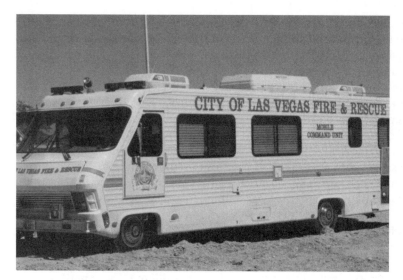

FIGURE 6.48 ◆ Some departments have mobile command posts that can be moved quickly to an emergency. The command post is equipped with all required radio frequencies and adequate workspace. *(Courtesy of Las Vegas Fire & Rescue PIO)*

Once established, the IC should announce the location of the base and instruct all incoming company commanders to report to the base officer. The base officer will give them instructions as to where and how to park their apparatus. Apparatus normally is parked diagonally so it can be moved independently. Company officers should keep their personnel together at their apparatus while awaiting assignment. A good rule of thumb is to maintain two companies in base for every one in staging. Additional resources are ordered through the IC.

Command Post.   The first arriving chief officer will assume the duties of Incident Commander. This chief officer should establish the location of the command post at least 200 feet from the fire building. (See Figure 6.48.) The location of the command post should be relayed to the dispatch center. The report might go something like this:

> Dispatch from Battalion 5. Battalion 5 is assuming command of the fire at 3rd and Adams. The command post will be located in the parking lot between 3rd and 4th on Adams.

The basis for controlling a large fire will have been met with the establishment of the command post, the fire attack team, the lobby control, the staging area, and the base. Any expansion or augmentation of this organization will be made by the Incident Commander after receiving information from the fire attack commander that a need exists.

The first arriving chief officer will remain as the IC until relieved by a higher-ranking chief officer. When relieved of the command, this chief officer should assume the duties of planning chief and remain at the command post to assist the IC. There is a good chance that the first arriving chief officer is the one most familiar with the building and the strategy and tactics initiated. Therefore, sending this officer into the building would be a mistake.

The second arriving chief officer should be sent directly to the fire floor and assume command of that division. The third arriving chief officer should be assigned to logistics. The next arriving chief officers should be assigned to divisions or functional groups as needed.

Logistics.   The success of a high-rise fire depends on the ability to move personnel and equipment to the fire in a timely manner. The officer responsible for accomplishing this is known as the logistics chief. Logistics is not only the most important but also the most difficult position in a working high-rise fire.

Logistics is responsible for lobby control, stairway support, communications, getting air cylinders and hose to the fire floors, traffic control, scene security, and related duties such as keeping fire protection systems working. In addition, logistics is responsible for firefighter comfort, including food and water, and any firefighter medical needs.

In addition to the medical unit established in staging, another unit may need to be established in base, in lobby, or in both. It may also be necessary to establish a firefighter rehab area in each location to ensure that firefighters remain hydrated.

The replacement of air cylinders places an immediate burden on logistics. Companies moving from base to lobby to staging should take as many cylinders as possible with them. It has been found that after the first wave of companies arrive on the fire floors, there are a sufficient amounts of hose, fittings, and forcible entry tools available for their use. However, air cylinders are in short supply.

The use of 45- or 60-minute air bottles should be considered. These bottles allow for a longer concentrated attack or search. However, the physical toll and fatigue of the firefighters must also be taken into consideration. Using higher-capacity bottles permits firefighters to work longer and depletes firefighters' strength and endurance sooner. That is one reason why rehab and medical screening need to be available in staging and all officers must monitor their members closely.

ACCESS OPERATIONS

Gaining entrance into a high-rise building is generally not a problem. The building is open during business hours and security personnel will normally meet the first arriving unit during nonbusiness hours. However, there are times when the building is locked and no one is on the premises to meet incoming units. In these instance, forcible entry will probably have to be made not only into the building but also into each of the different occupancies within the building because they will most likely be individually locked.

Before forcing entry, make sure that a fire department exterior lockbox is not available. If one is available, it is usually in a location that requires a ladder to reach it.

If forcible entry is necessary, it should be done in a manner that will result in the least amount of damage. At times the task can be accomplished by the use of two pry tools working together to force open entry doors. It may also be possible to cut the bolt between the doors or between the door and the floor with the use of the rotary saw.

If it is not possible to force a door, it may be necessary to break glass to gain entrance. The entrance doors are normally made of extremely thick glass, which makes them difficult to break. In addition, they are very expensive to replace. An attempt to break the glass in these doors should be made only as a last resort. A better choice is the glass panels adjacent to the entry doors. These panels are generally made of thinner glass and are easier to break.

Consideration for gaining entrance should not be limited to the front entrance. At times, the building has a basement with a stairway leading from the outside into this area. Gaining entrance at this point is normally quicker and easier than through the front; however, the initial action within the building commences in the lobby. This makes it better if entry can be made directly into the lobby area.

On some high-rise buildings it may be possible to raise a ladder and gain entrance through an upper level. This is not possible, however, in many of the newer buildings because the windows are not operable.

If fire is showing on arrival of the first unit, priority should be given to gaining entrance as quickly as possible.

RESCUE OPERATIONS

Rescue operations in high-rise fires are considerably different than in other types of fires. For example, a fire in a dwelling generally places all occupants in immediate danger. This is particularly true when the fire occurs at night. However, this is generally not true at high-rise fires. Only a few, if any, of the occupants are in immediate danger when the department arrives.

Occupants of high-rise buildings can be placed in danger in the event of a fire by three primary methods. Those in the vicinity can be directly exposed to the **flame.** This normally occurs in residential buildings in which the fire is confined to a single unit. Others in adjacent units may be exposed if the fire has a pathway for extension.

flame   The visible and colorful portion of a fire.

The most likely and by far the greatest threat to occupants from the fire is exposure to **smoke** or other products of combustion. Smoke may travel up elevator shafts, stairwells, and other unprotected vertical openings. It may also extend through air handling systems in older buildings that lack automatic shutdown devices. The newer buildings are designed with such devices as compartmentalization, pressured stairwells, and pressured elevator vestibules to restrict smoke travel, but these measures are not always successful.

smoke   One of the primary products of combustion.

The third threat is that of panic. Panic can occur when occupants know or suspect that a fire exists. The fire does not have to be in the immediate vicinity to cause irrational behavior.

As with fires in most other buildings, the best means of saving lives and facilitating rescue operations may be an aggressive attack on the fire. This, together with proper ventilation methods, may provide the highest degree of safety to occupants. However, it does not eliminate the need to conduct an adequate search if the fire is of any size whatsoever.

Search Procedures.   People can be found in high-rise buildings at all times of the night and day. It is important during fire operations that all occupants be accounted for. A complete search of a high-rise building is a long and tedious task and normally does not have to be made except when the fire is extensive. It is time-consuming and requires a considerable number of personnel to do the job effectively. However, a limited search generally is required if the fire is of any size whatsoever. Search procedures should be organized, directed, and coordinated by a responsible officer.

The search should be organized on a priority basis. First priority should be given to the immediate area on the fire floor. Second priority should be given to the floor above the fire. The search should then continue, floor by floor, both above and below the fire until a complete search of the area deemed necessary has been made.

Search efforts should be conducted through the use of search teams, each headed by an officer. The use of companies as search teams is ideal if a sufficient number of companies is available. Each search team should be provided with keys to gain access from the stairwell to individual floors and also room keys if such are available.

FIGURE 6.49 ◆ Avoid duplication of effort.

It is imperative that a strict accounting and documentation system be developed to ensure that every room on every floor is searched and that search efforts are not duplicated. (See Figure 6.49.) The method used by one of the larger departments is worthy of consideration. As the search team enters the floor from the door leading from the stairway, a large diagonal line is made with a piece of chalk on the door. The search team's identification number is placed below the diagonal line. A similar diagonal mark without the identification number is placed on the door leading to an office, dwelling unit, and so on. When a complete search has been made of the unit, the searching members close the door and place another diagonal mark on the door. This makes the original diagonal mark into an X. On completing the search of all rooms on the floor, the search team makes the X on the stairwell door and moves on to the next floor that needs searching.

Relocation of Occupants. Total evacuation of a high-rise structure during firefighting operations may be neither practical nor feasible. The primary effort should be devoted to moving those people who are threatened by the fire to a place of safety. This normally requires the movement of those on the fire floor and two floors above and below the fire floor. Occupants of other than these five floors should be instructed to remain where they are, unless of course there is a chance that they may be placed in jeopardy. Those on the fire floor and the two floors below the fire floor normally should be moved to the floor below the staging area. Those on the two floors above the fire floor should normally be moved to the third floor above the fire floor. In some cases, however, it may be necessary to move all those above the fire to the roof.

However, do not forget the lessons learned from the World Trade Center incident. If a severe fire is in progress in a high-rise building, every effort should be made to get as many people out of the building as possible. It was formerly thought by many firefighters that the chances of a high-rise building's total collapse were virtually nonexistent. The results of the World Trade Center incident have provided some serious reservations about keeping high-rise occupants inside the building while a severe fire is in progress.

Providing information and instructions to those in public areas within the building is simplified if the building is equipped with a well-designed communication system. These systems allow a fire department member to talk to occupants on individual floors or given areas of the building as the situation deems necessary. The system normally is controlled from the fire control room. The individual chosen to send the message should be one who is capable of projecting confidence. In cities where a number of the inhabitants speak a language other than English, it is wise to repeat the instructions in the appropriate language.

FIGURE 6.50 ◆ Tape similar to this can be used to set up a corridor to control evacuees from the building.
(Courtesy of Las Vegas Fire & Rescue PIO)

Controlling Evacuees. Regardless of the department's desire to keep occupants where they are, some will leave. There is no harm to this as long as they do not hamper firefighting operations. It is best to designate which stairway should be used by occupants for evacuation and then maintain control of this pathway. The important thing is to keep the traffic moving and not to let bottlenecks form.

Once occupants have left the building, they will have a tendency to stop and talk. This not only puts them in jeopardy from the possibility of falling glass but also can create a logjam that could extend back into the building and slow down the orderly evacuation of other occupants. To prevent this from happening, a corridor should be established that will lead occupants to an assembly area away from the building. It is a good idea to use fire line tape or rope to designate the corridor. Police officers and/or security personnel can be used to keep the evacuees moving. (See Figure 6.50.)

All evacuees should be logged in once they reach the assembly area. The information should include where they work or live within the building. If the individuals have no medical or other problem relating to the emergency, they should be free to leave the assembly area. It is important to instruct them on leaving not to return to the building. Whether they leave the scene is between them and their individual employers.

FIREFIGHTING TACTICS

The basic principle of fighting a fire in a high-rise building is no different from that of fighting a fire in any other type of building. The time-honored sequence of locating, confining, and extinguishing prevails. However, the big difference is time. It is not unusual for a department to take 15 to 20 minutes to reach the fire. This time factor needs to be taken into consideration when the first arriving officer requests additional companies and equipment.

If ascent has been made by way of the stairwell, the basic operations will begin when the reported fire floor is reached. As when entering any other potential fire area, the door should be felt and a check made for smoke. The door can be opened if indications are that it is safe to do so. If no fire can be found, the floor both above and below should be checked because a malfunction may have caused the alarm system to indicate the wrong floor.

If fire is suspected on the reported fire floor, the door should not be opened until a charged line is in place. The line may be connected to the outlet on the fire floor or the floor above or below the fire floor depending on the circumstances encountered. A 2½-inch line should be used as it is always easier to reduce the line than it is to increase it. The line should be equipped with a shutoff arrangement that permits the removing of the tip if it becomes necessary to extend the line.

It is good practice before opening the door to advance the hose line to ensure that the stairwell is not being used by occupants of the building for egress. The door cannot be closed once it is opened and the line is taken inside. This means that smoke and other products of combustion can escape from the fire area and quickly make the stairwell unsuitable for occupants. If occupants are using the stairwell, it may be necessary to delay entry until all evacuees have passed the fire floor.

If the door is hot to the hand, it may indicate that a potential backdraft condition exists inside. This condition could be caused by the shutting down of the heating and ventilation system, which might result in an oxygen deficiency in the fire area. In case of a hot door, the door should be opened with caution. If air rushes in when the door is opened, the door should be closed immediately and entry not be attempted until the area is ventilated.

Venting the fire floor of a high-rise building is not easy. Glass windows in the building may be tempered, heat strengthened, or laminated with a plastic coating or wire. Breaking these windows presents a challenge for firefighters. Some building codes mandate that a percentage of the windows at measurable distances be openable or tempered to break easily. These windows are usually marked in one corner.

One method is to break the windows with a ball and chain or with the use of a heavy object such as an axe attached to a drop line. The windows can be broken from two floors above the fire floor. The Incident Commander should be notified before any glass is broken so that the area below can be cleared. Breaking glass is a hazardous situation at a high-rise fire. Falling glass can be projected away from the fire building as much as 200 feet. At least this great a distance should be cleared of personnel and equipment prior to breaking the glass.

It is important to take into consideration the direction and speed of the wind prior to breaking any windows. Winds at higher levels can be considerably greater than they are at ground level. The wind pushing in on the firefighters on the fire floor can adversely affect their ability to advance hose lines. However, it may be necessary to break or remove glass in order to provide a place from which to work if the building is equipped with inoperable windows. Charged lines should be in position on the floor above the fire floor and the floor from which the windows will be broken prior to breaking any glass. The objective of these lines is to prevent the fire from leaping out the broken windows on the fire floor and extending into upper floors. The fire attack team should be ready to move in with charged lines once the windows are broken.

Breaking glass on an upper floor is a risky procedure. It is possible that the vacuum or winds could take a firefighter out a window. It is good practice to secure the firefighter who is breaking the glass with a safety rope prior to the glass being broken.

The potential for a flashover is always present in a high-rise fire. One of the contributors to this potential is the extensive use of plastics and other synthetics in the furnishings and decorations. Firefighters entering the fire area should be alert for this possibility constantly. Some of the indicators of a possible flashover are a freely burning fire and an accumulation of extensive heat and/or smoke at ceiling level. The smoke may sometimes remain at the ceiling level or it may bank down. The company officer should consider that the potential is extreme if firefighters are driven to the floor by extensive heat from above. It is acceptable at this point to project a stray stream across the ceiling to reduce the built-up heat.

Elevators are the quickest method of moving personnel and equipment to upper floors if it has been determined that they are safe to use. Normally their use is restricted to later arriving companies as the decision for their use is based on information relayed to the Incident Commander from the initial attack team. If an elevator is used to take an attack team aloft, the team should unload two floors below the fire floor and proceed to the fire floor by way of the stairwell as a precautionary measure against the fire meeting them when the elevator doors open. If any elevator terminates at a level at least five floors below the fire floor, it can generally be assumed that this elevator is safe for firefighters to use.

MOVEMENT OF PERSONNEL AND EQUIPMENT

Movement of personnel and equipment to the fire is accomplished through a relay operation. The relay points are the base, the lobby, and the staging area.

The Base.   The base is the collection point for incoming companies and equipment held in reserve for use in the staging area. Equipment or personnel requested from the staging officer is sent from the base to the lobby control officer.

Stairwell Support Group.   The most effective method to transport members and equipment to staging from the lobby is through the use of elevators. An important strategy is to ensure that the elevator vestibules are free of fire. Until the vestibules are cleared and elevators are usable, a system to ferry equipment may have to be to be established using the stairways. The stairwell support group is used for this purpose or for the task of moving equipment from the roof to the staging area when equipment is delivered to the roof by a helicopter.

Lobby control will deliver the equipment to the designated stairwell entrance at ground level where it will be received by the stairwell support group. This group consists of personnel stationed on every other floor from the ground floor to the staging area, with an officer supervising every four or five people. One member picks up the equipment at the ground floor entrance to the stairwell and carries it to the third floor landing. He or she then returns to the ground level to pick up another load. Another member picks up the equipment at the third floor and carries it to the fifth. This process continues until the equipment reaches the staging area hallway. At this point it is picked up by a member from the staging area and carried to the proper storage area.

Stairwell support members shall have all their personal protective equipment and SCBA with them. However, this equipment may be removed while equipment is being moved in order to reduce fatigue. Supplemental lighting may be required in the stairwell for the safety of personnel.

The Staging Area.   The staging area is the collection point for the equipment and personnel that will be used on the fire. It normally is located two floors below the fire floor. A general principle exists that no one is to move from the lobby control to the staging area empty-handed. A determination as to the minimum amount of equipment and the minimum number of personnel to be held in reserve is made early during the emergency. One procedure used for personnel is that there will be two fire attack units in the staging area for every fire attack unit presently fighting the fire.

FIGURE 6.51 ◆ Firefighters taking equipment to the staging area at the First Interstate Bank high-rise fire. *(Courtesy of Rick McClure, LAFD)*

One of the units should be equipped and ready to replace the unit fighting the fire. The second unit is one that was just relieved and is replenishing bottles and then taking a rest. (See Figure 6.51.)

A medical unit and a responder rehab unit also should be assigned to the staging area. Although they may not be needed, it is best that they be available. Another medical unit set up to provide first aid should be established in the lobby.

Normally the equipment supply list will include air bottles, hose, fittings, breathing apparatus, forcible entry tools, salvage equipment, pike poles, ladders, and an oxygen unit. The supply is replenished as it is dispatched to the fire. Empty bottles that have been used on the fire are also stored in the staging area.

The used equipment is separated from that which is available for dispatch. If the staging area is large, one end can be used for used equipment and the other end for fresh equipment. If the area is relative small, the two groups should be separated so that they can be readily identified. Requests to replenish supplies or personnel are made to the base.

RAPID INTERVENTION TEAMS

The rapid intervention teams inside the building report directly to the operations officer. One team should be available at the base of each stairway on the floor below the fire. While on this floor, each team should make a survey of the floor plan as it will most likely be the same as that on the fire floor. The teams should monitor radio messages constantly so as to keep aware of what is occurring. If a team is sent to the fire floor to assist firefighters, another team should be dispatched to replace the team that left.

MAINTAINING RECORDS

It is important that records be kept at each stage of the relay. The base officer should log in all companies and equipment that report to the base. All entries should include time. He or she also will log in all requests for equipment or personnel received from the staging officer and log out what was sent to staging.

The same logging in and out procedure continues with the lobby officer.

The staging officer logs all requests made for equipment and personnel. In addition, he or she logs in all equipment or personnel received from the lobby control officer and all sent to or received from the attack forces.

AIR SUPPORT

Those departments having helicopters normally include this equipment in the plans for fighting fires in high-rise buildings. Helicopters can be used effectively for reconnaissance, delivering equipment, and rescue. (See Figure 6.52.) At times the use of helicopters is the only effective means of rescuing people who have been trapped above the fire floor. If endangered, these people generally will move to the roof. From here they can be rescued by helicopters, providing the heat and smoke do not make flight operations impractical.

Sometimes it is easier to move equipment from the roof to the staging area than it is to move it from the ground floor to the staging area. This is particularly true if the fire is high in the building and the building is equipped with smoke towers that permit a downward movement of supplies. In these instances it may be advisable to move a stairway support group to the roof to facilitate the transportation of equipment.

Helicopter operations are placed under the direction of an air operations director who reports to the operations commander. The helicopter pad should be located at least half a mile from the incident site. Operations at this location will minimize the confusion caused by the noise and rotor downwash from the aircraft.

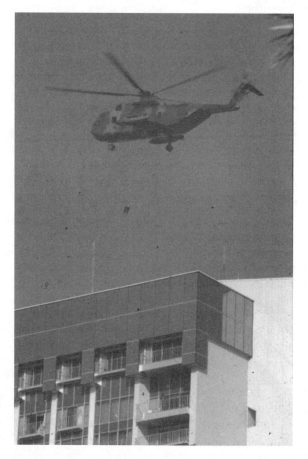

FIGURE 6.52 ◆ Helicopters can be extremely useful on high-rise fires.
(Courtesy of Clark County, Nevada, Fire Department)

WATER SUPPLY

Modern high-rise structures have fire pumps rated to pump to the highest floors. These systems use domestic water and water tanks for supply. The systems are designed to provide adequate water for the sprinkler system and, through the use of pressure reducing valves (PRVs), provide effective fire streams. However, these pumps may be ineffective or malfunctioning for a multitude of reasons. To protect against this possibility, it is important that supply lines be laid into the exterior outlets of the building's system with the apparatus operator prepared to pump if it becomes necessary.

The lines attached to the pumper should be loaded and the apparatus operator should set the pump pressure below the building system pressure and monitor the layout for water flow. The IC must be notified and the pump pressure increased to the building system pressure if the apparatus operator notes water flow from the apparatus. It may be necessary for the fire department to establish an auxiliary water supply if the amount of water used on the fire is excessive. The stairwell support should be used to accomplish this task. It should be noted that this is a timely and labor-intensive task.

When establishing an auxiliary water supply, base will supply water to the entrance of the stairwell. Stairwell support will lay a sufficient amount of hose from ground level either to staging or, if necessary, to the fire floor. The hose should be moved to the side of the stairwell to remove tripping obstacles. It is worth noting that it is much easier to lay the hose down from staging or from the fire floor than working up.

ADDITIONAL CONSIDERATION FACTORS

The Building Engineer.   No one knows more about the building than the building engineer. This makes this individual an ideal person to contact early in the emergency. The building engineer is usually available to first arriving officers during business hours. Although there may be a tendency for the fire attack team to take the engineer with them to the incident location, he or she will be utilized better by remaining available to lobby control or the IC. The engineer's experience with the HVAC system and other building systems may prove valuable in combating the emergency. If the fire occurs during nonbusiness hours, the engineer should be contacted to report to the command post to provide assistance.

Radio Communications.   The capability of radio communications in a high-rise structure may be problematic. The large amount of metal and concrete in the building's skeleton may interfere and block transmissions. If radios are used in repeat modes, they may have to be switched to direct mode. In addition, other means of communication, such as cellular phones, and hard-wired or built-in emergency systems, should be established. However, first arriving companies must monitor the radio communications and make adjustments if the system fails.

Elevators.   All elevators must be checked for occupants early in the emergency. In some situations, the elevators may not return to ground level. The elevator group may have to locate the elevators to ascertain their status. If the elevators are determined to be safe, they are available for transporting members and equipment to staging.

Ventilation.   The migration of smoke and toxic gases throughout the structure may present a greater hazard to occupants and firefighters than the actual fire. Smoke normally moves from lower floors to upper floors by way of vertical shafts, poke through construction, and heating and air-conditioning shafts. Additionally, smoke may "stratify" on one or more floors. Therefore, a ventilation plan must be included into the strategic thinking. Consideration should be given to building construction, HVAC control, and exterior weather conditions. All these factors will affect vertical and horizontal ventilation efforts.

Salvage.   When a significant amount of water is used for fire extinguishment, it can be expected that a large amount of the water discharged by fire streams will run off and potentially create water damage to contents unaffected by the fire. Salvage operations should be an integral part of the fire suppression operations.

MULTIPLE-ALARM FIRES

Adequate establishment of the basic command system by the first alarm companies provides the pathway for smoothly moving into the organization required for coping with a major fire.

A second alarm assignment is normally requested if the first arriving officer observes any smoke coming from the building when the company first arrives on the scene. Any request for personnel and equipment beyond the second alarm will be initiated by the Incident Commander based on information received from the first arriving officer when the officer has reached the reported fire floor and determined the extent and intensity of the fire.

The standard operating procedures (SOPs) of a department should include the method to be used for the buildup. A suggestion for a smooth procedure follows:

+ The Incident Commander is responsible for all operations at the incident. The safety officer, the liaison officer, the information commander, and a rapid intervention team leader all answer directly to the Incident Commander.
+ The operations chief is responsible for all firefighting operations. The staffing officer, the air operations director (if air support is available), and all division commanders are under this individual's command.
+ The logistics chief is responsible for logistic support. The lobby control officer, the base commander, the stairwell support unit, and communications are under this individual's command.
+ The planning chief is responsible for the documentation of the incident and keeping the Incident Commander apprised of the status of the various phases of the operation. This individual gathers and analyzes information and assists the Incident Commander in planning strategy for the overall operation of the incident.

Such movement should be accomplished on a priority basis. An example for the utilization of the chief officers responding to a multiple-alarm assignment follows:

The first arriving chief officer assumes the duties of the Incident Commander.
The second arriving chief officer is assigned the duties of operations chief.
The first arriving higher-ranking chief officer relieves the Incident Commander (first arriving chief officer). The relieved chief officer is assigned the duties of planning chief.

Any additional command level personnel will be assigned at the discretion of the Incident Commander.

Strategic Goals.   The immediate goals when the first unit arrives at the fire are:

+ Locate the fire.
+ Establish a fire attack team.
+ Establish a command post.
+ Establish a base station.

   ◆ Establish a staging area.
   ◆ Establish and maintaining lobby control.

If it is determined that a fire is in progress, the following additional goals are established:

   ◆ Request additional companies and equipment that may be needed.
   ◆ Set up a full-blown Incident Command System.
   ◆ Set up rapid intervention teams.
   ◆ Set up first aid stations.
   ◆ Move personnel and equipment to the staging area.
   ◆ Stop the extension of the fire.
   ◆ Extinguish the fire.
   ◆ Provide for ventilation as needed.
   ◆ Provide for salvage operations.
   ◆ Commence rescue operations.
   ◆ Relocate occupants to a place of safety.
   ◆ Provide for the control of evacuees.
   ◆ Set up a record-keeping system.

DEMOB/REENTRY

After the fire has been extinguished and the building searched, the final phase can begin. The final phase includes the determination of the cause of the fire, an overhaul plan, a demobilization plan, and a reentry/housing plan.

Regardless of the number of floors above and below the fire that the Incident Commander deemed necessary to be searched during the progress of the fire, a complete search of the building after the fire is an important operation that should not be neglected. The following article that appeared in the *Joplin Globe,* Joplin, Missouri, on October 19, 2003, illustrates the value of the search. It is recommended that fire officials review the final report on this fire when evaluating their SOPs on high-rise firefighting procedures.

On Scene Scenario

Downtown Chicago Fire Kills 6

CHICAGO—Eight people remained hospitalized early Saturday, some in serious or critical condition, after a fire at a downtown office tower that left six dead.

Trapped government workers frantically dialed 911 Friday as they tried to make their way through smoke-filled staircases and hallways, officials said. Hours later, 13 were found unconscious in the stairwell and on the 22nd floor, six of them dead.

The bodies weren't discovered until after the fire was brought under control Friday evening and firefighters started searching the 35-story Cook County administration building floor by floor, authorities said.

Any high-rise fire will create numerous hardships to the community. These hardships include lost jobs, business interruptions, lost revenues, and displaced persons. An effective plan should be developed to deal with and minimize these issues.

The primary responsibility for determining the cause of the fire rests with the IC. Due to the size and significance of most high-rise fires, it is wise to bring in technical experts. These may include local arson investigators or members from the Bureau of

Alcohol, Tobacco, Firearms and Explosives. The IC should ensure preservation of the scene until the necessary investigations have been made.

The development of a demobilization plan should be started early during the emergency. The IC should consult with Planning and Operations to consider the remaining needs of the incident and the needs of the community to determine which units should be released. Consideration should be given to overhaul, reentry of occupants into the building, and security needs. The companies can be released in controlled stages.

Criteria should be developed for allowing occupants to reenter the building after the fire situation is completely controlled. These criteria may restrict reentry to escorted inhabitants to obtain essentials and to entry to the fire floor.

INCIDENT COMMAND SYSTEM CONSIDERATIONS

Of all the different types of building fires, a high-rise fire is probably the only one that will ever come close to setting up a complete Incident Commander System. With the exception of the very large cities, cities in general do not have the personnel nor are they equipped to fully staff a high-rise fire incident. Of course, it goes hand-in-hand that only the very large cities have buildings of sufficient height that will demand an extensive use of personnel and equipment. Figure 6.53 is a sample of an ICS organization chart for a high-rise fire.

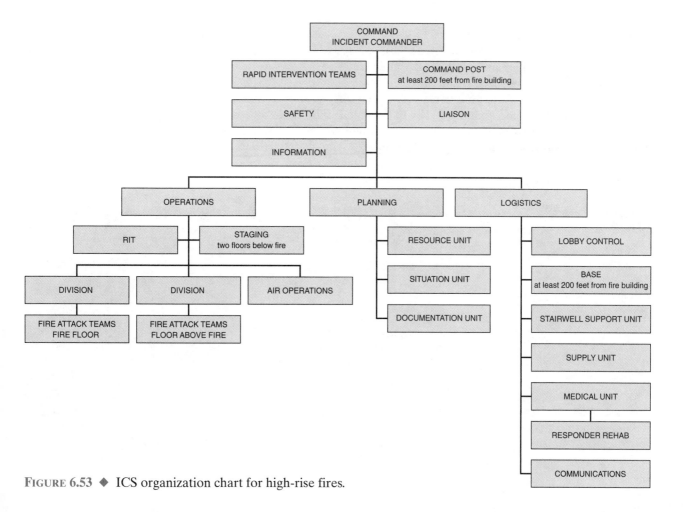

FIGURE 6.53 ◆ ICS organization chart for high-rise fires.

FIGURE 6.54 ◆ One of the many stairways where smoke penetrated. *(Courtesy of Clark County, Nevada, Fire Department)*

A little after 7:00 A.M. on November 21, 1980, a fire broke out in the Deli, a restaurant in an unsprinklered area on the casino level of the MGM Grand Hotel/Casino in Las Vegas. A flashover occurred in this area and extended out into the main casino area. The entire casino area was involved in fire within six minutes of the discovery time. The fire spread from the casino area out the west portico door on the casino level, where it was met by the arrival of the first fire department personnel.

The fire in the casino area was fed by flammable furnishings, wall coverings, PVC piping, glue, fixtures, and plastic mirrors hanging on the walls. The heat and smoke from the fire spread rapidly through the seismic joints, the elevator shafts, and the stairways and through the 21 residential floors of the hotel. Except to a minor extent,

FIGURE 6.55 ◆ An example of the damage in the casino area. *(Courtesy of Clark County, Nevada, Fire Department)*

FIGURE 6.56 ◆ The fire traveled from the deli, across the casino, and burst out the main entrance where it was met by the first arriving fire department units.
(Courtesy of Clark County, Nevada, Fire Department)

the fire itself did not extend above the casino level. However, the extension of the heat was so intense that it activated the automatic sprinklers in the lobby area adjacent to the elevator shafts on the 26th floor. The 26th floor was the top floor. The rapid extension of the heat and smoke trapped people in stairways and elevators. Most of the fatalities occurred in the casino area and on the 16th floor and above, with the majority found between the 20th and 25th floors.

The fire turned out to be the second largest loss of life hotel fire in U.S history, claiming 85 lives and injuring 778 guests and seven hotel employees. Fourteen firefighters were hospitalized, mostly from smoke inhalation, and nearly 300 firefighters showed signs of smoke and carbon monoxide inhalation.

FIGURE 6.57 ◆ This damage was possibly in the casino or nearby corridor.
(Courtesy of Clark County, Nevada, Fire Department)

The location of the fire and the type of occupancy in which it is found are two of the many factors that contribute to the differences between fires. This chapter explained and highlighted some of the variables found in various types of buildings and occupancies that influence the fire problem. Fires can and will occur in any phase of construction and will occur without warning for a multitude of reasons. Firefighters must understand why fire behaves differently in a "partition" fire from a fire that begins in an apartment building still under construction. High-rise fires and even chimney fires present their very own challenges to responding fire units. Firefighters need an understanding what they may expect in these different types of fires.

Review Questions

Attic Fires

1. What size lines should normally be used to attack attic fires?
2. What is the guideline as to the amount of water to use to extinguish attic fires?
3. When should salvage operations be commenced?
4. What is an effective method of extinguishing the fire if it is in a large finished attic with a gable roof and access from below is difficult?

Basement Fires

1. What is one of the first things that should be done on arrival at a basement fire?
2. Where should the first lines be placed?
3. What factor has the biggest influence on determining the best method of attack?
4. What procedure should be used if it becomes necessary to open the floor for ventilation?
5. What is a good practice to use when cellar devices are discharging water into the fire area?
6. What is the value of high-expansion foam at cellar fires?
7. What can be used as a last resort to extinguish the fire?

Chimney Fires

1. What is the best method to use if the fire is contained within the chimney?
2. What practice should be followed if water is used?
3. How can the chimney be cleaned out if it is stopped up?
4. What would smoke in the attic generally indicate if the fire were in the chimney?

Partition Fires

1. What methods should be used to detect partition fires?
2. If it is necessary to open a wall to check for a possible partition fire, where should it first be opened?
3. How should water be used to extinguish a partition fire?

Dwelling Fires

1. What is the primary problem in dwelling fires?
2. Where should the first attack line be brought in on a single-room fire?
3. What firefighting tactics should normally be used on a well-involved, one-story dwelling fire?
4. To what location should the first lines normally be advanced in multiple-story dwelling fires?

Garage Fires

1. What two construction changes have increased the problem in garage fires?
2. How is the attack on a garage fire generally made?
3. What procedure should be used to protect a one-story dwelling when the fire is in the attached garage?
4. What procedure should be used to protect a two-story residence if the fire is in

the garage, which is attached at the first floor and under the second floor?

Mercantile and Industrial Fires

1. What type of streams are many times necessary for use in the initial attack?
2. What type of fire tactics are generally used?
3. How should fires in dust-collecting bins be extinguished?
4. How should fires in areas where molten metals are found be extinguished?

Places of Public Assembly

1. What is the big contributor to the loss of life in these fires?
2. What is one of the first actions that should take place by fire department personnel after arriving on the scene?
3. What action can be taken to help eliminate panic in the event the building is thrown into darkness?
4. What action should the Incident Commander take when responding to a fire backstage in a theater?

Taxpayer and Strip Mall Fires

1. What are the general construction features of these occupancies?
2. What is the primary problem in fires in these occupancies?
3. What action should be taken to stop the spread of fire in the attic?
4. What size lines are generally used to stop the spread of fire in the attic?

Church Fires

1. What should be anticipated if the fire is of any size at all?
2. What are some of the construction problems?
3. What action should be taken at the fire if the church has an unprotected passageway extending from the church to the rectory?
4. What can be done to extinguish the fire if it has entered the hanging ceiling or enclosed attic area?
5. What precautionary measure should be taken regarding the placement of apparatus that will be used to provide elevated heavy streams?

Fires in Buildings Under Construction

1. Fires in what types of buildings under construction appear to present the most troubles for firefighters?
2. What are some of the general firefighting problems on buildings under construction?
3. In reality, what is the only method of effectively fighting a fire in these occupancies?
4. What is the best method of protecting exposures?

High-Rise Fires

1. What should be the action of the first arriving officer at a high-rise fire?
2. What should be the action of the second arriving officer?
3. What are some of the factors that should be determined in the lobby?
4. Where should the staging area normally be located?
5. How is procedure used to move equipment from the base to the staging area?
6. Who is responsible for setting up the staging area?
7. Where should the base be set up?
8. Where should the command post be set up?
9. What action should the first arriving chief officer take?
10. What action should the first arriving officer superior to the first arriving chief officer take?
11. What is generally the best method of saving lives and facilitating rescue operations?
12. What is one method of eliminating the duplication of search efforts?
13. Normally, which people should be moved to a place of safety?
14. What should be done if air rushes in when the door on the fire floor is opened?
15. Where should the crew get off the elevator if it is used for taking a fire attack crew aloft?
16. What is the collection point for incoming companies and equipment?

CHAPTER 7 Fires in Mobile Equipment

Key Terms

Objectives

The objective of this chapter is to introduce the reader to firefighting tactics as they apply to mobile equipment. Car fires, truck fires, aircraft fires, small boat fires, and ship fires are explored. Upon completing this chapter, the reader should be able to:

- Describe the tactics used for extinguishing car fires, truck fires, aircraft fires, and small boat fires.
- Explain how a fire under the hood and a fire in the passenger area of a car should be attacked.
- Explain the principal problems that may be encountered when fighting a truck fire.
- Discuss how to extinguish engine fires in both piston engines and jet engines of aircraft.
- Explain what precautions should be taken when extinguishing wheel fires on aircraft.
- Describe the precautions that should be observed when approaching an aircraft crash.
- Explain what procedures should be used to save lives in an aircraft forced landing with no fire.
- Explain how a rescue team is formed and how the team advances to rescue occupants from a forced aircraft landing with the plane on fire.

◆ Explain how fires in the hold of a cargo ship should be fought.
◆ Describe what actions should be taken for a fire in the hold of a ship that started while at sea.
◆ Explain how a fire in a hold of a ship containing oxidizers should be fought.
◆ Explain the various methods available for extinguishing a fire in the engine room of a cargo ship.
◆ Describe how fires should be extinguished in the passenger areas of passenger ships.
◆ Explain the method that should be used to extinguish a fire on the deck of a tank ship.

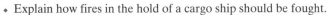

Mobile Equipment Fires

FIGURE 7.1 ◆ Car fires. Small lines generally are used to extinguish car fires. *(Courtesy of Las Vegas Fire & Rescue PIO)*

FIGURE 7.2 ◆ Aircraft crashes. Aircraft with a flammable liquid spill can occur in any community at any time. Foam has been used on this incident. *(Courtesy of Rick McClure, LAFD)*

(Continued)

Mobile Equipment Fires *(Continued)*

FIGURE 7.3 ◆ Ship fires occur in various locations on a ship and require different tactics in different locations and on different types of ships. *(Courtesy of District Chief Chris E. Mickal, New Orleans Fire Department—Photo Unit)*

◆ INTRODUCTION

Although the basic firefighting tactics are applicable to fighting fires in mobile equipment, some peculiarities apply to individual units that are worthy of consideration. One factor that applies to all is that of surprise. Fires occur in places where a few minutes before no problems existed. In most cases the fire department will be faced with a situation in which it is not exactly sure what is burning or how much of the material is involved. At times firefighters will be able to determine what materials are being transported prior to attacking the fire, but in most instances the attack will be made and the determination as to what was involved has to wait until after extinguishment has been completed. This in itself makes it mandatory that all attacks on mobile equipment fires be made with firefighters using full protective gear including SCBA. However, some basic tactics can be developed that will not eliminate all risks yet will reduce them to a minimum.

In the development of these tactics, it should be remembered that any fire in hazardous materials or where hazardous materials are involved or threatened requires a cautious and fully protected approach. In some cases it is better to evacuate and let the fire burn. If a hazardous material team is available, it should always be requested in these situations.

◆ CAR FIRES

At first glance, the extinguishment of a fire in a car appears to be a simple task. In should be kept in mind that car fires account for a large number of fatalities to occupants, and a number of firefighters have been severely injured as a result of not following proper safety precautions. (See Figure 7.4.) The primary problem for

FIGURE 7.4 ◆ Car fires occur in almost every community. They can be quite spectacular.
(Courtesy of Chris E. Mickal, New Orleans Fire Department—Photo Unit)

firefighters is the chance for a rapid spread of the fire due to a spill of gasoline, a rupture of the fuel tank, or a rupture of hydraulic fluid lines. The rupture of hollow drive lines, gas-filled shock absorbers, shock-absorbing bumpers, safety air bags, tires, and other units as well as the burning of toxic plastics are also major concerns. Adequate steps always must be taken to guard against these potentials. (See Figure 7.5.)

One of the first steps to take when arriving at a car fire is to park the apparatus in a safe location. Where the apparatus should be parked will depend on a number of circumstances. It is important to park in such a manner as to protect the firefighters from moving traffic. If the burning car is on an incline, it is best to park the apparatus uphill at least 150 feet from the car. This will prevent the car from rolling into the apparatus in the event the brakes give way. Regardless, it is a good idea to place a chock block under the

NO →   ← NO

FIGURE 7.5 ◆ As a precaution, do not attack the car from the rear or the front.

wheels of the car to prevent its movement during the extinguishment phase. Thought also should be given to protecting the firefighters in the event of a fuel tank rupture. It is good practice to park the fire apparatus to the front of the car, even if it means passing the vehicle on the way in. It is particularly important to follow this procedure if the fire is in the rear where the gasoline tank could become involved. Of course, if the fuel tank is in the front of the car, the apparatus should be parked to the rear.

At most car fires, the responding company will find smoke issuing from under the hood or from inside the body of the car. If the smoke is coming from under the hood, it is necessary to raise the hood to get at the fire.

If there is access to the interior of the car, the hood latch should be pulled to unlock it. This latch is normally found under the dash on the left-hand side. Care should be taken, however, when opening the car door to release the hood. It is possible that smoke and heat have penetrated the firewall and built up inside the car. There have been cases in which a backdraft occurred when the door to a car or van has been opened when all signs indicated that the fire had been confined to the engine area.

Sometimes the hood will not be released when the hood latch is pulled. This might occur because the release cable has been burned through. When this happens, it will be necessary to force open the hood to get at the fire.

If the hood latch releases the hood, it will also be necessary to release a second latch. This latch is generally found directly under the hood at the front. Gloves always must be worn for this purpose because the metal undoubtedly will be hot. Of course, the firefighters should already have gloves on, as each of them should be equipped with full protective clothing including SCBA.

A pike pole or similar tool can be used to pry open the hood if the metal is too hot to attempt to release this secondary latch. The fire generally can be knocked down with a small spray stream from a water extinguisher or a small line once the hood is opened.

Some officers prefer to knock the fire down before completely opening the hood. This will help prevent a backdraft if the conditions are ripe for one. Extinguishing the fire can be done by discharging CO_2 or dry chemical through a crack or by applying a fog stream from the ground level up into the engine area. Regardless of which method is used, there is a likelihood of a reflash once the hood is opened if a follow-up is not made with water to cool down the engine. If the fire under the hood is severe, the knockdown by one of these extinguishers followed by water usually is more effective and less hazardous than opening the hood and using water.

A fire under the hood generally will damage part of the electrical system. It is good practice to disconnect the battery any time the fire is in any part of the engine or in any electrical circuit. The air bags still may deploy even if the battery is disconnected (due to the reserve power). This can be done quickly with the use of a screwdriver and a pair of pliers.

Fires inside the car generally involve the upholstery. The water from an extinguisher or a small line normally can knock down these fires. There is, however, a possibility that the fire may have extended into the trunk if it is severe and involved the backseat. It is necessary to open the trunk under these conditions.

The trunk lock can sometimes be released from inside the car or if the keys are available by remote control. If not, the trunk will have to be pried open or a slide hammer used and the trunk checked for fire extension. A charged line should be in position when the trunk is opened, as it is impossible to determine what might be involved inside. There also is the possibility that backdraft conditions prevail. (See Figure 7.6.)

Backdrafts in vehicle fires are a possibility. Care always should be taken to avoid this possibility when opening up any part of the vehicle.

FIGURE 7.6 ◆ Engine company members waiting for water prior to approaching a well-involved station wagon fire. *(Courtesy of Rick McClure, LAFD)*

Occasionally fully involved cars are encountered. These should be approached with caution. It is good practice to open up on these fires with a straight stream from the maximum distance. The stream should first be directed at the gasoline tank to keep it from rupturing or venting under pressure. The tires should be hit next to prevent a similar action. The stream can be changed to a spray or fog stream as the car is approached. *Do not* make the initial attack from the rear of the vehicle or the front of the vehicle depending on the location of the gasoline tank. If it is not known where the tank is located, remember that most tanks are located in the rear of a vehicle. (See Figure 7.5.) Stay away from the hazardous area until assurance has been made that the tank has been cooled sufficiently to eliminate the threat of a rupture. Remember that some fuel tanks are made of plastic.

The life hazard in a car fire can be severe. A person trapped in a burning vehicle is a frightening sight. The only logical means of rescue is normally an aggressive attack on the fire. The victim could be unconscious or physically trapped. If unconscious, it may be simply a matter of releasing the seat belt and removing him or her. Seat belts, however, sometimes present a problem. It is good practice for firefighters to carry a knife that is capable of cutting the seat belt if necessary.

Vehicle extrication procedures have to be used to remove the occupants if they are physically trapped. There is always a chance of air bag deployment, and care should be taken during this operation to ensure that a reflash will not occur due to leaking gasoline. All potential sources of ignition should be removed, hot parts of the engine cooled, and a charged line handled by a firefighter kept available until all occupants have been safely removed.

◆ **TRUCK FIRES**

Truck fires in or around the engine or inside the cab should be handled the same as those found in cars. (See Figure 7.7.) Rescue practices are also similar. The problem with truck fires is the cargo. Firefighters are generally not aware of what is being carried.

FIGURE 7.7 ◆ Truck fires are not always found in upright positions. *(Courtesy of Rick McClure, LAFD)*

Signs indicating the presence of hazardous materials may be displayed on the truck, but it is possible that these hazards are present without the luxury of warning signs.

Fires in the cargo area of trucks always should be suspected as involving hazardous materials and approached as such. Full protective gear with SCBA is a must. It is good practice to attack a fire in the cargo hold of a truck from the upwind direction from a maximum distance using a straight stream for the initial attack. Thoroughly cool the fuel tank and tires as the advancement is made. Spray streams can be used as the fire is closely approached.

Be on the alert for a possible backdraft as the cargo space is opened. The manifest of what is carried should be checked prior to overhaul. It generally can be found in the driver's compartment. Standard overhaul procedures should be modified in accordance with what is found in the manifest. If available, a hazardous materials team should be requested if hazardous material are indicated in the manifest or suspected.

In some cases, depending on the evidence, it is best to let the fire burn and evacuate to a safe distance. This should be done if there is any doubt as to the safety of personnel or civilians.

◆ AIRCRAFT FIRES

Regardless of the size or location of the community or the size of the firefighting force, a fire department faces the possibility of having to fight a fire in an aircraft at any hour on any day. (See Figure 7.8.) The majority of all aircraft crashes have occurred on or within five miles of an airport. However, aircraft have crashed in some of the most unlikely places. (See Figure 7.9.) This section will cover some of the basic fires that occur in aircraft on the ground, together with those resulting from crashes or forced landings.

FIGURE 7.8 ◆ Airport units constantly train for that "big one."
(Courtesy of District Chief Chris E. Mickal, New Orleans Fire Department—Photo Unit)

ENGINE FIRES

Fires in engines of aircraft on the ground are normally not too serious. The procedure for handling the fire will differ, depending on whether it is in a piston or a turbine (jet) engine.

Fires in Piston Engines. **Piston engines** are internal-combustion reciprocating engines. Unlike the engines in an automobile, they are air-cooled to eliminate the weight of the heavy engine blocks that are typical of liquid-cooled engines. The housing around the engine is called a **nacelle.**

If the fire is confined within the engine nacelle, the pilot or maintenance crew should first try to extinguish it by using the aircraft's built-in extinguishing system. If this is not successful, the fire should be attacked using carbon dioxide. If deemed necessary, water spray or foam can be used to keep the adjacent parts of the aircraft structure cool.

piston engines Internal-combustion reciprocating engines.

nacelle The housing around an aircraft engine.

FIGURE 7.9 ◆ Aircraft crashes can occur in some unusual locations.
(Courtesy of Las Vegas Fire & Rescue PIO)

jet engines Jet engines draw air in through the front, compress it, mix it with fuel, and ignite it.

Fires in Jet Engines. **Jet engines,** also called **turbine engines,** draw air in through the front, compress it, mix it with fuel, and ignite it. The mixture is then exhausted out the back. Fires in the combustion chamber of jet engines are best controlled if the engine can be kept turning over by the crew. Extreme caution should be observed, however, when working around a running jet engine. Personnel should never stand within 25 feet of the front of, to the side of, or directly to the rear of the engine intake ducts. The suction created by some engines is sufficient to draw in a 200-pound person.

Personnel should also stand clear of the turbine plane of rotation area. This area is the path of flying parts in the event of engine disintegration. The area to the rear of the exhaust outlet should also be avoided for a distance of at least 150 feet. Exhaust temperatures reach approximately $3000°$ F at the outlet.

Extinguishment of fires that are outside the combustion chamber but within the engine nacelle would first be attempted by using the aircraft's built-in extinguishing system by the pilot or maintenance crew. If this is unsuccessful, then carbon dioxide should be tried; however, these extinguishers should not be used if magnesium or titanium parts are involved. In these cases it is best to allow the fire to burn itself out. Foam or water spray should be used to keep the nacelle and surrounding exposed parts of the aircraft cool while this is taking place.

WHEEL FIRES

Potential wheel fires should be approached with caution. The responding fire apparatus should be parked within effective firefighting distance from the aircraft but never to the side of the aircraft or in line with the wheel's axle. Debris is generally thrown to the sides but not to the front or rear if the tire should explode. Consequently, if it is necessary for firefighters to approach the aircraft, it should be done from a fore or aft direction. (See Figure 7.10.)

Smoke around the drums and tires of the wheels does not necessarily mean that the wheel is on fire. Overheating of brakes is not an unusual occurrence on aircraft. If the brakes are overheated, they should be allowed to cool by air only. On propeller-driven aircraft the cooling can be assisted if the crew will keep the engine running. This will direct fresh air onto the wheel. Firefighters should remain a safe distance and not approach the aircraft while this is being done.

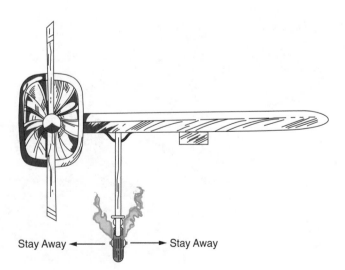

FIGURE 7.10 ◆ Remain clear of the danger area.

Stay Away ← | → Stay Away

It is possible for firefighters or members of the ground crew to use a smoke ejector to help reduce the cooling time in the event the aircraft's engines cannot be used for cooling or a jet aircraft is involved. Do not use water for cooling as the rapid reduction of heat in the wheels may result in an explosion. Carbon dioxide extinguishers are frequently used for cooling wheels.

An extinguisher can be used if there are actual flames in the wheel. A dry chemical extinguisher is recommended because it is less likely to chill the metal in the wheel parts. Water can be used if a dry chemical extinguisher is not available; however, it should be used with caution. Firefighters should protect themselves from a possible explosion by using the fire apparatus as a shield or by standing fore or aft of the wheel. The water should be applied in a fine spray and in short bursts of 5 to 10 seconds. At least 30 seconds should elapse between bursts. Water should be used only as long as visible flames are showing.

INTERIOR FIRES

Fires in the interior of an aircraft on the ground normally involve ordinary combustible material. It should be kept in mind that there is sufficient combustible material that gives off toxic gases on decomposition in the interior of most aircraft to warrant a standard procedure of wearing breathing apparatus to attack such fires, regardless of the size of the fire.

Fires originating when the aircraft is occupied are normally detected in the early stage. This generally results in the damage being restricted to a minimum. However, the fire will probably smolder for some time before being discovered if ignition takes place when the aircraft is unoccupied.

Fires discovered in the early stages normally can be extinguished by the use of a hand extinguisher. Small lines, however, will have to be employed if the fire is beyond the control of hand extinguishers.

The interior of a larger aircraft is similar to a corridor in an ordinary structure that contains a considerable amount of combustible material. This type of arrangement always presents the potential for a backdraft or a flashover. Consequently, extreme care should be taken when entering the cabin and while working on a fire once inside. It goes without saying that firefighters should always be equipped with full protective equipment, including SCBA. If the fire is small, it generally can be extinguished by making a direct attack with the use of small lines. The most stable point to initiate the attack on large commercial aircraft is through the exits over the wing. If it is decided that it is best to enter the aircraft to fight the fire, the positive pressure ventilation method of firefighting might prove extremely effective. Open the top of the aircraft or a cabin door some distance from the entry point and have firefighters with blowers precede those with 1½- or 1¾-inch hose lines. This should help remove the atmosphere of smoke and products of combustion and make it easier for the nozzle handlers to make entry.

The temperature inside the aircraft will probably be extreme if the fire is extensive. In these circumstances it might be best to use the indirect method of attack from outside the aircraft. Several lines should be used in a coordinated effort if such a selection is made. Where lines should be positioned can be determined by observing the conditions through the various cabin windows, together with the blistering of the external paint. Care should be taken so that the lines are not positioned in such locations that will result in forcing the fire into the uninvolved portion of the cabin area, providing that there are any.

It is possible that the fire can extend into the hidden areas in the ceiling, bulkheads, and overhead storage areas. These areas should be checked thoroughly once the fire is under control.

AIRCRAFT CRASHES

Aircraft range in size from those carrying one person to those transporting nearly 500 people. A crash involving an aircraft can occur in or near any city at any time. The magnitude of the problem facing firefighters will depend on a number of factors. Some of the primary factors are the seriousness of the crash, the size of the aircraft, the number of people aboard, and the personnel and equipment available to cope with the situation.

The primary objective in fighting a fire in an aircraft is the same as fighting one in any other occupancy—saving lives. However, the time frame is much more critical when aircraft fires are concerned. The primary reason for this is that flammable liquids are normally involved. This can result in an extremely intense fire that spreads rapidly.

The material contained in this section regarding aircraft crashes is not extensive but merely covers some basic essentials. It is intended to furnish a basic understanding of the problems involved and some general principles used to cope with them. It has not been written for those departments that have specialized crash crews nor does it cover all the varied situations with which the specialized crash crews are faced. Specialized crash crews are trained not only in the basics but also in the methods used to cope with the particular aircraft that operate from their field, whether they are private aircraft, military aircraft, or larger commercial aircraft. The knowledge and experience of the members of these crews goes far beyond what is covered in this chapter.

Aircraft crashes include those disastrous crashes in which all or nearly all of the occupants are killed on impact, those forced landings in which no fire exists on the arrival of the firefighting forces, those forced landings where the aircraft is involved in fire when the firefighting forces arrive, and variations of the three. (See Figure 7.11.)

FIGURE 7.11 ◆ Foam has been used on this plane that crash-landed in a residential area.
(Courtesy of Las Vegas Fire & Rescue PIO)

Certain precautions should be taken when approaching crash situations regardless of the type of incident. Constant thought must be given to the possibility of a flammable liquid spill existing that could find a source of ignition during the approach. Consequently, an approach from the windward side is preferred to one from the leeward side, and one from the top of the hill is preferred to one from the bottom of the hill if sloping terrain is involved. Drivers should also be on the watch for people who may have been thrown clear or escaped from the aircraft. This is particularly true when visibility is limited by smoke, fog, darkness, and so on, or if the response is made into an area where the ground cover is high. These factors also make it more difficult to avoid aircraft parts that may have been scattered over the area. It is good practice to lay protective lines once the apparatus has been positioned at the crash site, regardless of whether a fire is in progress on arrival.

Disastrous Crashes. A disastrous crash includes those in which all or a majority of the occupants are killed on impact. It may be one in which the aircraft dives into the ground at high speed or one in which impact is made at an angle that results in parts of the aircraft being spread over a wide area. The magnitude of the problem will depend on the size of the aircraft, the number of occupants aboard, and the population density of the area where the crash takes place.

Obviously, there will be no survivors if the crash is the result of a high-angle, high-speed dive. A large crater generally will be made and wreckage will be found both inside and outside of the crater. In some cases an explosion will take place and in others the crash will be followed by fire. If there is no fire, flammable liquid will be spread over a wide area, presenting the potential of a fire and/or an explosion. If there is fire in the crater, it may be advantageous to attack it with water and flood the hole. Any flammable liquid spread around the area should be blanketed with foam. It is good practice to cover the water in the crater with foam because flammable liquid will most likely be floating on top of the water. If the first arriving units do not have foam-producing capabilities, a request for such equipment should be made immediately.

The problem is extremely compounded if the aircraft hits at a low angle and crashes into structures. Gas lines generally will be broken and electrical lines severed. Fire normally will consume part of the aircraft and a number of surrounding buildings. It is imperative that an immediate call for help be made with the request including ambulances, police assistance, and a sufficient number of firefighting units to cope with the immediate fires and exposures and the potential for expansion. It is extremely important not to underestimate the magnitude of the situation. Most likely assistance will be coming from a considerable distance away and it will take a long time for additional help to arrive. The delay and resultant loss will be much greater if the situation is underestimated. The Incident Command System for major emergencies should be placed into operation immediately.

These disasters normally are of such a magnitude and compounded by so many variables that it is almost impossible to establish any guidelines as to what action should be taken first. Of course, the saving of life and safety to personnel will always take first priority.

Forced Landings—No Fire. Some forced landings in which no fire is involved are made with the wheels down; others are made with the wheels up. Although the objectives and procedures for handling the two are similar, the situation is normally more

FIGURE 7.12 ◆ Foam has been used to extinguish the fire in this aircraft crash. *(Courtesy of Rick McClure, LAFD)*

hazardous when a wheels-up landing is made. The reason for this is that there is a much greater chance of fracturing flammable liquid tanks or lines when the landing is made with the wheels up.

All the precautions listed for approaching an aircraft crash should be observed. The approach always should be made with the thought that fire could break out at any second. Consequently, lines need to be laid in such positions as to make any rescues and knock down the fire if ignition takes place.

Any flammable liquid spills should be covered with foam as soon as possible, providing of course that the responding vehicles have this capability. If not, requests for units having this capability should be made immediately. (See Figure 7.12.)

The desired results will have to be accomplished with the use of water if the first responding units do not have foam-delivery capabilities. Streams will have to be used to push the spilled liquids away from the aircraft to safe areas. Care should be taken to ensure that the spilled liquids are not pushed toward buildings or into sewers where disastrous results could occur if ignition were to take place.

Every attempt should be made to stop any leaking fuel. It may be possible to stop a leak by closing fuel valves; however, if necessary the lines can be crimped or the severed ends of the lines plugged. The firefighter performing this operation should be protected at all times by a hose line.

Occupants of the aircraft should be removed when it is safe to do so. The regular exits should be used if possible. If not, the second choice should be given to the emergency exits. It may be necessary to breach the outer skin of the aircraft to effect rescue if both the regular and emergency exits have been frozen shut by the impact of the landing. Hand or power tools can be used for this purpose.

Extreme care must be taken to prevent ignition if the spill was covered with foam and power tools or any spark-producing equipment is used for gaining entrance into the aircraft. It is possible that those firefighters carrying the cutting equipment to the aircraft could have broken the protective blanket. Any breaks in the protective blanket should be covered. Loaded lines advanced by firefighters should be in position during cutting operations as an additional precautionary measure.

Forced Landings with Fire. This type of emergency is the most difficult of all. Every member of the responding companies is aware of the critical position of occupants of the aircraft and the need for quick action if lives are to be saved. The situation has most likely been caused by a fracturing of flammable liquid lines or tanks that released flammable liquids that were ignited by sparks resulting from the impact of the aircraft with the landing surface. In many cases the fire starts before the aircraft comes to rest. The time frame is critical. Rescue efforts must be commenced within a few minutes if any success whatsoever is to be achieved.

The term rescue as used in the context of aircraft rescue and firefighting is slightly different from when it is used at a structure fire. With **aircraft rescues,** the term refers to the control of a life-threatening fire in the critical area during the time that it takes all physically able occupants to leave the aircraft on their own.

> **aircraft rescues** The control of a life-threatening fire in the critical area during the time that it takes all physically able occupants to leave the aircraft on their own.

The same precautions must be observed as for other types of crashes. The direction of the initial approach is critical. The wind direction, the terrain, and the exit facilities on the aircraft must be carefully considered. Additional help should be requested if it appears that it is or might be needed.

Although rescue is the most important factor, it must be kept in mind that a quick, direct attack on the fire may be the best means of accomplishing it. A maximum effort must be launched if the decision is made that this is the best possible method of saving lives.

It is usually best to make the initial attack with at least two lines using foam if rescue efforts are to be made while the fire is still in progress. The use of AFFF foam is recommended.

The most severely exposed area should be hit first and the lines directed outward to cover the remaining parts of the fuselage. (See Figure 7.13.) This action will cool

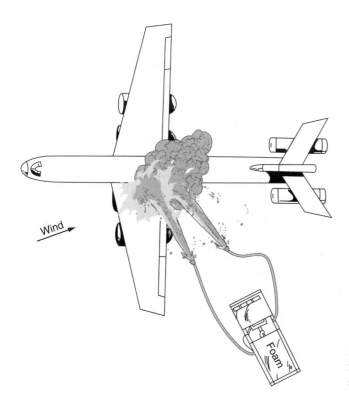

FIGURE 7.13 ◆ Direct initial lines to the most severely exposed area.

FIGURE 7.14 ◆ If possible, make approach from the windward direction.

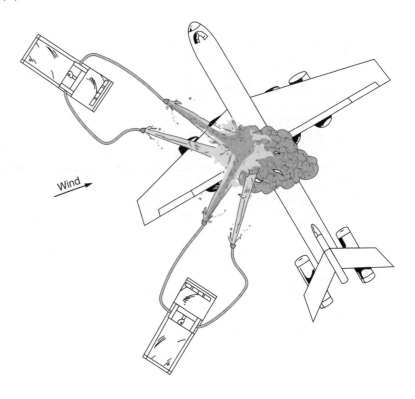

Wind

down the fuselage and reduce the heat exposure to the occupants. This protective measure should be continued with the objective of keeping the entire fuselage cool until rescue lines can be brought into play. As many lines as can be employed by the personnel and equipment on hand should be used to achieve the objective. (See Figure 7.14.)

If the initial responding units do not have foam-delivery capabilities, the same basic procedure should be followed using water. The use of water could result in ground fires spreading. This might place firefighters in a precarious position; however, keep in mind that there probably will be little chance of saving those inside the aircraft if the fuselage is not kept cool. It is also crucial that no firefighter should be placed at risk unnecessarily. Protective lines should be moved into position as a precautionary measure.

Forming of a Rescue Team.   If possible, the rescue team should be composed of four firefighters using spray nozzles on 1½- or 1¾-inch lines. The nozzles should be kept 12 to 24 inches off the ground and should be interlaced so as to form a solid mass of water as advancement is made on the aircraft. (See Figure 7.15.) Firefighters are provided better protection if they work inside their lines.

Forming of the team should be done as quickly as possible; however, it is most likely that the firefighters will not be positioned and lines loaded all at the same time. As each individual is positioned, his or her nozzle should be turned to a straight stream in order to help keep the fuselage cool until the entire team is formed.

It is better that the rescue team work from the windward side of the fire. This will expose the team to less radiant heat, ensure better reach for the streams, and

FIGURE 7.15 ◆ The proper positioning of rescue team members. Nozzles should be 12 to 14 inches off the ground and the streams interlaced.

ensure better visibility as the wind to their backs will blow the smoke away from the team.

With this arrangement of the rescue team, the two firefighters in the middle will perform the rescue work while the two outer firefighters provide them protection from the heat once the rescue exit is reached. All firefighters should be provided with full protective equipment, including SCBA. If available, additional protection can be provided if the four advancing firefighters are used only for protection while two additional firefighters who will actually perform the rescue operations follow behind the rescue team. These firefighters should also be equipped with spray nozzles on 1½- or 1¾-inch lines and use them in the same manner as members of the advancing line. Their lines can be used to extend the width of the protective screen while helping to protect the initial four firefighters as the advancement is made. If an additional firefighter is available, he or she should follow the two to provide additional protection.

If foam is available, it is possible to put it to good use in a coordinated attack. Foam lines can be used to blanket the area behind the advancing rescue team. This will provide an additional protection for the firefighters pushing toward the exit doors.

This type of protection is most effective when the rescue team consists of only four firefighters and the fire is being fed by a flammable liquid leak. Under such conditions there is a constant threat that the fire could get behind the advancing rescue team. Of course, backup lines are normally positioned to prevent this, but a protective blanket of foam behind the advancing team provides a greater degree of security.

It is possible for a small amount of fire to appear around the feet of some of the rescue team members if the team advances too rapidly. The natural tendency of the foam operator is to direct his or her stream at the fire when this happens. This action should be avoided as it will most likely splash fuel onto the firefighters. It is better to direct the foam stream at the buttocks and legs of the firefighters and allow the foam to roll down onto the fire to extinguish it.

Strategic Goals

- Determine the best approach to the fire after considering the wind direction, the terrain, and the exit facilities on the aircraft.
- Commence rescue as soon as possible.
- Make the initial attack on the fire with at least two foam lines if rescue will be attempted while the fire is still in progress.
- Hit the most severely exposed area first and then direct the lines outward to cover the remaining parts of the fuselage.
- Lay as many lines as possible to continue the cooling of the fuselage.
- Move additional lines into place to protect the firefighters keeping the fuselage cool.
- Form a rescue team of at least four firefighters using 1½- or 1¾-inch lines.
- Add additional firefighters to the rescue team, if available.
- Provide protection for the rescue team by laying a blanket of foam behind the team as it advances.

Incident Command System Considerations

- Incident Commander
- Safety officer
- Search and rescue group
- Rapid intervention team
- Staging officer
- Exterior divisions
- Medical unit

◆ SMALL BOAT FIRES

The potential for a small boat fire exists in any community or fire district that has within its jurisdictional boundary a body of water of any size. Sometimes the small boats are brought to the body of water on a periodical basis; at other times they are stored permanently on the water (in most cases in marinas or next to private docks). The marinas will vary in size from those housing only a few boats to those housing hundreds. The storage anchorages usually consist of one or more main floating walks that extend out for varying distances into the body of water. Extending out from these main floating walks are **finger piers** to which the boats are secured. Many of the boats are protected from the weather during nonuse by canvas covers that extend from one end of the boat to the other. This configuration presents the constant potential for a rapid spread in the event one of the boats becomes well involved with fire. The potential for a large loss is also present as some of the boats are valued into the thousands.

> **finger piers** At a storage anchorage for small boats, there are floating walks that extend out for varying distances into the body of water. Extending out from these main floating walks are finger piers to which the boats are secured.

Getting water to the fire may be a problem if the boat is tied up at a pier some distance from the shore line. Some of the larger marinas have hydrants adjacent to the main walkways and water available along the walkways. At other marinas it will be necessary to take draft to supply the hose lines. **Drafting** is a means of taking water from a nonpressure source such as the ocean or other open body of water.

> **drafting** Taking water from a nonpressure source such as the ocean or other open body of water.

Fires in the cabin or superstructure area should be attacked by using fog or spray streams. A single 1½- or 1¾-inch line normally will be sufficient for a fire in a small boat. Two lines generally will be required if the fire is of any size in a small yacht. A fire in a small yacht should be attacked from both sides so as not to push the fire onto an adjacent boat.

Many fires on small boats start around the engine or in the bilge area. The **bilge** is the lower part of a ship's hull or hold. The primary cause is the ignition of flammable liquid vapors trapped in the bilges. This is generally ignited when an attempt to start the engine is made. The fire usually starts with an explosion. Carbon dioxide is very effective if the fire is fairly well contained under the floor plates and has not advanced too far. Class A foam is effective if the fire is extensive because it has the ability to knock down the fire quickly while forming a seal to help prevent reignition. Water in the form of fog or spray streams is also effective; however, it presents the problem of possibly sinking the boat and spreading burning fuel onto the adjacent waterway. Thought should always be given to the protection of surrounding boats.

Regardless of the method used to extinguish the initial fire, there will probably be glowing embers in seat cushions, mattresses, and other combustible materials. These items will have to be thoroughly overhauled. There is also the possibility of a considerable amount of unburned gasoline in the bilge that will have to be siphoned into a sturdy container. It may be necessary to call for a vacuum truck if the spill is extensive. Caution should be taken to ensure that wiring is disconnected from the batteries and shore lines prior to commencing these operations. It may be necessary to pump out the boat to keep it from sinking if too much water is used.

Occasionally it will be found that a small boat on fire has been set adrift prior to the arrival of the fire department to keep it from damaging adjacent boats. In such cases the boat should be secured prior to hitting it with hose lines. If this is not done, the hose lines may push the boat into adjacent exposures or out of the reach of firefighters.

bilge   The lower part of a ship's hull or hold.

◆ SHIP FIRES

Fires aboard ships are considerably different from fires in structures. The problems are more unique. The uniqueness presents firefighters with situations that are not common to their day-to-day firefighting efforts. The primary difference is heat conductivity. A serious fire in a windowless brick room with fire-resistive floors and ceiling would expect to be confined to the room. A similar fire in a similar-sized area aboard ship would present the risk of fire spreading to adjacent compartments through bulkheads, deckheads, and decks. **Bulkheads** and **deckheads** are similar to walls and ceilings except they are made of metal. This presents the additional problem of conducted heat.

A second difference with ship fires is buoyancy. It can be expected that when master streams are used to extinguish a serious fire in a structure, large volumes of water will be found running from the floors and out the doorways. The same amount of water used on a ship fire might result in the ship capsizing as all the water would be held inside the ship. The ship's pumps might be able to dispose of part of it, but not at the rate that it is discharged by hose lines.

The third difference is the restricted capabilities for ventilation. This results in heat and products of combustion being held inside, compounding the problem of locating and extinguishing the fire.

bulkhead   Similar to a wall except it is made of metal.

deckhead   Similar to a ceiling except it is made of metal.

FIGURE 7.16 ◆ *Cargo ship Harpoon.*
(Courtesy of Los Angeles Maritime Museum)

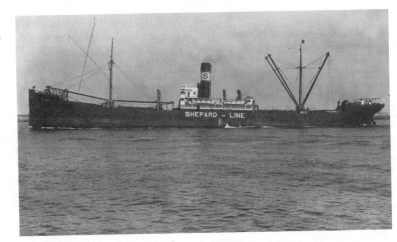

holds   Where the cargo is carried on cargo ships.

bow   The front portion of a ship.

stern   The rear portion of a ship.

lower deck   The bottom deck of a ship's hold.

'tween deck   The middle of three decks in a ship's hold.

upper deck   The upper of three decks in a ship's hold.

escape hatch   An escape hatch consists of an iron ladder inside a metal tube. A small door from the tube opens into the ship's hold at each of the various deck levels.

port   Left, looking from the stern toward the bow of a ship.

starboard   Right, looking from the stern toward the bow of a ship.

hatchways   The openings in the deck of a ship through which the cargo is loaded and unloaded.

coamings   The raised sides of the hatchways of a ship at deck level.

hatch covers   The hold of a ship is secured by placing metal support beams across the hatchways and attaching them to the coamings. Hatch boards (referred to as hatch covers) are then laid on the beams to cover the opening.

These three differences from structure fires are common to all ship fires, but there are sufficient differences in the construction of the different types of ships making it impossible to establish guidelines for firefighting tactics that apply to all ships in general. Even the tactics used for fighting a fire in the engine room of a cargo ship are different from those of fighting one in the engine room of a tanker. However, knowledge of the combined tactics for fighting fires in cargo ships, passenger ships, and tankers is probably applicable to most types of commercial ships.

CARGO SHIPS

Cargo ships vary in size, shape, and construction features; however, some similarities are common to most. Most ships are approximately 10 times longer than they are wide. (See Figure 7.16.) The cargo is carried in **holds.** There generally are a minimum of five holds, numbered from the **bow** (front) to the **stern** (rear). For descriptive purposes, consider that the ship has five holds.

Hold number 1 will probably be divided into two levels, whereas the other holds will be divided into three. The three levels are referred to as the **lower deck,** the **'tween deck,** and the **upper deck.** Each of these holds most likely will be equipped with an escape hatch. An **escape hatch** consists of an iron ladder inside a metal tube. A small door from the tube opens into the hold at each of the various deck levels. The escape hatch may be enclosed in a small deck house on the main deck. The hold may also be equipped with ventilators, one on each side (**port**—left; **starboard**—right, looking from the stern toward the bow).

The openings in the deck through which the cargo is loaded and unloaded are known as **hatchways.** The hatchways have raised sides at deck level, which are referred to as **coamings.** Coamings vary from a low level to about three feet high. The hold is secured by placing metal support beams across the hatchways and attaching them to the coamings. Hatch boards (referred to as **hatch covers**) are then laid on the beams to cover the opening. Two or three heavy tarpaulins are then placed on top of the hatch boards and fastened down to the outside sides of the coamings.

To a firefighter, a hold may be visualized as a three-story warehouse that measures approximately 60' by 60'. Similar warehouses are connected to opposite ends of the first warehouse with the ½-inch steel plate walls of the first warehouse acting as a common wall between the warehouses. Each of the warehouses is filled with miscellaneous merchandise that may contain flammable or hazardous materials. None of the merchandise is skidded and there is no aisle space. The only access to each warehouse is through a 30' by 30' hole in the roof. Any additional holes that need to be cut for access or ventilation will be restricted to 15 feet from the roof on two sides of the warehouses.

The machinery space and fuel tanks probably will be found between hold number 3 and hold number 4. This is near the center of the ship. The superstructure will be located above the machinery space. The superstructure houses the crew and officers' quarters, the galley, and the eating areas. The decks above the main deck in the **superstructure** are referred to as the **shelter deck,** the **boat deck,** the **bridge deck,** and the **flying bridge.**

To the aft of hold number 5 is the fantail. The **fantail** is a deck house with extra crew quarters. Below the crew quarters will be found the steering engines, the carpenter shop, the bos'n's locker, and miscellaneous areas. (See Figure 7.17.)

The ship is compartmentalized in such a manner that one or two compartments can be damaged and become flooded without sinking the ship. For example, between each hold and between the machinery space and holds number 3 and 4 are watertight transverse bulkheads. There are no interconnecting doors or openings between the holds and none from a hold into the engine room.

Standard equipment on most ships includes both a steam system and a CO_2 or halon system that is capable of flooding individual holds or the engine room. Although halon is an excellent extinguishing agent, it is also environmentally damaging to the earth's ozone layer. Because of this damaging effect, the material is no longer manufactured. However, it may be found in some older fixed fire systems. Controls for these shipboard fixed fire systems are normally located in the passages

superstructure The decks of a cargo ship above the main deck.

shelter deck The first deck above the main deck in the superstructure of a cargo ship.

boat deck The second deck above the main deck in the superstructure of a cargo ship.

bridge deck The third deck above the main deck in the superstructure of a cargo ship.

flying bridge The fourth deck above the main deck in the superstructure of a cargo ship.

fantail A deck house on the aft of a cargo ship that contains extra crew quarters.

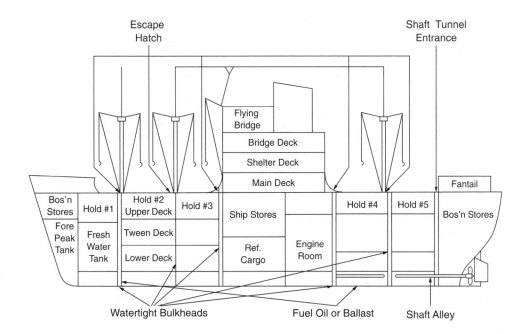

FIGURE 7.17 ◆ The general layout of a cargo ship.

in the superstructure somewhere below the bridge. The controls are generally well marked.

Most ships are equipped with standpipe systems having an attached 1½-inch linen hose. These systems are fed by the ship's fire pumps; consequently, they are not charged unless the fire pumps are operating.

Some areas of some ships are protected by automatic sprinkler systems. These systems have manifold inlets on the deck that can be used by the fire department to boost the pressure in the system.

A word of caution is in order regarding the use of the CO_2 system to flood the machinery space. A warning bell normally sounds in the machinery space when this system is activated. Any firefighter in the machinery space at the time this warning bell sounds should get out as soon as possible as the system is capable of completely flooding the machinery space in about two minutes. Firefighters should be aware that the CO_2 injected into the machinery space rapidly dissipates the oxygen. Anyone trapped in this area could die due to the lack of oxygen.

Fires in the Hold. It is common practice for apparatus responding to a report of a fire aboard a ship that is tied up to a wharf to drive directly onto the wharf. Smoke coming from the ship will indicate that lines probably will be needed. Consequently, the first arriving pumper should take a position close to the scene so that any equipment needed will be accessible and unnecessarily long lines will not have to be laid. Most wharfs are void of hydrants. This generally makes it necessary for the pump operator to prepare to draft.

When preparing to draft, the suction hose should be dropped between the fender log and the wharf. A **fender log** is a log that is attached to the wharf at the water level to prevent approaching ships from damaging the wharf. Placing the suction hose between the fender log and the wharf will help prevent damage to the suction hose in case a boat comes alongside.

fender log A log that is attached to the wharf at the water level to prevent approaching ships from damaging the wharf.

The first arriving officer should take charge of the fire and go aboard the ship to size up the problem. It has been found best for the officer to wear a white hat rather than a helmet. The white hat appears to be a symbol of authority in marine tradition. This makes it easier for the officer to obtain needed information and the cooperation of the crew.

One of the first tasks of the responding officer will be to determine the location of the fire, what is burning, and the extent of the fire. This information generally can be determined by talking with the ship's captain or first mate. The first mate will have a **manifest** of the ship's cargo. The manifest will provide the information as to what is probably burning. The manifest also will indicate what materials are being carried in adjacent holds. It is important to determine whether any hazardous materials are being carried and, if so, their location. The characteristics of any chemicals being carried that are not familiar to the responding officer should be determined from the fire department's dispatch center. The first mate will also be able to provide a chart showing the layout of holds, bulkheads, ventilators, passageways, and other factors that are important to the successful extinguishment of the fire. Additionally, the first mate will be able to provide information regarding the firefighting facilities aboard the ship. The ship will probably have a CO_2 system with the capability of flooding the hold. It is important to find out whether this system was used prior to the department's arrival and also if the ship is being fumigated.

manifest The manifest indicates what materials are being carried on a ship and in what holds the material is stored.

Once the necessary information is obtained, the Incident Commander should notify the dispatch center of the extent of the fire and what additional equipment, if any, will be needed. The first alarm response probably will include fire boats. If the information gathered indicates that special equipment carried on the boat companies will be needed, this information should be relayed to the boat companies so that they can have the equipment laid out on deck on arrival. It will be necessary to detail several members of the land companies to help secure the fire boat lines, haul up equipment, and get a Jacob's ladder into position if it will be needed.

A **Jacob's ladder** is a rope ladder with wooden steps. It is kept coiled up, which makes it easy to put over the side of the ship so that members of the boat company can climb aboard. Its construction is similar to the chain ladders developed for escape from the second floor of a dwelling.

It is best for the Incident Commander to plan on taking full advantage of all firefighting systems aboard the ship such as the CO_2, steam, halon, and foam systems. It is important, however, that department hose lines be laid out and wetted as sometimes the ship's equipment is unreliable, particularly on some foreign ships. Hose lines should be laid from both the land companies and the boat companies.

Getting lines aboard is not always an easy task. Those from land companies are generally taken up department ladders that are raised from the wharf or hauled up by ropes that have been lowered from the ship's main deck. Lines from fire boats are generally hauled aboard by ropes lowered from the main deck. Most of the equipment from land companies is carried up the gangplank or up department ladders. Most of the equipment from the fire boats is brought aboard by lowering ropes and hauling it aboard. If ladders are used to take hose lines or equipment aboard, the ladders should be checked periodically as the rise or fall of the tide could affect their security. A ladder should not be tied to the ship because the ship's movement could damage it.

The hatch can be opened and the covers over the ventilators removed once hose lines are in position and charged. Removing the hatch covers will provide addition information as to the extent of the problem. Usually the admittance of air into the hold will result in considerable smoke and heat being given off. However, if no fire is visible, it means that the first thing that will have to be done is to commence the removal of the cargo.

It is important that water not be used until visible flames are encountered. Use of water at this stage not only will increase the loss but also will make the task of removing the cargo more difficult. Water will help break open any cardboard containers in the hold and certain cargo will swell, making it extremely difficult to remove.

The ship's gear, a barge crane, or a railroad crane can be used to help remove the cargo. If conditions in the hold are not too smoky or hot, it will probably be possible to load the cargo on lift boards and remove it with the use of the cranes. Longshoremen may or may not assist firefighters in this task. If conditions are such that people cannot work in the hold, then it will be necessary to remove the cargo by the use of a clamshell bucket with teeth.

Water can be used on the fire as soon as flames become visible or if the heat becomes excessive. The use of water, however, should be limited to that necessary to control the fire so that removal of the cargo can be continued. The fire may be deep in the hold or possibly in the 'tween decks. The only solution is to continue to remove cargo until the fire is reached.

Occasionally the fire may be so severe as to cause hot spots to develop on the main deck or along the sides of the ship. These spots can become red hot if left unattended. A hot spot developing on the deck should be kept cool by a constant flow of water over it. The hot spot indicates that the main body of fire is probably directly below it. It may be possible that a hole can be cut in the deck and a spray nozzle

Jacob's ladder A rope ladder with wooden steps used aboard ships. It is kept coiled up and lowered over the side to permit someone to climb aboard.

FIGURE 7.18 ◆ Hot spots have developed on the side of this ship. Large amounts of water are being used to keep the side cool. *(Courtesy of District Chief Chris E. Mickal, New Orleans Fire Department— Photo Unit)*

inserted to knock down the main body of fire. The firefighter inserting the nozzle should be equipped with full protective gear, including SCBA, and should be continuously wetted down with water as the steam and heat coming from the hole will be severe. The water inserted through the hole should be restricted to the amount necessary to effect control. The nozzle should be kept in the hole in the event of a flare-up, but it is best that the seat of the fire be reached by removing cargo. (See Figure 7.18.)

Water should be played onto the side of the ship to cool down any hot spots developing there in order to prevent the plates from buckling. If a hot spot develops, it is good practice to cut a hole in the side of the ship at the hottest spot and insert a nozzle, providing of course that the hot spot is located well above the water line. A hole cut in the hull too close to the water line could cause a problem if the ship develops a list. As with cutting a hole in the deck, the personnel operating this nozzle should be wetted down continuously with a spray stream to protect them from the heat and steam coming out of the hole. It may be advantageous to increase the size of the hole to permit personnel to enter if it appears that the seat of the fire can be reached from this location. If the hole is increased to an adequate size, it may be possible to remove sufficient cargo through the hole to develop a good working area inside the hold. If this is attempted, it should be remembered that safety to personnel always should be given first consideration.

If the fire is severe and extensive, it may be necessary to use a large amount of water to gain control. This procedure, as well as that of flooding a hold, should be used only as a last resort. The overall effect is total loss to the entire inventory of the hold, but perhaps more important is the possibility that an unequal absorption of water by the cargo could cause the ship to list and possibly capsize. It should be kept in mind, however, that the responsibility for the stability of the ship is not that of the fire officer but rather that of the ship's officer who is in charge. The ship's officer should constantly be kept aware of the amount of water that is being directed into various parts of the hold. The Incident Commander should request the ship's officer to advise him or her if the ship is likely to get into a dangerous condition of stability.

Thought should be given early in the fire to the possible extension of the fire to cargo in the adjacent holds by conduction through the bulkheads. The cargo in the adjacent holds should be removed if there is any indication that such a possibility exists. This operation should be commenced as soon as possible as it may take 24 hours or more to empty the hold. After the cargo is removed, it is possible to cut a hole through the bulkhead to gain entrance into the fire hold; however, this is not always advisable. Many times large crates, machinery, and the like may be encountered once the hole is cut. Such conditions would limit the effectiveness of operations. It is always wise to check the manifest first to see whether it can be determined as to what would be found on the fire side. Of course, if hot spots appear on the bulkhead, it would be wise to cut a hole as this would permit direct access to the seat of the fire. All the precautions taken to provide safety to personnel while cutting a hole in the side of the ship should be taken if a hole is cut in the bulkhead.

Extinguishing a fire in the hold of a ship is not an easy task. Every piece of cargo in the hold must be removed. Cargo that has been burned should be placed in metal railroad cars or loaded onto trucks, wetted down, and hauled away to a place of safety. Cargo that has not been burned should be placed in a selected area of the warehouse where it can be checked for salvageable value. A fire watch should be established as a precautionary measure against hidden fire even if there is no indication of fire in this material.

Firefighters who assist in removal of the cargo are subjected to unusual conditions. They should always wear breathing apparatus during this operation even if heavy concentrations of smoke are not visible. Heated materials give off a variety of toxic fumes that are not always apparent. In addition to this, much of the work is tremendously difficult. Consequently, it is important that arrangements be made to relieve these personnel at regular intervals. It is possible that it might take several days to extinguish and overhaul a hold fire completely. In such instances, fresh crews should be brought in as often as possible.

The successful extinguishment of a fire in a hold requires careful planning and hard work. If only ordinary combustible materials are involved, the process includes five basic parts:

1. Keep the sides of the ship and the ship's deck wetted down to prevent buckling of the plates.
2. Remove the cargo until the main body of the fire can be located and extinguished.
3. Separate the burned from the unburned cargo.
4. Wet down and overhaul the burned cargo.
5. Segregate and establish a fire watch for the unburned cargo.

Hold Fires Starting at Sea.   If a fire starts in the hold while the ship is at sea, the crew will generally cover the ventilators and secure the hold into as nearly an airtight condition as possible. Crew members then will flood the hold with CO_2, halon, or steam and head for the nearest port. The result of their action is that the fire will continue to smolder as a sufficient amount of air normally is trapped in the hold to maintain this type of combustion.

Ships coming into a harbor with a fire reported in the hold should not be allowed to proceed to a berth until the conditions have been examined by a fire officer. If the fire officer determines that it is safe to do so, the ship should be directed to proceed to the berth that will provide the best facilities for effecting extinguishment. If the cargo is explosive or considered too dangerous, the ship should be anchored in a safe location away from congested areas. In this case, all firefighting will have to be done by boat crews, and the cargo will have to be loaded onto barges for transporting to a safe location. If the cargo, together with the conditions found, indicates that the risk is too high to attempt to fight the fire, the ship should be placed in shallow water in a safe location where it can be flooded. A safe location is one in which there will be no loss of life or excessive damage in the event of an explosion.

Hold Fires Containing Oxidizers.   Fighting a fire in a hold containing oxidizers such as ammonium nitrate, Chile nitrate, and sodium chlorate is more a matter of knowing what not to do than it is knowing what to do. This is one type of fire in which doing the wrong thing can be disastrous.

Ammonium nitrate, sodium nitrate, and similar materials are oxidizers. Oxidizers provide their own oxygen when heated; consequently, fires in these materials cannot be extinguished by excluding the oxygen by battening down the hatches and covering the hold ventilators. Thus, when these materials are involved in a hold fire:

1. *Do not* batten down the hatches.
2. *Do not* cover the hold ventilators.
3. *Do not* flood the hold with steam.
4. *Do not* flood the hold with CO_2.

The most effective method of fighting fires in which these materials are all or part of the cargo is to apply large quantities of water. The hatch and the ventilators should be kept wide open so that pressure is not allowed to build up. Serious consideration should be given to towing the ship to a **safe location** in shallow water if the fire is severe or if the amount of material involved is extensive. The application of large quantities of water should be continued during the towing process. The hold should be flooded once the ship is secured in shallow water.

Fire in the Machinery Spaces.   Fires in the machinery spaces can be caused by a number of things such as an improperly placed oil burner, a broken oil line, removal of the wrong oil strainer cover, dry firing of the fire box, and other acts of carelessness. The problem usually remains small if the fuel oil pump is shut down immediately; however, it can become serious if this action is not taken. A number of things have to be done in the event of a fire. Most of them require action on the part of the crew. The induced and forced ventilation of the engine room must be stopped, the fuel pump stopped, the main engines stopped, and the skylights over the engine room opened.

Serious fires in the machinery space can present some unusual problems to firefighters. The most logical statement that can be made regarding these fires is that there is no one best method for extinguishing them. There are too many variables for any hard and fast rules to be made. However, several methods have been successfully used that can be tried.

The engine room of a ship is like a basement several stories high. The normal entrance into the engine room is through a watertight door on the main deck level. The first lines normally will be laid to the front of this entrance if there are indications that there is a well-involved fire in the engine room. Usually the first attempt is to advance a line down into the engine room from this location; however, in most cases the extreme heat and smoke will prevent advancement. If this happens, it will then be necessary to try to make entrance into the engine room through the shaft alley.

The **shaft alley** is a fairly large tunnel that houses the propeller shaft. It extends from the engine room to the propeller. Entrance is made into the shaft alley through

safe location This is in reference to a ship that enters a harbor when the cargo is explosive or considered too dangerous. The ship should be anchored in a safe location away from congested areas.

shaft alley A fairly large tunnel that houses the propeller shaft of a ship. It extends from the engine room to the propeller.

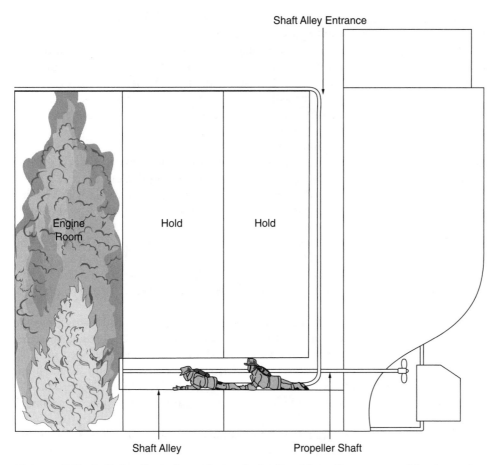

Shaft Alley Entrance

Engine
Room

Hold

Hold

Shaft Alley

Propeller Shaft

FIGURE 7.19 ◆ Bring lines through the shaft alley. Note: Some cargo ships have the superstructure and engine room in the aft section of the ship. With this arrangement, access to the engine room is not available from the shaft alley.

the **shaft tunnel hatch.** This hatch is found in the aft of the ship. This vertical, tunnel-like opening extends from the main deck to the shaft alley. (See Figure 7.19.)

Firefighters entering the shaft tunnel hatch should always wear breathing apparatus. This makes it a tight fit but breathing apparatus is necessary because there is no assurance that the shaft alley is clear of smoke and toxic fumes. If the space is too small, it probably will be necessary to abandon the turnout coat while advancing the line. The turnout coat should be put on, however, before entrance is made into the engine room. A 2½-inch line and a foam line should be extended down the shaft alley, into the shaft alley, and advanced to the watertight door to the engine room. It is a good idea to equip the 2½-inch line with a spray nozzle.

The watertight door to the engine room can be opened from both the engine room and the shaft alley. Some of the doors swing out and some open vertically. It is easy to determine which way the door will open.

The door should be opened slowly and cautiously. Gloves should always be worn because the handle may be hot. It may be necessary to cool down the handle before it is operated. If it is possible to enter the engine room, then it is usually better to use the

shaft tunnel hatch A vertical, tunnel-like opening that extends from the main deck of a ship to the shaft alley.

2½-inch line to control the source of the flammable liquid and cool down the heated metal. The fire can then be extinguished by using either water or foam.

It is well to remember that there probably will be fuel oil tanks located behind the bulkheads that are adjacent to the engine room. The fuel oil in the tanks may become heated if the fire is extensive. This usually presents no hazard unless the vents become clogged. If the tanks are full and the riser in the vent pipe is short, there may be some fuel oil forced out of the tank through the vent. The amount probably will be small and the problem can be easily controlled.

If entry cannot be made into the engine room from the shaft alley, a secondary method can be used. In fact, some departments may prefer to try this method prior to attempting the method previously outlined. With this method, all doors leading into the engine room should be closed, including the one from the shaft alley. The engine room ventilators and skylights should be covered. A 1½- or 1¾-inch line should be used to keep the ventilators and skylight housing cool. Another 1½- or 1¾-inch line should be advanced through the shaft alley to the opening into the engine room. Two 1½- or 1¾-inch lines equipped with spray nozzles should then be lowered approximately 15 to 20 feet into the engine room through the skylight housing. It will be necessary to remove the skylight tarpaulin while this is being done, but the tarpaulin must then be replaced.

Water should be supplied to the two lines as soon as they are lowered into the engine room. It probably will take 30 to 45 minutes before the fire is extinguished and the area cooled sufficiently for the line in the shaft alley to be advanced into the engine room. The covers should be removed from the ventilators and the skylights opened prior to the shaft alley line being advanced. If this method is successful, the 1½- or 1¾-inch line will be needed only to extinguish any glowing embers in Class A material.

A third method is to attempt to extinguish the fire by using the CO_2 system. If this system is employed, it must be done within a very short time after the fire has been discovered. Delay in operating the system may allow the metal in the immediate vicinity of the fire to reach a temperature above the ignition temperature of the burning oil. If this happens, there undoubtedly will be a reignition of the oil once the CO_2 is dissipated.

The boiler fires should be extinguished by the crew and the auxiliaries stopped prior to activating the system. The ventilation system should be shut down, all openings into the engine room closed, and all personnel ordered out of the engine room.

Once the engine room is flooded with CO_2, it should not be opened (except in an emergency) for some time. It is necessary to give the burning substances a chance to cool below their ignition temperature. It is good practice to take a 1½- or 1¾-inch line into the engine room and cool down the hot metal once the room is opened.

PASSENGER SHIPS

Fires aboard ships that carry passengers can present a serious life problem if the passengers and crew are aboard. It is possible that the passengers could panic and all run to one side of the ship. This action is capable of capsizing the ship. A smoky fire below decks can fill the passageways with heat and smoke, making it difficult for passengers to find their way to safety and making it difficult to locate the seat of the fire.

The passenger spaces on a large passenger ship are somewhat like a hotel but present many more problems. First of all, the ship's "hotel" is made of metal. There are hundreds of small rooms and some very large rooms served by a number of passageways. The rooms on the outside are equipped with portholes that can be used to ventilate the rooms but are too small to enter. The interior rooms have no direct access to the outside but are equipped with air ventilators. The headspace is less than in a room in a hotel. This results in the heat condition being kept closer to the deck (floor) in the event of a fire. All of these factors assist in complicating the problem.

The Incident Commander should contact the ship's officers to determine the location of the fire and the probability of what is burning when first arriving at a fire aboard a passenger ship. If possible, the Incident Commander should get a member of the ship's crew to guide the firefighters. Use of 1½- or 1¾-inch lines will usually be sufficient to knock down fires in passenger staterooms, the galley, the paint locker, and so on.

Fires in the passenger spaces may be on, above, or below the main deck. It is best that the initial lines be brought in from the main deck when the fire is on or above the main deck. Care must be used when taking these lines aboard so as not to block passenger gangplanks. The gangplanks may be needed to clear passengers from the ship. It is usually best to raise aerial ladders to the bow or stern or both for the purpose of advancing these lines.

If the fire is below the main deck, it may be best to enter the ship through the cargo port on the side of the ship. These doors are at approximately wharf level and are connected to the wharf by a wide gangplank.

The age-old rule of "never opening a line until the fire is seen" may have to be abandoned when advancing lines aboard ship. It may be necessary to use spray or fog streams to act as an absorbing shield to enable firefighters to reach the seat of the fire. This action will cause little damage because the bulkheads and decks in the passageways are made of metal.

Firefighters should always wear breathing apparatus and always work in pairs. The passageways are long and visibility is usually very poor. This makes it difficult for those who are not familiar with the ship to find their way around. Consequently, firefighters should not drop their line and search for the fire as the hose line may present the only logical means for them to find their way out in the event they get lost.

All sections of the ship below, above, and to each side of the fire should be checked once the fire is knocked down. A thorough check should be made of the heating and ventilating systems. These systems extend through all decks—often through combustible partitions, false ceilings, closets, and so on. The fire could extend through these systems to inaccessible spaces. A rekindle after the ship has sailed could seriously endanger the lives of the passengers and crew.

TANK SHIPS

A tank ship (tanker) is constructed differently from a cargo ship. Although there is no standard tank ship, certain features are found on most of them.

On most tankers the bos'n and paint locker are found in the bow. Behind these usually is found a dry cargo hold used to hold miscellaneous ship's gear. Under the cargo hold are two **deep tanks** that are used to trim the ship. These tanks do not ordinarily carry oil. From the deep tanks aft are a number of storage tanks for carrying the

deep tanks Tanks that are located under the cargo hold of a cargo ship that are used to trim the ship.

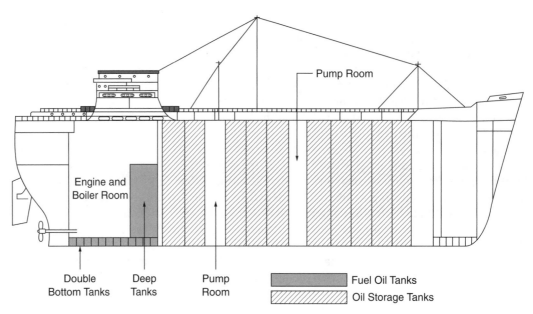

FIGURE 7.20 ◆ The layout of a tank ship.

liquid cargo. The number of these tanks will vary according to the size of the ship and its carrying capacity. The machinery spaces are located aft. The pump room will be located somewhere between the deep tanks and the machinery spaces, sometimes directly in front of the engine room. The engineer's and crew's quarters are directly above the machinery space, and the captain's and mates' quarters are located amidships under the bridge. (See Figure 7.20.)

It should be noted that the engine room is located in the stern rather than amidships. This means that there is no shaft alley that can be used to gain entrance into the engine room in the event of a fire in this location. Fighting a fire in the engine room will have to be done by using one of the other methods described for fighting a fire in the engine room of a cargo ship.

Fires in the galley, paint locker, crew's quarters, and so on aboard a tank ship are handled the same as a fire in similar spaces aboard any ship.

The most serious problem of a fire on a tanker is involvement of the flammable liquid or the potential involvement of flammable liquid. Consequently, there are certain steps that should be taken once it has been determined what is burning, its location, and the extent of the fire. The first is to establish a fire area. All tanks, hatches, and ports should be closed within the fire area. Ullage hole covers, Butterworth plates, tank tops, ventilator ducts, and all other vents not already closed should be closed and secured throughout the ship. Additionally, all electrical circuits in the vicinity of the fire should be de-energized. If any loading or unloading operations are taking place, or if any tanks are being cleaned, these operations should be closed down, all valves closed, and all hose disconnected.

Fires in Cargo or Bunker Tanks.   The first attempt to extinguish fire in these tanks should be made by excluding the air and suffocating the fire. This requires that all openings into the tank be closed. It is best that the ship's crew be used for this purpose because crew members are familiar with all openings to the tank. After the tank has been made airtight, a follow-up that can be taken is to utilize the ship's steam-smothering or CO_2 system. The steam-smothering valves on branch lines to pump rooms, cofferdams, or tanks not on fire should be shut down prior to activating the system.

In some instances it may not be possible to close off all openings to the tank. In these cases foam probably will be the only effective agent for extinguishing the fire. Water can be used sparingly from spray or fog nozzles to precool the liquid, but should not be used once the application of foam has commenced. It should be remembered, however, that water should not be applied if heavy materials such as asphalt or tar are involved, as this could result in a very severe and drastic explosion.

Fires on the Deck.   Fires on the deck can be the result of a tank overflowing or they can be fed by a broken transfer hose or leaking pipeline. If the fire is being fed from a supply source, the flow fuel should be stopped if at all possible. Regardless of the source of the fire, all tank openings throughout the vessel and the pump rooms should be closed as quickly as possible.

If possible, fires on deck should be attacked from the windward side. This direction of attack will have a twofold benefit. First, it will carry the flames, smoke, and fumes away from the attacking firefighters. Second, it will help carry the extinguishing agent into the fire.

Normally the best extinguishing agent will be foam. AFFF is preferred. The foam should be played against the nearby vertical structures and be allowed to run down and flow smoothly over the fire. It also can be applied to the deck ahead of the fire.

Foam is not particularly effective on flowing liquid fires as the movement of the liquid will prevent the formation of an airtight foam blanket. Under these circumstances, if the supply of fuel cannot be shut off and it is burning at its source, portable CO_2, dry chemical, or water fog probably will be effective when applied directly on the fuel as it emerges from the opening. It is good practice to play foam ahead of the flowing fuel to form a dam and thereby prevent the spreading of the fire. Water from spray or fog nozzles can be used to cool the surrounding structures of the vessel and to protect the firefighters using the foam. However, the water should be used in such a manner as not to affect or break up the foam blanket or dam.

◆ INCIDENT COMMAND SYSTEM CONSIDERATIONS

The Incident Command System for ship fires is different from that for structure fires. First, the Command position normally will be a unified command rather than an individual Incident Commander. As a minimum, the unified command generally will consist of the fire department, the vessel, and the Coast Guard. While the fire department will have the primary decision powers for firefighting operations, the chief officer representing Command no doubt will need the advice of the ship's officer and the Coast Guard on several issues.

FIGURE 7.21 ◆ Ship fire ICS organization.

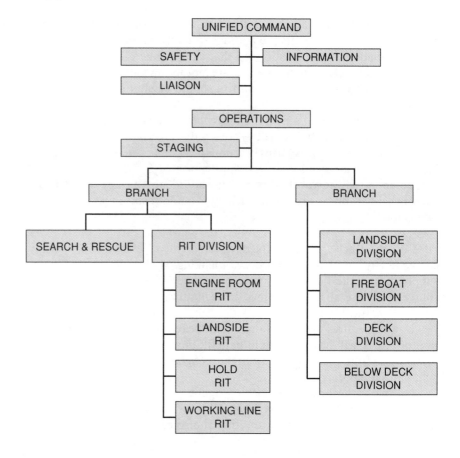

The rapid intervention team concept will also be expanded on a serious ship fire. Several teams will be required depending on where firefighters are operating. There should always be a RIT (or RIC) stationed at the entrance to any location where lines have been taken. There should also be a RIT based on shore. All of the RITs should be under the command of a RIT officer who answers to a branch chief. A sample Incident Command System organization for a ship fire is shown in Figure 7.21. The actual organization used on a ship fire will depend on the type of ship involved and the location of the fire or fires.

◆ **SUMMARY**

Whether a car fire or an airplane incident, firefighters may encounter fire emergencies that are not routine situations. Small boat fires may seem like an easy task to handle, but firefighter safety and the utilization of the Incident Command System still must be included in any given incident action plan. Who will be the Incident Commander and the location of the command post as well as having a rapid intervention team ready, are as important for these types of fire emergencies as they are for others. The nature of these fire incidents dictates that firefighters must be as trained and equipped to handle any mobile fire emergency as they would any structure fire in their district.

Review Questions

Car Fires

1. What is the primary problem facing firefighters in a car fire?
2. What is one of the first precautionary measures that should be taken at a car fire?
3. How should a fire under the hood of a car be extinguished?
4. How should the attack be made on a fully involved car fire?

Truck Fires

1. What is the primary problem involving truck fires?
2. What tactics should be used to fight a truck fire?

Aircraft Fires

1. What tactics should be used to extinguish a fire in a piston engine of an aircraft when the fire is confined within the engine nacelle?
2. How are fires within the combustion chamber of jet engines best controlled?
3. What method should be used to extinguish a fire outside of the combustion chamber of a jet engine but within the engine nacelle if the fire cannot be extinguished with the aircraft's built-in extinguishing system?
4. What area should be avoided by the apparatus driver when responding to a wheel fire in an aircraft?
5. On large commercial aircraft, where is the most stable point to initiate an attack on a fire inside the aircraft?
6. What are some of the precautions that should be taken when approaching an aircraft crash?
7. From what direction is it generally best to make the approach to an aircraft crash?
8. What precautionary measures should be taken before cutting into an aircraft for the purpose of making rescues?
9. If rescue efforts are to be made while an aircraft is on fire, what is usually the best method of making the initial attack?

10. When making an initial attack on an aircraft for the purpose of making a rescue, which area of the aircraft should be hit first with the foam lines?
11. Which firefighters in the rescue team perform the rescue work?
12. How can foam lines be used to protect the members of the rescue team?

Small Boat Fires

1. How should a fire in the cabin or superstructure area of a small yacht be attacked?
2. What extinguishing agents are effective if the fire around the engine of a small boat is contained under the floor plates and has not extended too far?
3. What action should be taken in a small boat fire if there is a considerable amount of unburned gasoline in the bilge after the fire is extinguished?

Ship Fires

1. What are the three differences between a fire aboard ship and one in a structure on land?
2. Where are the controls for the steam and the CO_2 systems generally found on a cargo ship?
3. Where should the first responding pumper to a cargo ship fire with smoke showing generally be spotted?
4. What is the first job of the first arriving officer at a cargo ship fire?
5. How are hose lines taken aboard for a fire in the hold of a cargo ship?
6. What action should be taken when hot spots develop on the deck of the ship when there is a fire in the hold?
7. When should the hold of a ship be flooded if there is a fire in the hold and the hold contains regular cargo?
8. What are the five basic parts to the successful extinguishment of a fire in a hold that contains ordinary combustible materials?
9. What action should be taken by the Incident Commander regarding a fire coming into the harbor when the fire started in the hold while at sea?

10. What three techniques *will not* extinguish amononium nitrate fires?

11. What four things should *not* be done when fighting a fire in a hold that contains oxidizers?

12. To what point should the first lines normally be laid for a fire in the engine room?

13. Where should entrance to the engine room be made if it cannot be made through the door on the main deck of a cargo ship?

14. Where is the entrance into the shaft alley made?

15. What method should be employed to extinguish the fire if entrance into the engine room cannot be made through the shaft alley?

16. What should the Incident Commander do when first arriving at a fire aboard a passenger ship?

17. What is the best method of bringing lines aboard ship if the fire is below the main deck on a passenger ship?

18. What should be the first attempt to extinguish a fire in a tank on a tank ship?

19. What method should be used to extinguish a flowing liquid fire on the deck of a tanker?

Special-Interest Fires

8 CHAPTER

Key Terms

Objectives

The objective of this chapter is to introduce the reader to firefighting tactics as they apply to flammable liquid fires, flammable gas emergencies, electrical fires, silo fires, wildfires, brush fires, and pier and wharf fires. Upon completion of this chapter, the reader should be able to:

- Explain the tactics used to extinguish flammable liquid spill fires.
- Explain the tactics used to attack vertical storage tank fires.
- Describe the tactics used to attack horizontal storage tank fires.
- Explain the tactics used to control breaks in flammable liquid transportation lines.
- Explain the tactics used to extinguish fires in oil tanks.
- Discuss the control tactics used on LPG emergencies.
- Quote the basic principle of fighting fires in combustible gases.
- Explain the meaning of the term BLEVE.
- Explain the methods of extinguishing electrical fires.
- Discuss the use of water on electrical fires.
- Describe the methods of fighting fires in various types of silos.
- List the classifications of wildfires.
- Define origin, head, fingers, rear, flanks, perimeter, island, spot fires, slopover, flare-up, blowup, firestorm, green area, and black area as related to wildfires.
- Discuss the various factors affecting fire behavior at wildland fires.
- Discuss the various fuel types common to wildland fires.
- Discuss the weather factors that affect wildland fires.
- Define relative humidity, local winds, sea breeze, offshore breeze, surface winds, gradient winds, and gravity wind.
- Define slope, aspect, chute, saddle, and canyons as related to wildland fires.
- Explain the tactics used on direct attacks on wildland fires.
- Explain the tactics used on indirect attacks on wildland fires.
- Explain the difference between backfiring and burning out on wildland fires.
- Discuss air operations at wildland fires.
- Discuss the mop-up procedures used on wildland fires.
- List the 10 standard firefighting orders for wildland fires.
- List the 18 "watch out" situations for wildland firefighting.
- Discuss the organization chart for the Incident Command System for wildland firefighting.
- List the factors that need to be considered in fighting brush fires.
- Explain the tactics used to protect a structure at a brush fire.
- Discuss the tactics used for extinguishing a brush fire.
- List some of the personnel considerations for use at brush fires.
- List some of the apparatus considerations for brush fires.
- Explain the difference between a dock, a pier, and a wharf.
- Explain the best method of attacking the underside of a wharf fire.

Special-Interest Fires

FIGURE 8.1 ◆ Flammable liquid fires. This is a fire in a large storage tank in a refinery. This type of fire requires the use of many heavy streams. *(Courtesy of Rick McClure, LAFD)*

FIGURE 8.2 ◆ Flammable gas fires. A flammable gas fire in a transmission line caused by an earthquake. *(Courtesy of Rick McClure, LAFD)*

(Continued)

FIGURE 8.3 ◆ Wildfires—West Glenwood, Colorado, June 11, 2002— An extensive backfire has been set by the Flathead Hotshot crew on Horse Mountain. *(Photo by Andrea Booher/FEMA News Photo. Courtesy of FEMA)*

FIGURE 8.4 ◆ Brush fires. A fast-moving brush fire. *(Courtesy of Rick McClure, LAFD)*

◆ INTRODUCTION

The title of this chapter is somewhat misleading. It almost sounds as though the firefighting practices to be discussed are of less importance than those outlined in previous chapters. This is not the case. Many of the fires covered in this chapter are extremely challenging to firefighters. In fact, a number of firefighters have been killed and others injured during the violation of some of the basic safety principles outlined in this chapter. The information contained here should be thoroughly understood by those involved in the practice of firefighting—not only for the purpose of doing a more effective job in fire extinguishment but also for the purpose of establishing practices of personal safety.

FIGURE 8.5 ◆ Silo fires.
(Courtesy of Code 3 Images/Steven Townsend)

◆ FLAMMABLE LIQUID FIRES

From a technical standpoint, a **flammable liquid** is a liquid that has a flash point below 100°F. Those liquids having a flash point of 100°F or more are referred to as **combustible liquids.** However, for the purpose of discussing firefighting principles, both will be considered as flammable liquids. Oil, however, will be excluded from this definition. (See Figure 8.6.)

flammable liquid   A liquid that has a flash point below 100°F.

combustible liquid   A liquid that has a flash point of 100°F or more.

FIGURE 8.6 ◆ Flammable liquid fires in refineries offer different types of challenges for firefighters.
(Courtesy of District Chief Chris E. Mickal, New Orleans Fire Department—Photo Unit)

As discussed in Chapter One, flash point refers to the temperature at which a flammable liquid will start giving off sufficient vapors to burn. Consequently, it can be said that vapors will be given off that will burn any time the flash point is below the ambient temperature. Those liquids having a flash point above the ambient temperature will have to be heated before they will give off vapors that will burn. In some cases, this concept is a factor that should be considered when determining the way in which a fire in a flammable liquid will be attacked.

Another concept of flammable liquids that has an impact on firefighting procedures is whether the flammable liquid is **miscible.** If it will mix with water, it is miscible; if it will not, it is not miscible. A good portion of the flammable liquids, including gasoline, encountered in firefighting are not miscible. These liquids are referred to as **hydrocarbons.** The overall effect is that if water is used as the extinguishing agent on a hydrocarbon fire, the flammable liquid will float on top of the water and continue to give off vapors once the fire is extinguished.

Those flammable liquids that are miscible are referred to as **polar solvents.** Water mixed with polar solvents will increase their flash points and reduce the hazard. However, conventional foams cannot be used to extinguish fires in these liquids because the liquid will destroy the foam. Alcohol is an example of a polar solvent. Polar solvents require special foams for extinguishment.

miscible Capable of being mixed with water.

hydrocarbons Those flammable liquids that are not miscible.

polar solvents Those flammable liquids that are miscible.

SPILL FIRES

One characteristic of a flammable liquid spill that finds a source of ignition is that the resultant fire will propagate over the entire spill area at an amazing rate of speed. Consequently, special precautions should be taken when approaching a flammable liquid spill fire or, for that matter, a flammable liquid spill that has not yet become ignited. If possible, approach the fire or spill from the windward side and from the uphill side if the terrain is sloping. This is particularly important if ignition has not yet taken place. The vapors capable of being ignited may be invisible, which could place the apparatus in the vapor area or possibly in the spill area itself if approach is made from another direction. It does not take much imagination to visualize the potential result if ignition occurs while the apparatus is so located.

When referring to a flammable liquid fire, the size of the fire is determined by the surface area involved, not the amount of fuel available to burn. If the surface area is small, the fire is small. If the surface area is large, the fire is large.

If the fire is small and not being fed from a continuous source, it may be possible to extinguish it with a few shovels of loose dirt. If dirt is not available, or the fire is too large to handle in this manner, a carbon dioxide or dry chemical extinguisher may prove to be effective. As a general rule, a handheld carbon dioxide extinguisher can be used on fires up to about 100 square feet of area. The dry chemical extinguisher is preferred in either case because it not only is capable of knocking down the fire quicker but also has the advantage of being effective at a greater distance from the fire.

Fires larger than 100 square feet should, if possible, be attacked with foam. The most useful foam for hydrocarbon flammable liquid fires is AFFF. Polar solvents require special foams.

The attack should be made by directing the foam ahead of the fire and allowing it to spread over the fire. If vertical surfaces are available behind or to the side of the fire, the foam can be projected onto these surfaces and allowed to run down and out onto the fire. (See Figure 8.7.)

An even more effective method of knocking down a large flammable liquid fire is the combined use of dry chemicals and foam. Of course this method is limited to those departments that have the capability of discharging large quantities of dry chemicals

FIGURE 8.7 ◆ Foam being used on a flammable liquid spill from an overturned gasoline truck.
(Courtesy of District Chief Chris E. Mickal, New Orleans Fire Department— Photo Unit)

on a fire. The attack on the fire should be made initially with dry chemicals, which could possibly result in complete extinguishment of the fire. This attack needs to be followed immediately by completely covering the flammable liquid with foam to prevent a reignition. Failure to follow through with the foam blanket could result in a more critical situation than the original fire.

If foam is not available, it may be necessary to extinguish the fire by using water. Water can have a dual effect, depending on the flash point of the material involved. The water may have the effect of cooling the liquid below its ignition temperature, or it may be used to separate the liquid from the flame. A general rule is that those liquids having a flash point above 100°F can possibly be extinguished by cooling, whereas those having flash points below 100°F cannot.

Attacking a spill fire with water will result in the fire being pushed away from the direction of attack. For example, if the fire is attacked from the north, the fire will be pushed toward the south. It is important that the safest direction for pushing the fire be determined prior to commencing the attack. If exposures are involved, care must be taken to see that lines are laid to protect them prior to the attack being initiated. It should also be remembered that almost all hydrocarbon liquids are lighter than water. Consequently, any unburned liquid will float on top of the water after the fire has been extinguished. This could present the potential hazard of a reignition.

One of the better methods of making the attack on the fire is through the use of at least two spray nozzles on 1½-, 1¾-, or 2½-inch lines backed up by one or more similar protective streams. (See Figure 8.8.) The size and number of lines required will depend on the size and extent of the fire. The attack should be made from a direction that will push the fire to the least hazardous area.

A careful analysis must be made prior to the attack as to the amount of water available. The attack should not be started if there is not a sufficient amount of water available to complete the job. If extinguishment cannot be completed, the overall effect would most likely be a spread of the fire, which would intensify the overall problem.

In some cases, the most effective method of handling a spill fire is to let it burn itself out. This method could most likely be used in those instances in which there is doubt that the water available can do the job and no exposures are involved.

In many situations, the spill fire is being fed from a continuous supply source. The magnitude of the problem will be affected by what is burning, the size of the fire, the potential for its spread, and the number and type of exposures. The solution to the problem will include the need to protect the exposures, protect the supply, stop the spread, and extinguish the fire.

The exposures could include a variety of types, including storage tanks or other flammable liquid facilities. (Exposure protective tactics for storage tanks and other flammable liquid facilities will be discussed later in the chapter.) Standard procedures for protecting exposures should be employed for other types of exposures. The key is to use a sufficient number of lines to complete the task effectively. In some cases it may be necessary to use hose lines to divert the fire to a safer direction of extension.

Shutting off the fuel supply to the fire should be given early consideration. Shutting off the flow generally requires that a valve be closed or a hole plugged. This action requires a coordinated effort by several firefighters. The team assigned the task of shutting off the flow should consist of a minimum of three firefighters. The two outside firefighters should advance 1½- or 1¾-inch lines equipped with

FIGURE 8.8 ◆ A good method of attacking spill fires.

spray tips. The inside firefighter is usually an officer. This person is responsible for making the shutdown. When advancing, the inside person maintains the integrity of the team by keeping one hand on the inside shoulder of each of the nozzle handlers. (See Figure 8.9.) The advance should be made from the direction that would best accomplish the task while providing the maximum safety to the firefighters. The wind direction, the terrain, and other factors that will influence the potential spread of the fire and the safety to personnel should be kept in mind when making the decision as to the direction of attack. It is good practice to provide a backup line or two as a safety factor for the advancing shutoff team.

Stopping the spread of the fire caused by flowing fuel can become a tricky job. The ideal is to dam up the area directly in the path of the flow direction. The heat given off by the fire is normally too intense to permit working close to the fire's edge. In these instances it is best to work some distance ahead of the fire and to use sand, dirt, or other noncombustible material to provide a dam. It is a good idea to request several loads of dirt or sand through the dispatch center if a sufficient amount is not at the scene of the emergency. It may be possible that the flow can be shut off before the requested material arrives on the scene; however, it is also possible that the source of the flow cannot be stopped. It is much better not to need the material than to suffer the consequences of the fire spread. Incidentally, if the

FIGURE 8.9 ◆ A typical approach to shut off a valve.

fire is located close to a road, it is not a bad idea to place a firefighter there to watch for truckloads of sand or dirt that might be in transit to a construction site. Confiscating this material could prove to be extremely beneficial to the control of the flow problem.

If it becomes impossible to stop the flow while attempting to shut off the supply, it may be necessary to divert it to a safer area. Shovels can be used ahead of the fire to provide flow paths, or hose lines can be used to change the direction of the flow.

Extinguishing the fire can be done simultaneously with protecting the exposures, stopping the flow, and stopping the spread, or at times it may be necessary that extinguishment be delayed until some of the other requirements are fulfilled. Regardless, extinguishment operations should be conducted as previously discussed.

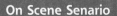

TANK FIRES

The problems of fighting fires in flammable liquid storage tanks will vary depending on the standard variables of weather, tank size, material burning, and so on. Each of these must be considered when making a size-up of the fire. However, when consid-ering the problem of exposures, there is a considerable difference in the hazard when the exposures are horizontal storage tanks than when the exposures are vertical stor-age tanks.

Generally, a fire threatening a flammable liquid storage tank is the result of a spill from a storage tank or transportation line becoming ignited, or from a fire in an adjacent tank. Consequently, the problem is initially the same regardless of whether the exposed tanks are vertical tanks or horizontal tanks.

The problem primarily will be one of protecting the exposures and extinguishing the spill fire if the fire is confined to a **diked area.** Normally, the exposure problem will be eliminated once the spill fire is extinguished. However, the problem of protecting the exposures may be continuous if the fire is being fed by a break in a supply source or as a result of a fire in an adjacent tank. In case of a break in a tank, the tank feed-ing the fire in the diked area is itself an exposure.

diked area An area around a storage tank that has been constructed to contain any spill from the tank.

If the liquid has pushed its way out of the diked area, or no diked area is involved, the problem becomes one of protecting the exposures and extinguishing the spill fire. In some cases, the spill area will include pipe trenches.

The extinguishment of spill fires that expose storage tanks follows the same basic principles for extinguishing spill fires in which storage tanks are not involved. The basic difference is the extinguishment of fires in the tanks themselves and the protection of exposures.

Many times the spill fire is so large that it is impossible to extinguish it with the foam carried on a single engine company. In this case it may be better to cool the tank or exposure until a sufficient amount of foam is on the fire ground to extinguish the fire. If it is necessary to depend on the foam carried on engine companies, then the sit-uation may require several engine companies to accumulate a sufficient amount of foam to do the job.

Vertical Storage Tanks. Fires may occur in vertical tanks or the tanks may be exposed to spill fires. The best method of extinguishing the fire in the tank will depend on what is burning, the size of the tank, and the construction of the tank top. If the roof of the tank is gone, it may be possible to extinguish the fire by using water if the tank is small and a hydrocarbon liquid is involved. This is particularly true if the entire surface area of the fire can be covered with water fog at the same time. Care has to be taken, however, to limit the amount of water because too much water could cause the tank to overflow. If a considerable period of time has elapsed between the ignition of the fire and the application of water, it will be necessary to cool down the inside exposed surface of the tank to prevent reignition.

If the tank is too large to extinguish the fire by using water, the application of foam will prove most effective. If the tank has its own foam-extinguishing system, it should be used. If the system is not available on the tank, then foam applicators will have to be employed. Although it is possible to apply the foam through handheld hose lines, in most cases this is neither practical nor feasible.

In many instances it is not possible to extinguish the fire and it has to be left to burn itself out. Occasionally it is possible to start discharging the fuel in the tank to other storage tanks through the tank's distribution system. This will cut down the burning time and reduce the overall fire loss. The firefighting problem becomes one of protecting the exposures and keeping the burning tank cool whenever a fire is allowed to burn itself out.

It is important to understand the construction of a vertical tank to protect it adequately as an exposure. Vertical tanks usually are designed so that the weakest part of the tank is the roof. The roof of the tank will normally give way under a pressure buildup within the tank. This action will send the roof skyward. The distance it travels will depend on the degree of the pressure buildup and the size of the roof. This action could rupture adjacent tanks and expand the overall problem. It always presents a serious hazard to firefighters.

Records show that on a number of occasions the base of the tank rather than the roof releases on a pressure buildup. This action sends the entire storage tank into the air, causing it to react as a missile.

If the tank has a dome roof and is burning at the vent, it is usually a simple task of extinguishing the vent fire. This may be accomplished by using dry chemicals or water in spray or fog form.

Fortunately, most of the modern vertical storage tanks have floating roofs. Fires in these tanks generally occur in the seal area between the tank and the floating roof. These fires generally can be extinguished through the proper use of light water or foam. Firefighters must be extremely cautious if an attempt is made to extinguish these fires with the use of water. It is possible to sink the floating roof. This will usually result in a simple fire developing into a fully involved tank fire with its magnitude of problems.

The hazard to an exposed tank depends to a considerable degree on the amount of fuel in the tank. The hazard is much greater to a partially filled tank than it is to a full one. The principle involved might be illustrated by the use of a paper cup holding water. The water in the cup can be heated to its boiling point from an outside source without damaging the cup as long as the heat is applied to an area of the cup containing liquid. The water absorbs the heat and the cup remains intact. However, once the water boils away, the fire will heat the vapor in the cup, the cup will absorb the heat, and the cup will ignite.

Fires in flammable liquid storage tanks may burn for several days if allowed to burn themselves out. Large amounts of water should be applied to the burning tank and to all exposed tanks during the entire burning period. If may be necessary to set

FIGURE 8.10 ◆ Tank car fires should be considered as horizontal storage tanks. *(Courtesy of District Chief Chris E. Mickal, New Orleans Fire Department— Photo Unit)*

up portable monitors both for protective streams and to cool the tank on fire. This not only relieves the strain placed on firefighters but also makes it possible to continue to protect the exposures in the event that a staffed position has to be abandoned temporarily due to an unexpected flare-up or other emergency.

Horizontal Storage Tanks.   For the purpose of this discussion, the term **horizontal storage tank** includes tank trucks and railroad tank cars as well as permanently installed tanks.

Fires seldom occur in horizontal storage tanks. Occasionally, fires will occur at the vents, but extinguishment of these fires is usually a simple matter of cooling down the tank. This action will reduce the pressure buildup in the tank and close the pressure relief valve. The most pressing problem with horizontal storage tanks occurs when they are directly exposed to a fire. The exposure problem to these tanks is much more critical than it is to vertical tanks. (See Figure 8.10.)

The weakest part of these tanks is the ends. There is a possibility that the ends will give way if pressure builds up within the tank faster than it can be released. The tank literally becomes a rocket when this happens if the released fuel is ignited as it escapes. The tank will be projected a considerable distance, with fire flaming out the rear end. Two examples are worth noting.

A number of years ago in Kansas City, Kansas, the end of a horizontal tank gave way while firefighters were fighting an extensive flammable liquid fire. The reaction caused the tank to break from its mounting and to rocket 94 feet ahead. During its progress it went through a 13-inch brick wall and also knocked down another wall of a brick service station building. The fire shooting out the rear of the tank cost the lives of five firefighters and one civilian.

Another incident occurred in Pennsylvania—this one involving a tank truck. When the ends of the tank let go, the remaining parts of the tank and truck were projected approximately 470 feet down the road where the tank smashed into a stone retaining wall. Parts of the truck cab and running gear were torn loose during the progress. After hitting the wall, the tank tumbled and rolled over the highway and through a crowded intersection. It came to rest approximately 900 feet from its original position, leaving 11 people dead in its path of travel.

horizontal storage tank   A term used to include tank trucks and railroad tank cars as well as permanently installed tanks.

A horizontal tank can become exposed to a fire as a result of a fuel spill or from a fire in an adjacent tank or structure. If it is due to a fuel spill, the spill fire can be extinguished by covering it with foam. However, if the spill fire is being fed by a break in the tank or from another source, top priority will have to be given to stopping the leak and pushing the fire away from the exposed horizontal tank. Tactics explained in the section on spill fires can be used for this purpose. Application of foam will have to wait until the leak is eliminated if the source of the leak is the horizontal tank. The reason for this is that the water used for the advancement of the shutoff team will wash away the protective foam blanket. It may be possible to use water to wash the spill to a safe area.

It is important that water be used to cool a horizontal tank any time it is being exposed to fire. It normally will take a minimum of 500 GPM applied to an exposed side to prevent a pressure buildup. Either handheld lines or heavy streams or both can be used for this purpose.

A basic principle that should not be ignored when playing water on an exposed horizontal tank is *never direct the water from the ends of the tank*. To do so would place firefighters in extreme jeopardy if an end of the tank let go. (See Figure 8.11.)

It is possible that a tank truck or railroad tank car can be involved with fire in an area where there is an insufficient supply of water to keep the tank cool, and a sufficient amount of foam is not available to extinguish the fire. In these cases it will be necessary to evacuate the area. This means that everyone within a distance of at least 3,000 feet in all directions should be removed. All available firefighters and police officers need to be used to accomplish this task. If any amount of water whatsoever is available and if time permits, it may be wise to set up a nonstaffed portable monitor to help keep the tank cool while the evacuation is taking place. This may eliminate the potential explosion, which would allow the fire to burn itself out harmlessly.

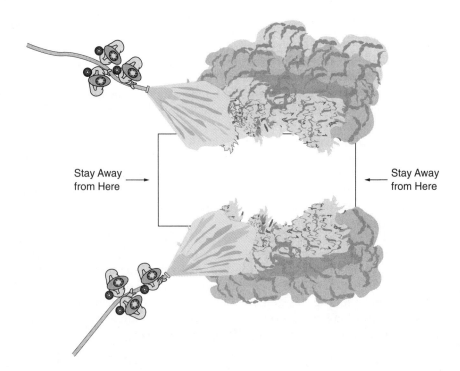

Stay Away from Here →

← Stay Away from Here

FIGURE 8.11 ◆ Never direct the water from the ends of the tank.

FIRES RESULTING FROM BREAKS IN TRANSPORTATION LINES

Occasionally flammable liquid spill fires are fed by breaks in transportation lines. At times these are above ground, but in most cases they are underground lines that have been broken by construction equipment. The extent and seriousness of the fire will depend primarily on the area covered by the spilled liquid prior to ignition and the rate at which the spill is being fed. The basic procedure for attacking spill fires should be used to extinguish the fire. The problem then becomes one of stopping the flow. The problem is somewhat simplified if a valve is readily available that can be closed to stop the flow. The method of making an approach to the valve was previously explained.

Unfortunately, in most cases when the break is in an underground line the nearest valve for shutting down the line will be located at a remote location and will have to be shut down by the company owning the line. Determining who owns the line is at times difficult, as several pipes used for transporting flammable liquids may run down the same street. In these circumstances it will be necessary to relay information to the dispatch center as to what type of liquid is involved and the distance that the pipeline is from the curb or another identifiable reference point. This information will help the dispatcher determine the owner of the line. In the meantime, there is some action that can be taken that will help minimize the problem.

If the transportation line is small and the flammable liquid flow from the line is minimal, a single 2½-inch line can be advanced to the break. For larger transportation lines and greater flows, two or more 2½-inch lines may be required. Firefighters advancing the lines should be protected at all times by large quantities of water from spray streams. Additional lines should be used to channel the escaping liquid to a safe location. It may be necessary to build dams to confine or control the flow. Once the lines have been advanced to the break, the nozzles should be opened and the lines pushed deep into the pipe. Where possible, positive shutoff nozzles should be used to eliminate the possibility of the nozzle accidentally being shut off after the line is inserted into the pipe. The objective is for the water to push the flammable liquid back into the pipe to stop the flow. Only water will be discharged from the break in the pipe if the operations are completely successful. The lines should be secured in place, all firefighters removed to a safe distance, and the operation continued until the Incident Commander has been assured by a responsible representative of the company owning the line that the pipe has been shut down and all flammable liquid removed from it.

The preceding procedure is risky. It is similar to moving in to shut off a valve feeding a flammable liquid fire; however, in this case it is on a much larger scale. If there is any doubt whatsoever in the mind of the Incident Commander that the risk is too great, it is best not to extinguish the fire, to protect the exposures, and to let the fire burn until the source of the spill can be shut off by owners of the pipeline.

OIL TANK FIRES

Fires in oil tanks present a fearsome spectacle, even to experienced firefighters. They usually are accompanied by dense black smoke intermixed with orange flames. It should be remembered, however, that this heavy black smoke actually offers protection as it screens radiant heat that might otherwise ignite other tanks or buildings in the area.

The two primary problems of oil tank fires is protecting the exposures and extinguishing the fire. Normally the initial lines are laid to protect the exposures. Once water has been played on an adjacent tank to keep it cool, the stream must be applied continuously. Intermittently cooling of an exposed oil tank could cause it to breathe, with the potential result of pulling the fire inside the tank.

Most oil tank fires occur in refineries or in similar types of installations. The first thing the Incident Commander should do when arriving at the fire is to check with the plant authorities to determine the type of oil burning. Most refineries have their own fire brigades and have predeveloped plans for coping with fires. They also generally have the equipment that will prove most effective for extinguishing the fire. Cooperation between the fire department and plant authorities usually produces effective results.

In most instances, the most effective extinguishing agent for oil fires is foam. If possible to do so, it is a good idea to spray water over the burning surface of the oil prior to applying the foam. It is a good practice to spray a small amount of water on the oil and then wait to see what happens. If no slopover occurs, the cooling can continue. A **slopover** is a small spill of oil out of the tank. If there is a slopover, it should be allowed to subside and then more water applied. This can be repeated until the oil is cool. This precooling of the oil often will reduce the amount of foam required for extinguishment and cut down the extinguishing time.

Most large storage tanks have their own permanently installed foam systems. These systems should be used, if possible. However, many times these installations are destroyed by an explosion that precedes the fire. In this case it will be necessary to use portable foam applicators. Most of these applicators are hydraulically raised and equipped with a gooseneck nozzle, which can be placed over the top of the tank and the foam applied. Almost all refineries maintain these applicators as standard equipment. If two applicators are available, the second one should be set up on the opposite side from the first. The foam should be applied for a considerable period of time in order to form a thick blanket over the hot oil. The extinguishment time can be reduced if it is possible to use hand lines from a ladder or elevated platform to knock down the hot spots once the fire has subsided.

In some cases an oil tank fire may be encountered where portable applicators are not available. It may be possible in these instances to apply the foam from a ladder pipe or elevated platform. The foam stream should be projected to the far side of the tank and allowed to flow back gently over the surface. If the equipment is available, it might be possible to cut down the extinguishing time by using two ladder pipes or elevated platforms from opposite sides of the tank.

Theoretically, the effectiveness of the foam blanket can be improved by keeping the outside of the tank cool by the use of hose lines while the foam is being applied. The cooling action should reduce the rate of vaporization of the oil, which would have the effect of limiting the amount of vapors available to burn. The overall result would be less heat available to work on breaking down the foam blanket. Regardless of whether the theory is true, no harm can be done by keeping the tank cool if the operation can be performed safely while the foam is being applied.

Some oil fires can be extinguished by water alone. Certain viscous oils will mix with water and form an emulsion that acts as a type of foam. Care should be taken when using this method, for it is possible to cause a slopover. To avoid a slopover, or to reduce it to a minimum, the water should be applied lightly first over a small area of the fire and then shut off to see what happens. Once frothing takes place and appears to be capable of being controlled, the water should be applied over the entire surface until the froth forms.

One of the greatest dangers to firefighters at an oil tank fire is the possibility of a boilover. A **boilover** is the expulsion of the contents of the tank by the expansion of

slopover (tank)   A small spill of oil out of an oil storage tank.

boilover   The expulsion of the contents of an oil tank by the expansion of water vapor that has been trapped under the oil and heated by the burning oil.

water vapor that has been trapped under the oil and heated by the burning oil. Fortunately, most oils are not capable of boiling over, as they lack the basic characteristics required for this to take place.

The basic requirements for oil to be capable of boiling over are a high viscosity, a wide range of boiling points, and the presence of water and heat. **Viscosity** is the ability to flow. A liquid with a low viscosity will flow easily. A liquid with a high viscosity flows relatively slowly. If water is trapped and heated below an oil with a low viscosity, the water will merely bubble up through the oil. However, if the oil is sufficiently viscous, the steam formed under the oil will lift the entire body of oil and discharge it from the tank.

Crude oil has all the requirements for a boilover. The higher the viscosity of the crude, the greater the chance of a boilover. A boilover should be anticipated any time a tank of crude has burned for more than a few minutes. A boilover can occur suddenly and without warning and normally will be accompanied by heat and flames that will extend several hundred feet into the air. The best action that can be taken when a boilover is expected is to keep everyone several hundred feet away from the burning tank. It is also wise to ensure that all avenues of escape are kept open.

Two other methods of extinguishing oil tank fires are the subsurface injection of air and the subsurface injection of foam. The subsurface injection of air is referred to as the **air agitation method.** The air injected at the bottom of the tank forces cool oil from the bottom of the tank to the surface where it replaces hot oil. This has the effect of limiting the vaporization of the oil by reducing the oil's surface temperature.

Similar results are obtained by the **subsurface injection of foam.** The foam and the cool oil both rise to the surface. A foam blanket is eventually built over the surface to obtain complete extinguishment.

Both of these methods are normally applied by oil company personnel without the participation of fire department members.

viscosity Viscosity is the ability to flow. A liquid with a low viscosity will flow easily. A liquid with a high viscosity flows relatively slowly.

air agitation method A subsurface injection of air used as a means of extinguishing an oil tank fire.

subsurface injection of foam Refers to injecting foam at the bottom of an oil tank fire. The foam and the cool oil both rise to the surface. A foam blanket is eventually built over the surface to obtain complete extinguishment.

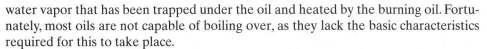

◆ FLAMMABLE GAS EMERGENCIES

The two flammable gases most commonly encountered by fire departments in emergency operations are natural gas and liquefied petroleum gas (LPG). The most commonly encountered LPGs are butane and propane or a mixture of the two. Both natural gas and LPG are referred to as fuel gases. A **fuel gas** is a flammable gas customarily used for burning with air to produce heat. Natural gas normally is stored and transported as a gas, whereas LPG generally is stored and transported as a liquid. One of the primary differences between the two is in their vapor density. (See Figure 8.12.)

Vapor density may be defined as the weight of a volume of vapor as compared with an equal volume of air. Vapor densities of less than 1.0 are lighter than air. Those greater than 1.0 are heavier than air. This means that gases having a vapor density of less than 1.0 will rise when they escape from their container, whereas those having a vapor density of greater than 1.0 will travel toward the ground. Natural gas has a vapor density of less than 1.0. Consequently, it will tend to rise when it escapes from its container. LPG, on the other hand, is heavier than air, having a vapor density of 1.5 or more. This means that escaping gas will travel to and hug the ground while it searches for a source of ignition. Both natural gas and LPGs are odorless; however, both are mixed with a material that gives them a distinctive odor that can be detected well below their explosive limits.

fuel gas A flammable gas customarily used for burning with air to produce heat.

vapor density The weight of a volume of vapor as compared with an equal volume of air.

FIGURE 8.12 ◆ A natural gas fire in a transmission line. The goal is to protect the exposures until the gas can be shut off. *(Courtesy of Las Vegas Fire & Rescue PIO)*

Firefighters respond to two basic types of emergencies involving flammable gases. One type is a leak of the gas and no fire. The other is where fire is involved. The latter primarily applies to fires in transportation lines and storage equipment. For the purpose of firefighting, tank trucks and railroad tank cars are considered as storage equipment.

Nonfire emergencies involving LPG are much more serious than those involving natural gas; however, the procedure for handling the two is basically the same. The primary objective is to keep the escaping gas out of contact with people, out of structures, and particularly out of contact with an ignition source. These three activities are coordinated with the attempt to shut off the source of the leak. The most commonly accepted method used to accomplish the contact objectives is to direct, dispense, and dilute the gas by the use of spray streams.

The extent of the critical area is normally much larger for LPG leakage than it is for natural gas. Natural gas is lighter than air and will tend to travel upward when it escapes from its container. Consequently, it normally will not travel too great a distance in search of an ignition source. However, LPG is heavier than air and will travel greater distances while physically maintaining a mixture with air. When it escapes, it will produce a heavy fog-like vapor cloud that assists in identifying the problem area. It is important to remember that the ignitable area extends well beyond the limit of the vapor cloud. In fact, it is good practice to consider that ignitable gases may extend a distance of up to 200 feet beyond the vapor cloud. This potential should be taken into consideration when eliminating ignition sources. (See Figure 8.13.)

The primary danger area for potential ignition is normally downwind from the source of the leak. However, the ignition potential exists on all sides. Initial action includes removing all people and shutting off all flame sources such as pilot lights, oil burners, and so on. Other potential ignition sources such as operating engines and electrical motors should also be eliminated. It should be kept in mind that a running engine on a fire apparatus could also be a hazard. It is possible for a diesel engine to "run away" due to a mixture of LPG and air being taken into the engine and compressed rather than the engine compressing nothing but air. If the spill is widespread, it may be wise to shut down all electrical power in the area.

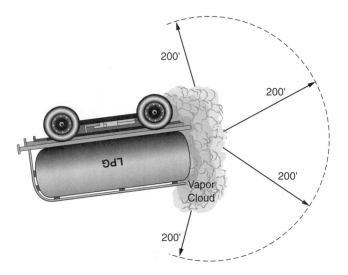

FIGURE 8.13 ◆ Consider that ignitable gases may extend out up to 200 feet.

A couple of important points should be kept in mind while directing, dispersing, and diluting the gas, removing the people, and eliminating the sources of ignition. LPG is much more hazardous and will spread out over a much larger area than is expected. For example, a container holding propane will produce approximately 270 cubic feet of gas for every cubic foot of liquid. If the gas finds an ignition source, the result will be a very violent explosion, with the fire involving the entire vapor area all the way back to the source of the leak. This could be a short distance or it might extend over several thousand feet. Consequently, it is good practice always to operate as though ignition could occur and in a manner that the resultant damages will be minimized if it should happen. Lines should be laid that will be needed to protect the exposures, and personnel should not be placed in jeopardy while performing their duties.

In many cases closing a valve can stop the leak. However, in a large number of situations the valve needing closing will be in the vapor area. Personnel should never be sent into the vapor area to shut off a valve unless a sufficient number of protective streams are provided to ensure maximum protection in the event that ignition occurs while they are performing the operation. The personnel assigned this task always should wear full protective gear, including breathing apparatus.

One last caution regarding LPG spills. It is good practice to approach the reported spill from the windward side whenever possible. Apparatus should be parked in the center of the street, a safe distance from the reported spill. Parking in the center of the street assists in eliminating the apparatus as a source of ignition, as the spilled LPG will probably travel and remain in the gutter area if the leak is not too extensive.

FIRE INCIDENTS

A basic principle of fighting fires in combustible gases is *never extinguish the fire unless the source of the leak can be stopped.* (See Figure 8.14.) Although there are situations that will require deviation from this principle, it is good practice always to consider the principle prior to taking other action. The principle is based on the concept that extinguishing the fire prior to eliminating the leak will normally create a much larger hazard than the fire itself. The escaping flammable gas will spread over a wide area, with the potential result of an explosion that could do extensive damage and possibly cost a number of lives.

FIGURE 8.14 ◆ Never extinguish the
fire unless the leak can be stopped.

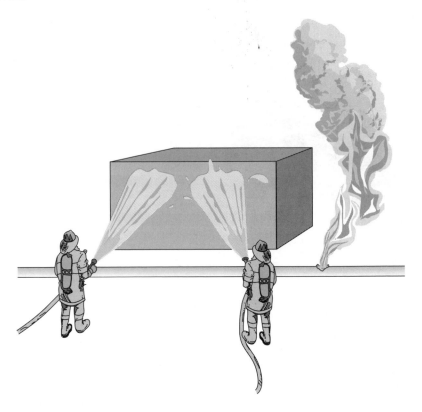

Small natural gas fires in buildings generally can be extinguished by shutting off
the nearest valve. If a fire occurs at the meter, it will be necessary to shut off the valve
at the street, as the valve at the meter controls only the flow of gas on the discharge
side of the meter. If a fire occurs in the street as a result of a break in the main transmission line, the general practice is to protect the exposures and let it burn until gas
company personnel have shut off the flow to the line.

Minor fires at the relief valve of small LPG storage tanks can be extinguished by
cooling the tank with water. This will relieve the built-up pressure within the tank that
caused the relief valve to open, and the valve will close.

The most challenging flammable gas fires occur in LPG transportation tank
trucks. The basic tactics of handling these fires apply to railroad tank cars and permanently mounted storage tanks; consequently, the discussion of firefighting operations
will be limited to tank trucks.

The initial action required when arriving at a large fire involving an LPG tank
truck is to protect the exposures and cool the burning tanker. This probably will
require large quantities of water and a number of hose streams. If sufficient water is
available, it may be best to set up heavy-stream appliances. Portable heavy-stream
appliances not only provide large quantities of water but also can be left in place to
continue the cooling action if it becomes necessary to evacuate the area suddenly.
Keep in mind that lines should not be set up at the ends of the tank, as the tank can
become a flaming rocket if one of the ends lets go.

If flames are impinging on the burning tank, it is imperative that water in large
quantities be applied to the impinged areas as quickly as possible. The overall objective
is to prevent a BLEVE from occurring. A **BLEVE** (boiling-liquid, expanding vapor
explosion) is a devastating effect caused by failure of the container. It is the result of

BLEVE   A boiling-liquid,
expanding vapor explosion.

impinging flames building up pressure within the tank faster than the spring-loaded relief valve can relieve the pressure. The overall effect is the release of the vapor and contents, which results in a mushroom-type fireball. The buildup of pressure is even more positive when the tanker is turned over. When the tanker is upright, the relief valve is in the vapor area. When the tanker turns over, the relief valve is in the liquid area and as a result is unable to relieve any vapor pressure buildup.

The BLEVE is generally a result of the combination of the pressure buildup and the loss of strength of the metal caused by the heat of the flame impingement. Failure of the metal usually will occur at the point of flame impingement. The buildup of pressure will cause a longitudinal tear in the metal, with the resultant release of large quantities of the pressurized LPG. Because the burning fuel is an immediate source of ignition, the result is a tremendous explosion accompanied by a huge fireball. The danger area for the radiated heat from the fireball and the concussion from the explosion could be as much as half a mile. People have been killed as far away as 800 feet as a result of flying fragmented metal. Anyone doubting the power of a BLEVE should take the time to research an incident that occurred in Kingman, Arizona, on July 5, 1973. In this incident, 15 people were killed, 13 of them firefighters.

A couple of factors regarding BLEVEs should be kept in mind when planning operations. Most BLEVEs occur when the containers are ½ to ¾ full of liquid. A tanker that is nearly full when the fire is initiated eventually will contain liquid within these proportions if a leak in the tank is feeding the fire. Records indicate that the time between the initiation of the fire and the BLEVE varies from a few minutes for small containers to a few hours for very large containers. However, the NFPA reports that 58 percent of the documented incidents in which pressurized containers failed from flame impingement occurred within 15 minutes. Remember that the time factor started when the impingement started, not when the fire department arrived on the scene.

The best preventive measure against a BLEVE is the playing of large quantities of water on the exposed tank. Sufficient cooling will keep pressure from building up within the tank and prevent the weakening of the metal. It is good practice to set up master streams for keeping the tanker cool and to withdraw personnel to a safe distance. This is particularly true if flames have been impinging on the tanker for any length of time when the firefighting forces first arrive. Although a BLEVE should always be anticipated, two positive warning signs must not be ignored. One is an increase in the pitch of the sound made by the escaping gas. This is generally a result of a rapid increase in the rate of vented gases. The other sign is a discoloration of the tank near the flame impingement area.

It should be remembered that the most dangerous situation involves flame impingement on the upper part of the tank, *above the level of the liquid.* Failure of the tank can occur rapidly in this situation because there is no liquid in contact with the tank in this area to partially absorb the heat.

If indications are that the fuel feeding the fire can be stopped by closing a valve, *and it is reasonably safe to attempt it,* then every effort should be made to do so. The same general precautions taken for shutting off a leaking source when there is no fire should be observed; however, where fire is involved, the number of lines used for protection of personnel normally will need to be increased.

Once the flow has been stopped, plans can be made to extinguish the fire. The best agent for fire extinguishment is a dry chemical, providing a sufficient amount is available to do the job. The initial discharge of the dry chemical can be started close to the leak or from some distance back. Several discharge lines should be used if the decision is to make a close approach. All nozzles should be directed at the source of the leak and

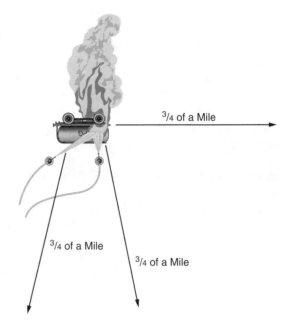

FIGURE 8.15 ◆ If necessary to abandon the fire, evacuate the area for at least ¾ of a mile.

opened simultaneously. This normally will knock down all the fire close to the source of the leak. Additional dry chemicals can be used to extinguish any remaining fire.

With the second approach, the dry chemical nozzles are opened up some distance from the fire. The firefighters handling the nozzles should be spaced across in a line. The objective is for the nozzle people to work in unison to produce a wall of dry chemicals as they advance on the fire. The nozzles should be operated up and down in a vertical alignment. This type of approach not only has the effect of quickly knocking down the fire but also provides a heat shield for the attacking nozzle handlers.

Occasionally fire departments will encounter LPG tanker fires in areas lacking sufficient water to fight the fire effectively. This might occur in rural areas or perhaps on freeways where the incident is remote from a good water source. In these cases it is usually wise to start an immediate evacuation of the area. If time permits, heavy-stream appliances can be set up and not staffed to use whatever water is available. It is possible that this action could prevent a BLEVE. The overall plan, however, should be to anticipate a BLEVE. Operations need to be directed to minimize the loss if the BLEVE does occur. It is good practice to evacuate the area for at least ¾ of a mile, and preferably 1 mile, around the fire. (See Figure 8.15.)

◆ ELECTRICAL FIRES

electrical fire A fire in energized electrical equipment.

For purpose of definition, an **electrical fire** is a fire in energized electrical equipment. Once the circuit has been de-energized in electrical equipment, the fire becomes a Class A fire or a Class B fire, depending on the material burning. It then can be handled in accordance with the material involved. (See Figure 8.16.)

The primary hazard for firefighters when responding to an electrical fire is the potential for electrical shock. Consequently, whenever possible, the electrical circuit to the involved equipment should be de-energized prior to an attack being made for the purpose of extinguishing a fire.

FIGURE 8.16 ◆ An example of a Class C fire. This fire will involve the kitchen quickly if not extinguished.
(Courtesy of Fire Cause Investigations, a division of System Engineering And Laboratories [SEAL])

Occasionally it is necessary to extinguish an electrical fire. Most electrical fires encountered by fire departments are small and can be extinguished by the use of a portable extinguisher. Extinguishers classified for use on electrical fires contain a nonconductive material. Dry chemicals and carbon dioxide meet this requirement.

Dry chemical extinguishers are effective for use on electrical fires but will require extensive cleanup at all electrical contacts after the fire has been extinguished. The multipurpose extinguisher also presents a potential corrosion problem if all the powder is not cleaned up soon after the fire has been put out. Consequently, it is best not to use it where the powder may get onto metal parts that are difficult to reach.

Carbon dioxide is noncorrosive and leaves no residue. This extinguisher should be given first choice when the fire involves sensitive electrical equipment. The use of dry chemicals on fires in sensitive electrical equipment such as computers could result in more damage than that produced by the fire itself.

In some cases it may be necessary to use water on de-energized electrical equipment. A good example is a fire in an electrical motor. Water can be used after the circuit has been shut off. The circuit generally can be de-energized by turning off a switch or pulling a plug. Electrical motors get very hot and it may be necessary to apply water for some time to cool it off. The water can do no additional damage because the windings on the motor will have to be rewound in any event.

The electrical wiring in many areas of some cities is run underground. Underground fires in manholes and street transformers are not uncommon where these systems are found. These fires present special problems to firefighters.

It is good practice when responding to a reported fire in underground equipment not to park the apparatus over a manhole cover, regardless of whether smoke is issuing from the opening. Fires in these systems have a habit of collecting flammable vapors and blowing a manhole cover high in the air when ignition takes place. As a precautionary measure against this potential, all people within a block of a smoking manhole or vault should be kept back a safe distance from all manholes. After a call for help from the electrical utility has been placed, all main service switches and fuses to all buildings within the block of the emergency should be pulled. It is a good idea not to use the bare hand to pull the switch, as the switch may be charged. If possible, a pike pole or axe can be used for this purpose. Tested, approved rubber gloves provide good protection when removing fuses.

Disconnecting the power provides some, but not positive, protection against the problem extending into buildings. As a precautionary measure against further extension into buildings, a constant check should be made of the electrical entrances into buildings until the Incident Commander has been informed by a responsible person from the electrical utility that the situation is under control.

Unless it is absolutely necessary for the purpose of making a rescue, underground vaults and manholes should not be entered. If entrance is necessary, firefighters should wear breathing apparatus and carry explosion-proof flashlights.

Water or other conducting liquids should never be poured down transformer vents. Discharging carbon dioxide into the vents can do no harm but there is no assurance that it will do any good. The best procedure is to keep everyone a safe distance away and not do anything unless requested by representatives from the electrical utility.

Occasionally fires will occur in electrical equipment that cannot be extinguished by portable extinguishers. In these instances it is best to let the fire burn and protect the exposures until all power to the electrical equipment can be shut off. The fire can then be attacked but particular care should be taken if high-voltage equipment is involved. Signs are normally attached to high-voltage equipment; however, in the absence of signs it should be assumed that the equipment uses high voltage.

If the fire department has been assured by a representative from the utility company that the equipment has been de-energized and a decision is made to extinguish the fire with water, there are still certain practices that should be employed as an additional safety factor.

The first precaution is for those handling the hose lines to stand on a dry spot. The second is to avoid the use of straight streams. The water should be discharged on the fire at the maximum reach of the stream.

◆ SILO FIRES

silos Large farm structures made of poured concrete, concrete staves, or steel.

staves Curved concrete blocks held in place by steel rings on silos.

ensiling A method of preserving green fodder.

forage (crops) Food of any kind used for feeding horses and cattle.

Silos are large farm structures made of poured concrete, concrete staves, or steel. Some of the older silos were made of tile blocks or wood. **Staves** are curved concrete blocks held in place by steel rings.

Silos are used for ensiling and storage of forage crops. **Ensiling** is a method of preserving green fodder. **Forage** crops are food of any kind used for feeding horses and cattle. Hay, corn, oats, and so on are examples of forage crops. Silos should not be confused with grain bins or grain elevators. Those structures are used to dry and store grains such as corn, wheat, and barley and have different firefighting problems from those of silos.

There are thousands of silos scattered around North America. Most of those in the United States are protected by volunteer fire departments, and the majority of silos are located quite a distance from the closest available fire company. The distance involved results in a delayed response to a reported fire in a silo; however, the time delay is generally not a problem. There is usually no hurry to extinguish the fire as it is generally well contained within the structure.

The primary cause of silo fires is the failure of farmers to follow proper forage storage procedures. The use of improper procedures can result in spontaneous combustion of feedstuffs after ensiling. Crops should be ensiled at 50 to 68 percent moisture content. A silo fire is likely to occur whenever silage or haylage is put up too dry. Fires are most likely to occur when silage is put up at less than 45 percent moisture content. Farmers should check the moisture content with a moisture meter or use an oven-drying method to determine the crop moisture content because the moisture content cannot be accurately estimated by sight or touch. Too low a moisture content can result in a fire while too high a content will result in seepage and silo damage.

Fires in silos can be controlled and extinguished effectively if they are promptly discovered and the correct extinguishing actions taken. Fire departments can take several steps prior to receiving a call of a reported silo fire that will make the entire operation proceed smoothly.

First, and most important, is that all fire department members be trained to recognize and identify the different types of silos. The information in this book should be helpful in achieving this requirement.

Second, all members should receive training on the action to be taken at silo fires. The chief officer who will respond to the fire should be present when the sessions on silo identification and the training of actions to be taken are presented and discussed.

If a fire company has any silos within its first-in district, inspections of the silos should be conducted, the silo identified as to type, and preliminary firefighting plans formulated. However, it must be kept in mind that modifications to the silo may take place between the date of the inspection and the date of the fire that might affect the preplanning.

Departments should obtain and have available the special equipment required for silo fires. The suggested equipment is:

1. A thermal imaging camera to help identify where hot spots are from outside of a silo.
2. A **high-temperature thermometer** (0–500°F) or other temperature-sensing device for use in locating hot spots in silage.
3. A specially designed probe to be used to lower the thermometer into the silage and injecting water directly into the silage. Figure 8.17 illustrates a suggested arrangement for the development of a probe.

high-temperature thermometer
A thermometer used for locating hot spots in silage.

Several steps should be taken by the fire company when it first arrives at the reported fire location. The fire official should seek and use the advice and assistance of the farmer and the silo technical experts in determining the type of silo involved and the action to be

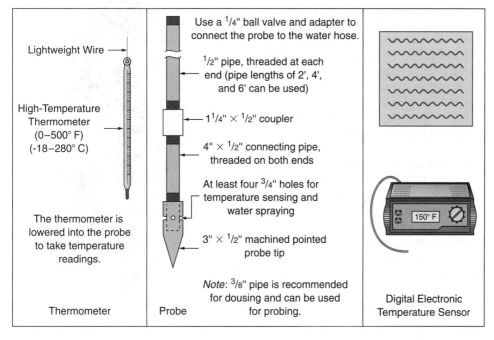

FIGURE 8.17 ◆ Thermometer, probe, and digital electric temperature sensor.
(Source: Extinguishing Fires in Silos and Haymows, NRAES—18)

taken. Don't be in a hurry. If the fire is in the silo it is well-confined. However, plan for the possibility of the silo collapsing by establishing and maintaining a collapse zone.

There are a couple of signs available to farmers that a fire exists within a silo. One is a burnt odor or burnt appearance of the forage used for feed. Another is smoke in the feed chute or smoke coming from the roof of the silo.

Once a fire starts in a silo, it presents the fire department with many challenges to which there are no simple solutions. Silo fires generally cannot be extinguished with conventional firefighting tactics used for other types of structures. Each fire is different, and each must be approached with the realization that disastrous results can occur if the proper action is not taken. A number of firefighters have been killed or injured while working on silo fires.

Extinguishing a fire in a silo requires the combined efforts and knowledge of fire officials, the farmer, and the silo technical experts. The primary variable that determines the correct action to take is the type of silo in which the fire exists. Therefore, determining the silo type is the first and most crucial step that should be taken by the responding fire company officer.

Fire companies may encounter a number of different silo types. Each requires different tactics to extinguish the fire and limit the risks to firefighters. It is therefore imperative that fire officials responding to silo fires be capable of correctly identifying the type of silo involved and have preplanned the corrective action necessary to cope with the situation successfully. Misidentification of a silo or not taking into account modifications of a silo can result in injury or death of firefighters.

TYPES OF SILOS

The most common silo is a vertical silo, also referred to as a tower or upright silo. These silos are the recognizable round cylinders found on farmsteads. There are approximately one million vertical silos in North America.

Vertical silos are subdivided into three general types: conventional, oxygen limiting, and modified.

conventional silo A conventional silo is usually equipped with a three-foot diameter chute that runs the full length of the height of the silo. The chute allows the silage to fall down into the barn, into a loading wagon or into a conveyer during unloading operations.

Conventional Silos. A **conventional silo** is usually equipped with a three-foot diameter chute that runs the full length of the height of the silo. The chute allows the silage to fall down into the barn, into a loading wagon or into a conveyer during unloading operations. The silo may be equipped with a hemispherical domed roof or no roof at all. (See Figure 8.18.) Figure 8.19 illustrates a dome-type top-unloading conventional silo.

It is possible that the fire is in the chute. After knocking down the visible fire in the chute, a close inspection should be made of the chute doors to determine whether the fire was confined to the chute or has extended into the silo or has extended from inside the silo to the chute area.

Fire Suppression. There is generally no hurry in extinguishing a fire in a conventional silo unless there is an exposure problem to an attached or nearby structure. These silos are not designed to be sealed and therefore present a limited threat of explosion.

After identifying the type of silo, one of the first steps that should be taken by the arriving fire unit is to extinguish any hot embers that may have left the silo. If it is possible to do so, the farmer should be instructed to move any livestock or machinery from the area and from nearby adjacent structures. The farmer should also be requested to have the silo operator and silo dealer respond to the emergency. The silo unloader machine should be raised as far as possible to minimize damage to its parts from heat and fire. This can be done by the use of a hand crank or electrical switch at

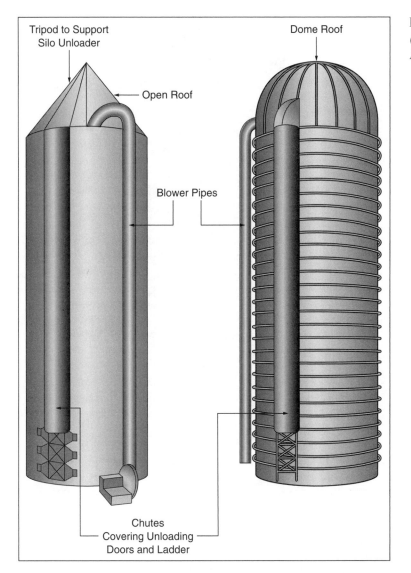

Tripod to Support
Silo Unloader

Open Roof

Blower Pipes

Dome Roof

Chutes
Covering Unloading
Doors and Ladder

FIGURE 8.18 ◆ Conventional silos.
(Source: Extinguishing Fires in Silos and Haymows, NRAES—18)

the base of the silo. The electrical power should then be disconnected to the unloader and all the electrical power to the silo should be locked out.

After these initial steps have been taken, a firefighter with full protective gear and full-body harness and lifeline should assess the fire from above the silage. Getting a firefighter above the silage can best be accomplished with the use of fire department aerial equipment if such has been dispatched to the emergency. If aerial equipment is not on the scene, or none is on the way, the firefighter can climb the silo's exterior ladder or chute if it is safe to do so.

Firefighters should take particular care in climbing silo ladders and always use a life belt and hook. The rungs of a chute ladder serve a dual purpose. They double as door handles and as rungs of a ladder. Consequently, they do not always provide a secure footing. Some may be broken, loose, or missing.

Climbing the chute ladder is very difficult if the firefighter is wearing SCBA gear. It is more effective to use a supplied-air respirator (SAR). A SAR includes a mask

FIGURE 8.19 ◆ A top-unloading conventional silo. *(Source: Extinguishing Fires in Silos and Haymows, NRAES—18)*

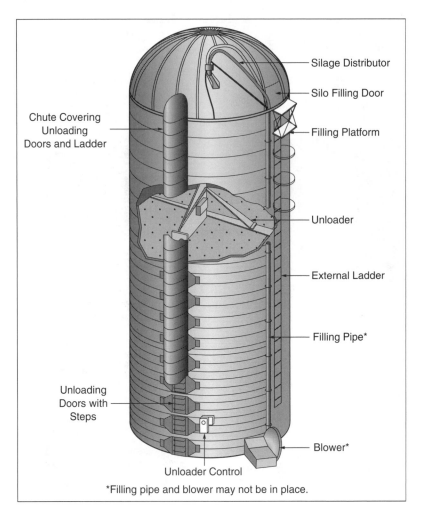

Silage Distributor

Silo Filling Door

Filling Platform

Chute Covering Unloading Doors and Ladder

Unloader

External Ladder

Filling Pipe*

Unloading Doors with Steps

Blower*

Unloader Control

*Filling pipe and blower may not be in place.

attached by a long air line to one or more large air cylinders located outside the silo. A SAR also has the advantage of not running out of air after 30 or 45 minutes.

Once the firefighter has arrived above the fire, he or she should extinguish any visible hot spots.

Fires in vertical conventional silos generally can be found around the perimeter of the silo, near the unloading doors, or in the top four to six feet of the silage. The fire has probably been burning for a considerable length of time and has most likely created hollowed-out cavities in the silage that will present a hazard to a firefighter entering the silo. Knocking down surface fires may be done from an aerial ladder or elevated platform, from the silo filling platform, or from the silo chute.

After all fire in the burning surface has been knocked down, a firefighter may enter the silo if it is positively safe to do so. It should be remembered that a silo is a confined-space structure and firefighters are required to follow OSHA's Standard 29 CFR Part 1910.146. This standard when applied to silo fires requires that:

- The firefighter use full personal protective equipment, including breathing apparatus and a harness and lifeline.
- Adequate rescue resources be present at the scene.
- All power to the silo be shut off and tagged out.

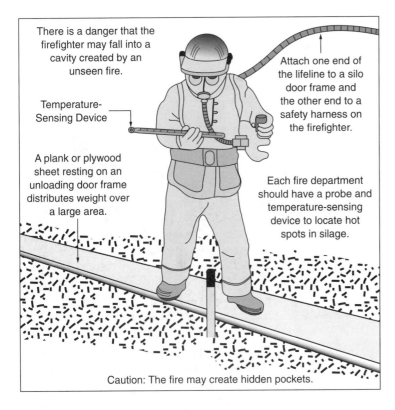

There is a danger that the firefighter may fall into a cavity created by an unseen fire.

Temperature-Sensing Device

A plank or plywood sheet resting on an unloading door frame distributes weight over a large area.

Attach one end of the lifeline to a silo door frame and the other end to a safety harness on the firefighter.

Each fire department should have a probe and temperature-sensing device to locate hot spots in silage.

Caution: The fire may create hidden pockets.

FIGURE 8.20 ◆ Probing for hot spots in silage.
(Source: Extinguishing Fires in Silos and Haymows, NRAES—18)

- Training and standard operating procedures be established and followed.
- A designated person must be stationed outside the silo to monitor the firefighter inside the silo.
- The air inside the silo must be monitored (IDLH).

Any firefighter entering the silo should be firmly secured with lifelines and stand on long boards or ladders. The body harness should be attached to a lifeline that is secured as high as possible to a door frame, a beam, or a silo hoop. (See Figure 8.20.)

The first action of the firefighter should be to obtain temperature readings. This is important as the key to extinguishing the fire is to determine its exact location or locations.

The firefighter should operate from wooden planks that are laid on the silage to distribute the firefighter's weight. A probe should be projected into the silage and a temperature-sensing device lowered into the probe. (See Figure 8.21.) How far the probe can be inserted into the silage depends on the condition of the silage and the length of the probe.

Temperatures below 140°F indicate no serious heating problems. Temperatures of 180°F or higher indicate that the silage will eventually burn and has lost its value as feed. It also indicates that action must be taken to cool the silage.

Water should be injected into the heated area; however, experience indicates that flooding of the silo with water will not extinguish the fire. (See Figure 8.21.) The water will follow the path of least resistance, which is down the interior wall of the silo and most likely will not reach any portion of the heated area. The best hope of reaching subsurface burning is to use penetrating nozzles. Unfortunately, it is usually impossible to extinguish the fire completely.

FIGURE 8.21 ◆ Inject water directly into the silage.
(Source: Extinguishing Fires in Silos and Haymows, NRAES—18)

Once the extent and location of the fire are established, small amounts of water are injected into hot areas via the probe.

Planks or Plywood

Not fully extinguishing the fire in a conventional silo is almost a norm. At times it may appear that the fire has been extinguished but it may reignite. The only safe procedure to ensure that the fire is out is for the farmer to empty the silo partially or completely. This may be a long affair, but it is best that a firefighter be present during the unloading to douse any hot spots that may mature. In such cases it will be necessary for the company officer to arrange for rotation of personnel.

oxygen-limiting silos
Oxygen-limiting silos are constructed of steel or poured concrete. The most common of these silos is the blue-colored Harvestore brand.

Oxygen-Limiting Silos. **Oxygen-limiting silos** are constructed of steel or poured concrete. The most common of these silos is the blue-colored Harvestore brand. The roof on oxygen-limiting silos is generally flatter than those on conventional silos. (See Figure 8.22.) Most of the roofs are equipped with two 18- or 24-inch openings. If the silo has not been modified, it will have no exterior unloading chute as the silo is unloaded from the bottom.

Fire suppression. Fires in oxygen-limiting silos present an extreme danger to firefighters as they are subject to exploding. An explosion can result if any air is allowed to enter the silo. The atmosphere inside the silo during a fire is an enriched carbon monoxide mixture with sufficient heat available to complete the fire triangle if oxygen is provided. Firefighters should consider the silo the same as they would an enclosed routine structure that is ripe for a backdraft. Therefore, nothing should be done to cause air to be drawn into the silo. In fact, it is highly recommended that firefighters allow the fire to burn itself out without taking any type of corrective action.

The farmer should be instructed to remove livestock and machinery from adjacent structures if it is considered safe to do so. It may be necessary to leave the silo closed for several weeks to allow the fire to burn itself out. The fire department's action during this period of time is to monitor the silo and adjacent areas regularly.

If the fire is still burning after three weeks, thought should be given to injecting liquid nitrogen or carbon dioxide into the structure to displace the oxygen and cool the fire. (See Figure 8.23.) It is best that this be accomplished by technical experts

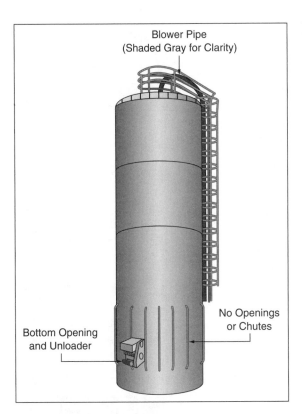

Blower Pipe
(Shaded Gray for Clarity)

No Openings
or Chutes

Bottom Opening
and Unloader

FIGURE 8.22 ◆ Oxygen-limiting silo.
(Source: Extinguishing Fires in Silos and Haymows, NRAES—18)

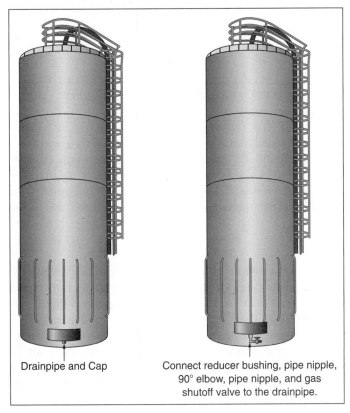

Drainpipe and Cap

Connect reducer bushing, pipe nipple, 90° elbow, pipe nipple, and gas shutoff valve to the drainpipe.

FIGURE 8.23 ◆ Connections for gas injection. *(Source: Extinguishing Fires in Silos and Haymows, NRAES—18)*

FIGURE 8.24 ◆ Modified oxygen-limiting silo with an access door added in the side wall.
(Source: Extinguishing Fires in Silos and Haymows, NRAES—18)

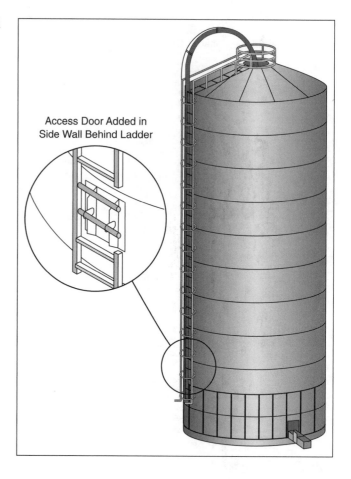

Access Door Added in Side Wall Behind Ladder

from the silo manufacturer with the fire department standing by a reasonable safe distance from the silo as a precautionary measure.

modified silo A modified silo is one that has been modified from its original design.

Modified Silos. A **modified silo** is one that has been modified from its original design. (See Figure 8.24.) Modification of a silo can change the tactics normally used for firefighting. An example of a modified silo is a conventional silo that has had the exterior chute replaced with a center unloading system. Another example is the redesigning of an oxygen-limiting Harvestore silo by modifying it for top unloading equipped with unloading doors and a center unloading design or exterior chute. (See Figure 8.25.)

Fire Suppression. Extreme care must be employed when a fire in a modified silo is encountered. One of the most important steps to be taken by a fire officer is to secure technical assistance prior to taking any action. In fact, it is recommended that a second opinion also be obtained. Firefighters have paid a high price in injury and death by using incorrect suppression techniques on modified silo fires. It is important to remember that there is nothing wrong with allowing a silo fire to burn itself out. In fact, it is recommended that any fire in a modified oxygen-limiting silo be handled using the exact precautions and procedures that would be used for a fire in an oxygen-limiting silo.

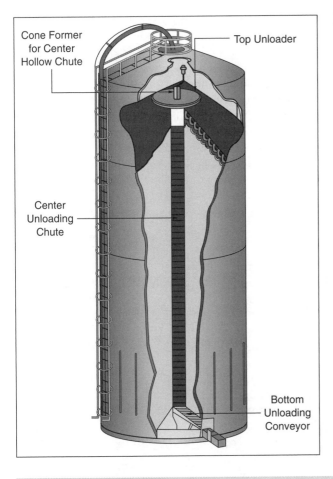

Cone Former
for Center
Hollow Chute

Top Unloader

Center
Unloading
Chute

Bottom
Unloading
Conveyor

FIGURE 8.25 ◆ Modified oxygen-limiting silo with top unloader and center unloading chute. *(Source: Extinguishing Fires in Silos and Haymows, NRAES—18)*

On Scene Scenario

The following incident was taken from an article written by Timothy G. Prather of the University of Tennessee and published in a National AG Safety Database bulletin.

Oxygen-limiting silos, such as Harvestore: [sic] silos, can explode if a fire is not handled properly. Two firemen were killed in Georgia on August 5, 1993 when they applied water and foam to a fire in an oxygen-limiting silo. The explosion blew the roof off, sending one fireman to the ground over 100 yards away and the other through the roof of the nearby metal building. Two firemen on the ground were injured by debris. The top 15 feet of the silo were severely damaged by the explosion and an adjacent silo dented by the debris.

◆ WILDFIRES

A **conflagration** has been defined as "a major building-to-building flame spread over a large area." Some people confuse a "group" fire with a conflagration. The primary difference between the two is that a **group fire** is confined to a small area whereas a conflagration spreads over a large area and crosses natural or man-made barriers.

conflagration A major building-to-building flame spread over a large area that crosses natural or man-made barriers.

The scenario of a conflagration that manifested itself over the last century where it involved narrow streets and closely packed buildings is today comparatively rare. The most common type of conflagration scenario that has appeared in recent years is what is called an urban/wildland interface. This scenario refers to forest fires or brush fires that spread to nearby buildings. It is interesting to note that Table 2.1.6 in the nineteenth edition of the *NFPA Fire Protection Handbook* lists the Oakland Hills (forest) firestorm, Oakland, California, October 20, 1991, as the fire causing the largest dollar loss in U.S. history in the year in which the fire occurred. However, the San Francisco earthquake and fire in 1906 and the Great Chicago Fire in 1871 are still listed as numbers one and two on an adjusted loss basis.

On Scene Scenario

During October and November of 2003, nine wildfires fanned by Santa Ana winds swept through southern California, causing several billion dollars in damage, killing 20 people including one firefighter, and destroying about 3,000 homes. The control of the fires required the services of over 14,000 firefighters from across the United States. The largest of these fires was the San Diego Cedar fire, which burned 280,278 acres and, destroyed 2,232 residences, 22 commercial properties, 566 outbuildings, and 148 vehicles.

Today, most conflagrations in the United States involve wildland fires. Additionally, approximately one out of every eight firefighters killed at a fire is killed at a wildland fire. The ratio was even higher during 2002. According to the U.S. Fire Administration report released September 30, 2003, "100 firefighters lost their lives while on-duty in the U.S. in 2002—nearly a quarter of those while fighting wildland fires."

Most firefighters assigned to the U.S. Forest Service and similar organizations such as the California Division of Forestry have been well-trained in wildland firefighting. However, the majority of firefighters assigned to fire departments in cities, counties, and similar jurisdictions have received little, if any, training in this area. These firefighters normally are referred to as structural firefighters. Despite their lack of training in this area, large numbers of structural firefighters are sent to participate in wildfire operations every year.

IFSTA (International Fire Service Training Association) recognized this problem a number of years ago and published a book to assist these firefighters in developing their knowledge of the problems and tactics involved. The title of this book is *Wildland Fire Fighting for Structural Firefighters,* fourth edition.

This section of this textbook also has been prepared to assist structural firefighters. The information is not of the depth required by firefighters regularly assigned to wildland areas.

CLASSIFICATION OF WILDLAND FIRES

Wildland fires are classified into three general types. In large fires, all three types are usually involved at the same time.

Ground Fires. The fuel for ground fires includes leaves, needles, small wood branches (less than ¼ inch), duff, muck, and peat. The smoke from these fires is minimal while the flame extends from approximately six inches to four feet. The spread of the fire is slow.

Surface Fires. The fuel for surface fires includes reproduction trees up to six feet in height, fallen trees, grass, brush, branches, and slash. The height of the smoke is medium to high, while flames extend from two to fifty feet. The fire can spread at a slow rate or very rapidly.

Crown Fires. The fuel for crown fires is brush and trees higher than six feet in height. The height of the smoke can be extreme. The rate of the flame spread is from fast running to extreme. A crown fire is one that spreads from the top of one tree (or shrub) to the top of another more or less independent of a surface fire.

WILDLAND FIRES TERMINOLOGY

Although structural firefighters are very familiar with the common terminology of a building, the language used when referring to wildfires will undoubtedly seem strange. However, it is important that those firefighters who wish to gain knowledge regarding wildland firefighting strategy and tactics be familiar with some of the common terms used to describe various parts of a ground fire pattern. Following are some of the most important to know. Knowledge of these terms is vital when discussing the fire with seasoned wildland firefighters and when receiving operational orders. (See Figure 8.26.)

Origin. The term **origin** refers to the area where ignition first occurred. The origin may be readily accessible or located in a remote area. When it is accessible, it is generally located next to a trail, road, or highway. Those fires started by lightning or by a careless camper may be in an inaccessible location.

origin   The area where ignition first occurs at a wildfire or brush fire.

Head. The **head** is that part of the fire that is spreading or traveling most rapidly. It is usually located at the opposite end of the fire from which the wind is blowing or the

head   That part of a wildfire or brush fire that is spreading or traveling most rapidly.

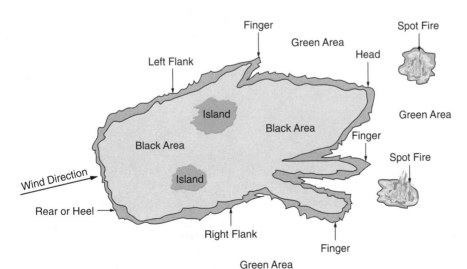

FIGURE 8.26 ◆ Parts of a wildfire.

FIGURE 8.27 ◆ West Glenwood, Colorado, June 16, 2002— Spot fires such as this one can quickly spread a wildfire or create a new one.
(Photo by Andrea Booher/ FEMA News Photo, Courtesy of FEMA)

upper part of a slope. (See Figure 8.27.) The head generally burns rapidly and intensely and is responsible for the most damage. At a large fire, there may be several heads. A good portion of the firefighting tactics used at a wildland fire is concentrated on gaining control of the head.

fingers Long, narrow strips of fire that extend out from the main body of a wildfire or brush fire.

Fingers. **Fingers** are long, narrow strips of fire that extend out from the main body of fire. If not controlled, fingers will form new heads. Fingers normally form in patches of mixed ground cover, some of it light and some heavy. They are also formed when the head is split by natural features such as bodies of water, rock formations, and so on. (See Figure 8.26.)

rear The side opposite the head at a wildfire or brush fire.

heel The side opposite the head at a wildfire or brush fire.

Rear. The **rear** is the side opposite the head. It is also referred to as the **heel.** It is generally located close to the origin of the fire and may be found to be burning downhill or against the wind. This is generally the easiest part of the fire to extinguish. Many times the initial attack on the fire is made at this point. (See Figure 8.27.)

flanks The sides of a wildfire or brush fire.

Flanks. The **flanks** are the sides of the fire. Looking from the rear of the fire toward the head, the left flank is to the left and the right flank is to the right. Fingers occasionally extend out from a flank. A shift in the wind can quickly change a flank into a new head. Early consideration should be given to controlling the flanks. (See Figure 8.27.)

Perimeter. The **perimeter** is the boundary of the fire. It is also referred to as the **fire edge.** It is constantly changing until the fire is suppressed. When the fire is suppressed, the edge of the burned-out area becomes the perimeter.

> **perimeter** The boundary of a wildfire or brush fire.
>
> **fire edge** The boundary of a wildfire or brush fire.

Island. An **island** is an unburned area located within the fire perimeter. Islands offer more fuel for the fire and are frequently burned out by the firefighters. If not burned out, they must be frequently inspected to ensure that no spot fires have started in the island. (See Figure 8.26.)

> **island** An unburned area located within the fire perimeter of a wildfire or brush fire.

Spot Fires. **Spot fires** are new fires that have been started by sparks or flying embers carried aloft by convection or winds from the fire area. They may form a new head or in some cases a major fire. The travel direction of a spot fire cannot always be predicted accurately. This fact could result in trapping personnel or equipment between the two fires. (See Figure 8.27.)

> **spot fires** New fires that have been started by sparks or flying embers carried aloft by convection or winds of a wildfire or brush fire.

Slopovers. **Slopovers** are fires that have extended over the control line or natural barrier into an unburned area. Slopovers are still connected to the main body of fire. They are different from spot fires, as they appear close to and adjacent to the barrier whereas spot fires can occur some distance ahead or to either side of the main fire.

> **slopover (wildfire)** A fire that has extended over the control line or natural barrier into an unburned area at a wildfire or brush fire.

Flare-Up. A **flare-up** is a sudden increase in the intensity or speed of the fire spread. It lasts for only a short duration and does not change existing plans.

> **flare-up** The sudden increase in the intensity or speed of the fire spread of a wildfire or brush fire.

Blowup. A **blowup** is similar to a flare-up but is of significant magnitude to force a change in existing control plans. A blowup is similar to a flashover in a structure fire. It may produce a convection column and other characteristics of a firestorm. A potential blowup might give some warning signs. Anticipate that a blowup is about to occur if a fire continues to grow rapidly in size or intensity despite firefighting efforts that normally would control it.

> **blowup** A situation that is similar to a wildfire or brush fire flare-up but is of significant magnitude to force a change in existing control plans.

Firestorm. Structural firestorms were the primary factor responsible for the destruction of German cities during World War II. A **firestorm** is the intense burning of fuel over a large area, which results in a huge convection column over the fire area. This convection column causes a low-pressure area at the fire center. This low-pressure area draws in fresh air in the form of strong winds from all directions. The winds created in some of the firestorms in Germany reached 100 MPH. (See Figure 8.28.)

> **firestorm** The intense burning of the fuel at a wildfire or brush fire over a large area, which results in a huge convection column over the fire area.

Wildland firestorms are similar in nature. The updrafts and strong winds can cause tornado-like firewhirls, the uprooting of vegetation, and can throw rocks and debris for a long distance through the air. The sign of a potential firestorm is a sustained high-intensity fire with a growing convention column.

Green Area. The **green area** is the unburned area adjacent to the involved area. It may consist of different types of ground cover. Some of the green area may have a high moisture content and therefore be fairly slow burning, while other sections may

> **green area** The unburned area adjacent to the involved area at a wildfire or brush fire.

FIGURE 8.28 ◆ A firestorm.

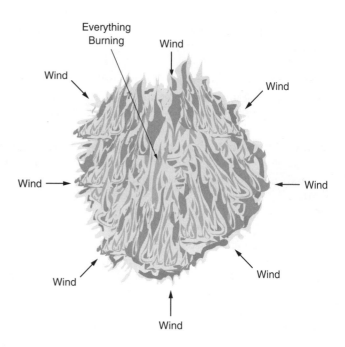

have a low moisture content and be highly flammable. The term green area is not, however, synonymous with a safe area.

black area The burn area of a wildfire or brush fire.

Black Area. The **black area** is the burn area. If the area is completely burned over with no islands, it is a relatively safe area during the progress of the fire. However, although the area is relatively safe, it is not always a comfortable environment as it may be very smoky with a large number of hot spots. The black area may also contain a large amount of aerial fuels, which were not burned by a surface fire. This condition leaves the possibility for a reburn.

FIRE CAUSES

Wildfires are a major cause of the fire loss in the United States. There have been more that 10,000 homes and 20,000 other structures and facilities that have been lost to wildfires since 1970. The leading cause of these fires has been incendiarism. Maliciously set fires have accounted for one-fourth to one-third of these fires.

Nationwide, lightning is the number-two cause; however, it is the number-one cause in the northwestern part of the country and in Arizona and New Mexico. (See Figure 8.29.)

FACTORS AFFECTING FIRE BEHAVIOR

While there is a common thread that runs through all of them, it can be stated confidently that the combination of factors affecting wildland fires is never exactly the same. Each setting is differently affected by the fuel, the weather, and the terrain. In fact, the burning speed and direction of a fire's travel can change drastically in a few seconds by a rapid shift in the wind direction and speed. The speed also will change radically as a fire approaches the base of a hill and starts up.

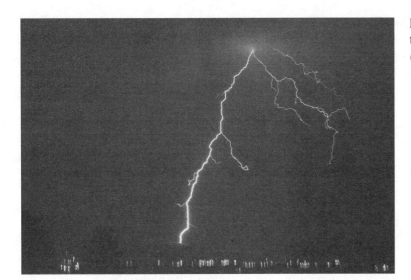

FIGURE 8.29 ◆ Lightning is one of the leading causes of wildland fires. *(Courtesy of Cathcart photo.com)*

The Fuel.   A **fuel** is a combustible or flammable material that is available for feeding a fire. Wildland fuels may be dead or alive. They are of many different types and variations, with varied fire-behavioral characteristics. Most people think of wildland fuels as grass, trees, and the like; however, the structures located in the wildland areas are also a source of fuel.

fuel (wildfire or brush fire)
The combustible or flammable material that is available for feeding a fire.

Fuel Types.

Grasses.   Grasses are annuals, which means that they are produced, grow, and die all in the same year. The pattern of this cycle generally repeats itself the following year. The season in which a fire occurs affects the burning characteristics of the material.

Brush.   There is a large variety of brush types found throughout the United States. Some are of small height while others grow quite large. Brush over six feet in height is classified as an aerial fuel. The burning characteristics of brush found in various parts of the country are quite different. For example, brush found in southern California is considered one the fastest-burning types in the world. A separate section in this book is devoted to brush fires.

Trees.   Some of the trees located in various parts of the country are deciduous trees, which means they lose all their leaves in the fall and grow new leaves in the spring. Some of the common deciduous trees are ash, birch, cottonwood, dogwood, maple, and some oaks.

The other classification of trees is evergreens. As the name applies, these trees do not lose their leaves but stay green all year. Some of the common types are redwood, spruce, hemlock, eucalyptus, cedar, and pine.

Slash.   A lot of material is left on the ground after logging operations are complete. This material may be a combination of tree limbs, stumps, and logs. The combination of these materials is referred to as **slash.** Slash is classified as light, medium, and heavy.

slash   The material that is left on the ground after logging operations are complete.

aerial fuels (wildfire) Those fuels that are physically separated from the ground.

Aerial Fuels. **Aerial fuels** are those fuels that are physically separated from the ground. The speeds with which aerial fuels will burn are generally considered as being in an inverse proportion to the horizontal distance between fuels. For example, the distance between trees will affect the rate and intensity of a fire in that environment. Other factors such as a hot, dry wind will also affect the rate and intensity of the burning of these fuels.

weather (wildfire or brush fire) The state of the atmosphere over the fire area.

The Weather. **Weather,** in regards to wildland fires, is the state of the atmosphere over the fire area. The weather is a critical factor in both the movement and the intensity of the fire but, more important, is a key factor affecting the safety of firefighters.

Weather is the most changeable factor affecting wildland fire behavior. It is extremely important that those firefighters participating in wildland firefighting operations be knowledgeable regarding the potential effects of weather changes on a fire's behavior. Each firefighter must keep himself or herself constantly informed of the current and anticipated weather in the fire area. The weather factors having the major impact on fire behavior are temperature, relative humidity, and wind.

lapse rate The rate of decrease of the temperature as the elevation increases.

Temperature: Air temperature is a condition that changes with elevation. Normally, the temperature decreases as the elevation increases. The rate of decrease is referred to as the **lapse rate.** The lapse rate varies with the moisture content of the air. Dry air has a higher lapse rate than moist air. The dry air lapse rate is about 5.5 degrees per 1,000 feet of elevation. This means that if the temperature at sea level is 100°F, it would be 16.5° less (3 × 5.5) at an elevation of 3,000 feet.

The moist lapse rate is about 3 degrees per 1,000 feet of elevation. As actual conditions are seldom 100 percent dry or moist, it is practical to assume that the lapse rate is approximately 3.5° per 1,000 feet.

stable air condition A condition in which the smoke will rise nearly straight up and then spread out nearly equally in all directions.

unstable air condition A condition in which the rising smoke will drift downwind.

temperature inversion A condition in which the temperature increases instead of decreasing when a given elevation is reached.

Whether the air is stable or unstable has an effect on how smoke from a wildfire will behave. For example, if the dry lapse rate is less than 5.5°, the air is considered to be in a stable condition. If it is greater than 5.5°, it is considered to be unstable. Under **stable air conditions,** smoke will rise nearly straight up and then spread out nearly equally in all directions. Under **unstable air conditions,** the rising smoke will drift downwind.

Normally, the lapse rate will remain steady as the elevation increases. However, at times it will reach an elevation at which the temperature increases instead of decreasing. This sets up a condition know as a **temperature inversion.** The result of a temperature inversion is that smoke from the fire will rise until it reaches the inversion level, at which it will stop rising and will flatten out. The temperature inversion is a common contributor to the smog conditions in many of the larger cities.

humidity The amount of moisture in the air.

relative humidity The amount of moisture in the air compared with the amount the air can hold at a given temperature.

Relative Humidity. **Humidity** refers to the amount of moisture in the air. The less moisture in the air, the greater the fire hazards. The amount in the air is referred to as the relative humidity.

Relative humidity is the amount of moisture in the air compared with the amount the air can hold at a given temperature. Clouds or fog will form when the relative humidity reaches 100 percent.

Relative humidity at ground level is influenced by such factors as elevation, slope, the time of day, and the season. For example, the relative humidity in southern California in certain seasons can get down to as low as 2 or 3 percent. Under these conditions, ground cover can become very dry and extremely flammable.

FIGURE 8.30 ◆ A wildfire being pushed by the wind. *(Courtesy of Rick McClure, LAFD)*

Wind. **Winds** are basically air in motion. One principle of wind is based on pressure. Air moves from a high-pressure area to a low-pressure area. The greater the pressure difference between the two, the stronger the wind. (See Figure 8.30.)

winds Air in motion.

As winds move across the earth's surface, they follow a path of least resistance. Their direction will be affected by such factors as rock formations, ridges, and smaller items such as trees. These obstructions to the wind's movement contribute to the formation of turbulence on the sheltered side of the obstruction.

Local Winds. **Local winds** are caused by heating and cooling patterns. For example, the sun's rays will heat the ground surface equally in a given area; however, some ground surfaces will absorb the heat while others will not. A grass-covered field will absorb heat slower than a plowed field. As a result, the heat of the plowed field will cause the air to rise. This rising air results in the formation of a local low-pressure area. Surrounding air then will move toward this low-pressure area. (See Figure 8.31.)

local winds Winds caused by heating and cooling patterns.

Sea Breezes and Offshore Breezes. Land masses heat faster than water. Consequently, due to solar heating, the pressure over the land during the heating period is less than the pressure over the water. This results in the movement of air from the water to the land. Near oceans, this is referred to as an on-shore or **sea breeze.** Sea breezes normally commence around noon and strengthen during the afternoon.

sea breeze A movement of air from the water to the land.

At night, the condition reverses itself. The land cools down faster than the water. This results in a low pressure forming over the water. As a result, the air will move from the land to the water. This is referred to as an **offshore breeze.**

offshore breeze The movement of air from the land to the water.

The overall result of the day and night conditions is that air moves from the ocean to the land during the daytime and from the land to the ocean at night. In between this change, a condition occurs whereby the pressures are the same. This results in an almost total calm. (See Figure 8.32.)

Surface Winds. Surface heating also influences air movement in hilly or mountainous terrain. The general principle involved is that heating causes air to move up slopes while cooling causes a downward movement.

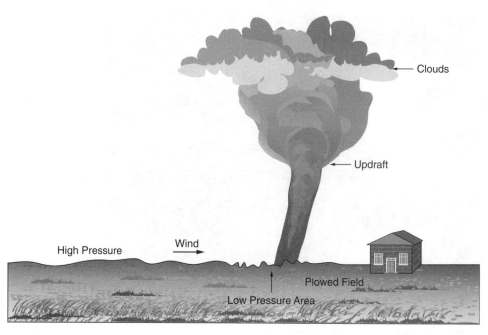

FIGURE 8.31 ◆ Local winds.

Gradient Winds. Large high-and low-pressure areas are generally expressed in a circle. In the Northern Hemisphere, winds typically move clockwise around a high-pressure area and counterclockwise around a low-pressure area.

gravity wind A downslope wind.

Gravity Winds. When an air mass moving toward a mountain reaches the mountain, it is forced upward. On reaching the ridge, it cascades down the other side. This downslope wind is referred to as **gravity wind.** As the air moves down, it compresses and heats. The result can be a strong, hot, dry wind. These winds are detrimental to wildland fire control. (See Figure 8.33.)

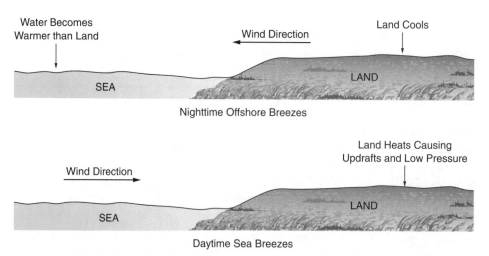

FIGURE 8.32 ◆ Sea breezes and offshore breezes.

Gravity Winds
Air Compresses
and Heats Up

Wind
Direction

FIGURE 8.33 ◆ Gravity winds.

Wind Effects on Fire Behavior.   Wind and fuel moisture are the two most important weather factors affecting wildland fire control. Wind is the most influential and variable of the two. It is also highly unpredictable. A wind can change its direction and speed rapidly.

Wind bends flames, dries out fuel, and brings in fresh oxygen to feed the fire. It picks up embers and sparks and carries them far in advance of the fire. An unexpected wind shift can change the progress and intensity of a fire rapidly. It is very important that the potential change of wind direction and speed be anticipated to help ensure firefighter safety and fire-control efforts.

Anticipating wind conditions is based on a number of factors. Predictions of the weather bureau should always be considered. Knowledge of the seasonal conditions in the fire area is also important. It is also valuable to be familiar with the generalizations of the daily weather cycles, recognizing that terrain will affect them. Some of the generalizations that should be considered are:

- The period from 10:00 A.M. to 6:00 P.M. is normally the most erratic. This time period, referred to as the "heat of the day," is the period in which the relative humidity is low, the fuel is dry, the temperature is high, and the wind is strong. These factors in combination are very unfavorable for fire control.
- The period from 6:00 P.M. to 4:00 A.M. is referred to as the evening and nighttime hours. During this period the air temperature cools down, the relative humidity generally rises, which tends to moisten the fuel, and the strength of the wind lessens. All of these factors are favorable for fire control.
- The period from 4:00 A.M. to 6:00 A.M. is the hours during which the fire activity is at its lowest.
- The period from 6:00 A.M. to 10:00 A.M. is the increasing period. The temperature begins to go up, the strength of the wind begins to increase, and the intensity of the fire will increase.

TOPOGRAPHY

Wildfires can occur on level ground or in mountainous or hilly areas. However, most of the largest and most difficult wildfires to contain have occurred in nonlevel environments. Although the factors previously examined applied to wildfires anywhere, those occurring in mountainous or hilly areas add the adverse conditions of terrain.

Slope.   **Slope** has a very pronounced influence on wildland fire behavior. For those firefighters knowledgeable in fire department hydraulics, slope is the same as grade. The **percent of a grade** (or the slope) can be defined as the rise in feet for every 100 feet of horizontal distance. For example, in estimating the percent of slope, a rise of 15 feet

slope   Slope is the same as grade. The percent of a grade (or the slope) can be defined as the rise in feet for every 100 feet of horizontal distance.

percent of a grade   The rise in feet for every 100 feet of horizontal distance.

in a 100 feet would be a 15 percent slope, and a rise of 40 feet in 100 feet would be a 40 percent slope.

There are several factors associated with slope:

1. In the absence of wind, fires normally move faster uphill than they do downhill.
2. On an uphill spread of fire, the flames are closer to the fuel and preheat the fuel much farther ahead of the fire than they would on level ground.
3. A fire burning uphill will create a draft that increases the rate of fire spread.

When the fire reaches the top of the slope (ridge) and starts down the other side, the reverse of the three factors takes place. This is a good place to stop the fire. However, it must be remembered that on steep slopes there is the possibility for burning material to roll down the slope and start new fires near the bottom. This could put any personnel working on the ridge in jeopardy. It is therefore important for personnel working in this area always to have an escape route.

aspect (wildfire or brush fire) Refers to compass direction.

Aspect. **Aspect** refers to compass direction. If a slope faces north, it has a north aspect. If it faces east, it has an east aspect. The importance of the aspect is related to the effect of the solar heating. For example, in the Northern Hemisphere (where the United States is located) a southern aspect slope receives more direct solar heat than does a slope facing in any other direction. This means that the air temperature is hotter and the moisture retention of the soil and vegetation is less on a southern slope than on any of the other slopes, and that a southern slope will be more susceptible to fires.

chute (wildfire or brush fire) A steep V-shaped drainage.

saddle (wildfire or brush fire) A depression between two adjacent hilltops.

Chutes and Saddles. A **chute** is a steep V-shaped drainage. A **saddle** is a depression between two adjacent hilltops. Chutes and saddles can have a drastic, adverse effect on a fire. They are capable of changing a relatively slow burning fire into a fast burning fire and can produce very erratic behavior. They are dangerous and can also be deceptive. Firefighters must be constantly on the alert and carefully evaluate what appears to be a harmless depression.

canyons Canyons can be thought of as topography chimneys. The heat, smoke, and fire gases can be channeled through canyons at a very rapid rate.

Canyons. **Canyons** can be thought of as topography chimneys. The heat, smoke, and fire gases can be channeled through canyons at a very rapid rate. A canyon acting like a chimney can increase the wind velocity and generate sufficient heat on one side to ignite combustible fuel on the other side.

direct attack (wildfire or brush fire) Refers to fighting the fire itself with the use of water, throwing dirt on it, using beaters, or building a fire line at the burning edge and throwing the material into the burned-out area.

FIRE SUPPRESSION METHODS

The two basic methods of attacking a wildfire are the direct attack and the indirect attack.

A **direct attack** refers to fighting the fire itself with the use of water, throwing dirt on it, using beaters, or building a fire line at the burning edge and throwing the material into the burned-out area. A **fire line** can be defined as a portion of a control line that is constructed near the fire's edge. It is made by clearing a trail of all vegetation. A direct attack involves attacking the fire at its edge or closely parallel to the edge.

fire line (wildfire or brush fire) A portion of a control line that is constructed near the fire's edge.

indirect attack (on a wildfire) The indirect attack is similar to the defensive mode on a structure fire. The indirect attack means applying control techniques some distance ahead of the fire to stop its progress. It generally consists of creating an area ahead of the fire that contains no fuel.

An **indirect attack** refers to building a fire line some distance from the fire's edge. As previously mentioned, a fire line is a portion of a control line. A **control line** may be defined as any natural or constructed barrier used to control a fire. The control line when an indirect attack is used may be wider than one at the fire's edge.

control line (wildfire or brush fire) Any natural or constructed barrier used to control a fire.

Direct Attack. The direct attack can be used on the head of small, light running fires involving ground or surface fuels such as grass, leaves, duff, field crops, and small brush. However, it also can be used on larger fires. When used on larger fires, the attack is generally concentrated on the flanks with the attack made from the rear of the fire. The attack is made working from the burned-out area.

Direct attacks are also made in the latter stages of a large fire, and on other fires in which the burning intensity and the heat and smoke are not too much of a detriment to firefighters working on the edge of the fire.

Direct attacks also are used to knock down flare-ups ahead of line construction activities and to knock down fingers that have spread from the edge of the main body of fire.

Where conditions permit, a direct attack is the preferred method of attacking a wildfire. A direct attack is an aggressive attack made at the fire's edge. Water is one of the most effective tools to use on a direct attack. The use of water additives mixed with water is widely accepted as an extinguishing agent for wildfires. Water additives reduce the surface tension of the water, which allows the water to penetrate fibrous materials faster and deeper than can be achieved with plain water. The most popular and what appears to be the most effective additive is Class A foam. The use of Class A foam reduces the amount of water required for an initial knockdown and has a definite effect on the reduction potential of a rekindle. It is estimated that the use of Class A foam is at least two times as effective as using plain water and that it allows firefighters to cover a larger area with a quicker knockdown than is possible with ordinary water.

One of the principal advantages of using a direct attack is that the area of the burn normally is kept smaller. Another advantage is that the firefighters work close to or in the burned area, which provides a safety zone in which to escape.

Probably the primary disadvantage of using a direct attack is that the firefighters are continually exposed to heat, smoke, and flame as the fire line is constructed directly adjacent to the fire's edge.

Indirect Attack.   An indirect attack is time-consuming and hard work. Using this technique, areas are sacrificed and the fire is allowed to grow. It involves withdrawing all fire suppression forces and equipment to other areas such as roads, trails, natural breaks, and constructed control lines. Then the hard work begins.

If a natural control line is not available, then a constructed line will have to be made. Both a natural line and a constructed line should be some distance from the main fire. There is no set rule as to how far from the fire a constructed line should be made. The decision depends on a number of factors such as how fast the fire is spreading, its intensity, the topography, and the type and amount of the fuel available to burn. (See Figure 8.34.)

The key to an indirect attack is the development of an effective control line. This line serves as a barrier to the spread of the fire in the event of a flare-up after the fire has been brought under control. The construction of the control line should be of sufficient width to prevent the flames from the fire from crossing over to the uninvolved area once it reaches the control line.

If possible, the line should be constructed in light fuels with full advantage taken of any natural barriers. Construction commences at a secure anchor point to prevent the fire from outflanking and possibly trapping the working crews. Regardless of the distance between the fire front and the control line, it is important that all unburned fuel between the two lines be removed. This is essential to creating an effective and safe control line.

Although firefighters assigned to an indirect attack are not working as close to the fire as those making a direct attack, indirect attacks are not completely safe from

FIGURE 8.34 ◆ Woodland Park, Colorado, June 11, 2002—Arkansas River. Thirty-one wildland fire crews work on creating a fire line on the southern flank of the Hayman fire in Colorado. *(Photo by Michael Reiger/FEMA News Photo, Courtesy of FEMA)*

risks. There is always a risk involved whenever firefighters are working in an area where there is unburned fuel between them and the main fire.

Water can be used in an indirect attack to construct a temporary wet line. The most effective method of accomplishing this is to mix the water with a penetrant. The most effective penetrant to use for this purpose is Class A foam. Class A foam wets the fuel more thoroughly but also less water is needed to complete the job. It should be kept in mind, however, that a wet line is a temporary line used to prevent ignition for a few moments. It is still necessary to burn or clear out any fuel between the control line and the fire in order to widen the control line.

Backfiring and Burning Out. Backfiring and burning out are similar operations but have different objectives. Both of them involve the intentional setting of fire to uninvolved fuels.

burnout (wildfire or brush fire) A method of widening a control line by eliminating fuel between the control line and the fire.

A **burnout** may be defined as a method of widening a control line by eliminating fuel between the control line and the fire. This reduces the fuel that the fire eventually has to feed on. Burning out also creates escape routes for firefighters and provides safety zones. Burning out around structures can be considered as a method of protecting an exposure. (See Figure 8.35.)

It is best that a burnout be started at the top of a slope. A burnout started at this point will burn downhill and against the wind. This creates a slow moving fire that will spread toward the main fire.

backfiring (wildfire or brush fire) The intentional setting of a fire to establish a defense perimeter for the fire or as an emergency safeguard for personnel.

Backfiring can be defined as a very aggressive offensive tactic that is used to stop the spread of a very intense wildland fire. It is a very dangerous and tricky operation

FIGURE 8.35 ◆ West Glenwood, Colorado, June 11, 2002—The Flathead Hotshot crew sets a burnout on Horse Mountain in the South Canyon drainage to stop the fire from consuming more forest land. *(Photo by Andrea Booher/FEMA News Photo, Courtesy of FEMA)*

performed under adverse conditions that should be done only by experienced personnel. (See Figure 8.36.) A backfiring operation should be done only with the approval of the operations chief or the Incident Commander. All personnel must be informed of where and when a backfiring operation is to be conducted prior to the first ignition.

FIGURE 8.36 ◆ West Glenwood, Colorado, June 11, 2002—The Flathead Hotshot crew has set a backfire to limit the consumption of more forest land. *(Photo by Andrea Booher/FEMA News Photo, Courtesy of FEMA)*

FIGURE 8.37 ◆ A fixed-wing aircraft making a water drop. *(Courtesy of Rick McClure, LAFD)*

AIR OPERATIONS

Air operations often are a part of every large-scale wildland fire. Support is provided by both fixed-wing aircraft and helicopters. Air operations are used for reconnaissance, transporting personnel and equipment, transporting smoke jumpers and helicopter rappellers to remote locations, and as a participant in both indirect and direct attacks on the fire. They also have been used effectively for purposely dropping burning material onto uninvolved fuels. (See Figure 8.37.)

Air reconnaissance is very important to the success of the fire. Information can be provided to the Incident Commander that is not available from any other source. Roads can be spotted, hot spots reported, and the Incident Commander can be kept informed of the extent and intensity of the fire in various areas.

Unfortunately, most Incident Commanders are not very familiar with the deployment of aircraft for firefighting. Consequently, it is important that an Incident Commander work very closely with the air tactical group supervisor. This supervisor is familiar with the turbulence problems associated with canyons, slopes, valleys, and so forth and can advise the Incident Commander when a decision is to be made regarding airdrops. (See Figure 8.38.)

Aircraft primarily are used for airdrops at wildland fires. Airdrops are used on both direct attacks and indirect attacks. On direct attacks they are applied directly to one of the edges at the head of the fire to reduce the rate of the spread. Both water and other suppressants are used for this purpose. Slowing the spread provides ground crews with more time to construct a control line around the fire and assists in the control and containment of that part of the fire.

On an indirect attack, airdrops of fire-retardant chemicals are made on uninvolved fuels in the path of the fire.

MOP-UP

mop-up The process of completing the extinguishment of a wildfire or brush fire. It entails finding and extinguishing all hot spots, particularly those close to the control line.

The mop-up should begin as soon as possible after the fire has been controlled. **Mop-up** is the process of completing the extinguishment. It entails finding and extinguishing all hot spots, particularly those close to the control line. The mop-up can mean the difference between success or failure of the entire fire control operation. More fires

FIGURE 8.38 ◆ Using a helicopter to make a water drop on a fire. *(Courtesy of Rick McClure, LAFD)*

have been lost because of poor or incomplete mop-up operations than for any other reason. (See Figure 8.39.)

Mop-up is a hard, dirty, and extremely exhausting job. While the ideal is to extinguish all hot spots in the burned area, this is almost an impossibility at a large fire. The amount of work that will have to be done depends on such factors as the size of

FIGURE 8.39 ◆ Los Alamos, New Mexico, May 4, 2002—Hot Shot members from Zuni, New Mexico, continue work on mop-up operations. *(Photo by Andrea Booher/FEMA News Photo, Courtesy of FEMA)*

the burned area, the type of fuel in the burned area, and of utmost importance the weather. A general rule is to mop up the entire area of small fires, if practical. On large fires, an effort should be made to extinguish all hot spots within 100 feet of the control line. Failure to do so could result in a rekindle and perhaps the start of another fire days after the original fire had been left.

Another general rule is to mop up the most threatening areas first. A search should be made for all smoldering hot spots. If a thermal camera is available, it can be of tremendous help. If water is available, make use of it wherever possible. A penetrant such as Class A foam mixed with the water will help prevent a rekindle. It is especially valuable in deep burning fuels. (See Figure 8.40.)

The job of mopping up is not complete when the handwork appears to have been finished. Patrol is an essential part of the mop-up procedure. Each firefighter who worked on the mop-up operation may be given a designated area to patrol. Any smoldering material found is to be extinguished or thrown back into the burned-out area and allowed to burn out. Patrols should continue until there is a reasonable assurance that the fire edge is completely out. To be able to have this assurance, it is possible that patrols will be required for several weeks.

FIRE CONTROL

The California Division of Forestry provides three definitions related to fire control. These definitions are:

fire containment (wildfire or brush fire) A fire is contained when it is surrounded on all sides by some kind of boundary but is still burning and has the potential to jump a boundary line.

fire controlled (wildfire or brush fire) A fire is controlled when there is no further threat of it jumping a containment line.

fire out (wildfire or brush fire) The fire is out when mop-up is finished and all crews can be released.

Fire containment. A fire is contained when it is surrounded on all sides by some kind of boundary but is still burning and has the potential to jump a boundary line. The boundary may be a *fire line,* which is a strip of area where the vegetation has been removed to deny the fire fuel, or a river, a freeway, or some other barrier that is expected to stop the fire. Hose lines from fire engines may also contribute to a fire being surrounded and contained.

Fire controlled. A fire is controlled when there is no further threat of it jumping a containment line. While crews continue to do mop-up work within the fire lines, the firefight is over.

Fire out. The fire is out when mop-up is finished and all crews can be released.

FIGURE 8.40 ◆ Los Alamos, New Mexico, May 4, 2002—Mopping up a wildfire is a dirty and exhausting job.
(Photo by Andrea Booher/FEMA News Photo, Courtesy of FEMA)

On Scene Scenario

During October 2003, a series of brush/wildland fires erupted in California. The primary contributing factors in these fires was dry weather, dry fuel, and the strong Santa Ana winds. The suspected cause of several of the fires was arson.

Although statistics at the time of this writing have not been confirmed, it is estimated that the destruction from the combined fires totaled 22 killed, including one firefighter, 3,600 homes destroyed, and nearly 750,000 acres of brush and timber burned (approximately 1,172 square miles). For comparison, the entire state of Rhode Island contains approximately 1,231 square miles.

FIGURE 8.41 ◆ San Bernardino, California, November 1, 2003—This aerial photo shows the patchwork of homes destroyed by wildfires in southern California.
(Photo by Andrea Booher/FEMA News Photo, Courtesy of FEMA)

FIGURE 8.42 ◆ San Bernardino, California, October 30, 2003—Twisted metal is all that remains of this home in Waterman Canyon that was destroyed by a wildfire.
(Photo by Kevin Galvin/FEMA News Photo, Courtesy of FEMA)

(Continued)

The most destructive of the fires was the San Diego Cedar fire, which tentatively destroyed approximately 850 homes and burned approximately 280,000 acres (approximately 437 square miles). It is estimated that this was one of the most destructive fires in the history of California.

FIGURE 8.43 ◆ San Bernardino, California, October 31, 2003— The evacuation shelter at Norton Air Force Base held over 3,000 evacuees following the wildfires in southern California. *(Photo by Andrea Booher/FEMA News Photo, Courtesy of FEMA)*

FIGURE 8.44 ◆ A small portion of the Simi/Val Verde fire. *(Courtesy of Rick McClure, LAFD)*

FIREFIGHTER SAFETY

Firefighter safety is a primary factor for consideration at every wildland fire. Unfortunately, too many firefighters have lost their lives at wildfires. Prior to the September 11, 2001, attack on the World Trade Center, the most firefighters killed at a single incident anywhere in the United States occurred at a wildfire. At the Great Idaho fire, which occurred in Idaho and Montana on August 20 and 21, 1910, 78 firefighters lost their lives. This occurred despite the knowledge shared by the firefighting profession that there is no fuel at a wildland fire, even structures, that is worth a firefighter's life.

No one will question the fact that wildland firefighting is extremely hazardous. Also, no one will question the fact that it is extremely important that every firefighter who actively participates in wildland firefighting operations should be thoroughly knowledgeable regarding the safety rules to be followed. Such knowledge should begin with a familiarity of the types of wildland fire behavior that kill firefighters. In a newsletter article by Retired Deputy Chief Vincent Dunn, Chief Dunn pointed out that a 10-year study of wildfires conducted by the National Fire Protection Association identified the following four types:

1. Firefighter fatalities often occur at the edge or perimeter of a large wildfire.
2. Firefighters often are killed when working in and around light fuels, such as grass, brush, or marsh weeds; they usually are not killed in large forest fires.
3. Unexpected wind shifts are a contributing factor in firefighter deaths.
4. A change in ground elevation increases the fire spread. This often kills firefighters on the side of a hill or mountain or on a cliffside.

The Ten Standard Firefighting Orders.   In 1957, a task force commissioned by Forest Service Chief Richard E. McArdle reviewed the records of 16 tragic fires that had occurred from 1937 to 1956. From this review came the 10 standard firefighting orders. Part of these orders was based on the successful "general orders" used by U.S. armed forces. Each of these orders, together with the "watch outs" that follow, form the bible for wildland firefighters. Every firefighter is expected to remember each of them and to use them to reevaluate fire suppression strategies and tactics.

It is also important for firefighters to understand the importance of the orders to their individual safety. In the analysis of the fire deaths reviewed in the study of the 16 tragic fires, which resulted in the orders, in every case in which a firefighter was killed while fighting a wildland fire, it was shown that one or more of the 10 standard firefighting orders had been ignored.

The Ten Standard Firefighting Orders

- *Fight fire aggressively, but provide for safety first.* This rule recognizes that firefighting is an extremely hazardous occupation. It cautions every firefighter to remember that no property is worth the life of a firefighter. Aggressive action is important to the overall achievement of the objective; however, safety has first priority.
- *Initiate all actions based on current and expected fire behavior.* A number of elements contribute to fire behavior, the three primaries being weather, topography, and fuels. All of these are considered in the original size-up of the fire. However, remember that a fire is not static. It constantly will move and grow until it is controlled. A continuous size-up requires that the fire's movements must be continuously anticipated. Every action taken at the fire should be based on what the fire is doing and what it is expected to do in the future.

- *Recognize current weather conditions and obtain forecasts.* Remember that the three most important weather conditions that affect fire behavior are wind, temperature, and relative humidity. The combination of these three factors determines how the fire will behave. A firefighter should use every tool at his or her disposal to gain information on the weather. If available, a portable/belt weather kit is capable of measuring wind speed and direction, relative humidity, and air temperature. Radios can be used to listen to the Weather Channel to obtain information on weather in the area in which he or she is working.
- *Ensure instructions are given and understood.* Every firefighter should make sure that his or her supervisor's orders are clear and precise. If there is a question or doubt, have the supervisor repeat and explain the order in plain language. A firefighter's life depends on knowing exactly what to do and what is expected.

Note that the first letter of the first four orders spells FIRE, and the first letter of the next six orders spells ORDERS.

- *Obtain current information on fire status.* It is important to know at all times what the fire is doing. Fires have a habit of changing very quickly. A small fire can become a large fire suddenly. It is important to know information such as: How fast and in what direction is the fire moving? Where is the fire perimeter? Are there any spot fires between you and the perimeter? If the required information is not available by observation, get it from someone who knows.
- *Remain in communication with crew members, your supervisor, and adjoining forces.* Keeping in communication with all those who are important to a firefighter's safety should be one of his or her first concerns. Good communications is essential to firefighters' safety. Communications on the fire ground can be by radio, by cellular telephones, face to face, by runners, or by any other reliable means.
- *Determine safety zones and escape routes.* A safety zone is an area that is unlikely to burn. An escape route is a way to get there. Safety zones include areas that have already been burned over and such locations as wetlands and lakes. Continually keep in mind the fastest way to get there. Consider the fact that it may be necessary to change an escape route as the fire progresses.
- *Establish lookouts in potentially hazardous situations.* The objective of a lookout is to keep the supervisor in touch with what is happening to the fire. It is important that the individual assigned to the lookout position recognize changes in the weather and in dangerous fire conditions such as spotting. Anything of importance that is observed by a spotter should be reported immediately to the supervisor. Portable radios are one of the best means for a spotter to keep in touch with his or her supervisor.
- *Retain control at all times.* It is important that the supervisor makes sure that his or her assignments and orders are understood at all times. It is also his or her responsibility to know at all times where those assigned to him or her are working. He or she must be alert constantly to make sure that all members of the crew are rested and capable of going to work.
- *Stay alert, keep calm, think clearly, act decisively.* Almost everyone, regardless of rank or experience, is capable of becoming confused and unsure of oneself at times on the fire line. There are just too many variables happening too fast for firefighters' minds to work properly when they are tired and nearly exhausted. If confused, do not make any decisions or give any orders that could place someone in jeopardy.

The Eighteen Situations That Shout "Watch Out." Rules of safety at wildland fires have been developed as a result of tragedies. Each of the 18 situations that shouts "watch out" has resulted in the death of at least one firefighter. The list identifies specific situations in which the risk is greater than normal. It is hoped that every firefighter assigned to a wildland fire will be aware of these warning signs and recognize that his or her life might depend on recognizing these situations. More important, any individual responsible for assigning crews to work areas should not make assignments where any item on the list exists.

1. *Fires not scouted and sized up.* For safety's sake, all members of a crew must be aware of where the fire is and what it is doing. A crew has been placed in danger whenever it is assigned to an area of the fire where the perimeter cannot be visually seen and the crew has not had an opportunity to make a size-up. Placing crews in such situations should be avoided.

2. *In country not seen in daylight.* Crews should not find themselves at night in areas where they have not had the opportunity of seeing the topography in the daylight. This places them in a situation in which they are not aware of important factors such as the density of the vegetation and the distance between points that might make a situation unsafe.

3. *Safety zones and escape routes not identified.* It is extremely important when working at wildland fires for every individual always to know the location of a safety zone and an escape route to get there. At any time a firefighter finds himself or herself in a situation in which he or she is not aware of either one of these locations, the firefighter should stop whatever he or she is doing until it is determined where these safety features are located.

4. *Unfamiliar with weather and local factors influencing fire behavior.* The microclimate and burning conditions of no two areas are exactly the same. Yet it is important for firefighters to be aware of these factors in areas where they are assigned to work. It should be kept in mind that different conditions require different strategies and tactics.

5. *Uninformed on strategy, tactics, and hazard.* Crews always should be briefed on the plan of attack and how their operation fits into the overall plan. Not having this information can place the crew in serious jeopardy. They should be particularly informed of factors such as airdrops or firing operations, which might affect them.

6. *Instructions and assignments not clear.* Crews should be given clear, precise instructions prior to being assigned to the fire line. Failure of a crew to obtain adequate instructions can result in inefficient operations and perhaps place the crew at risk. Once a crew is on the fire line, it is too late to have an order clarified.

7. *No communication link with crew members.* It is important that lines of communication be maintained up and down the chain of command. Without such lines of communication, critical and perhaps lifesaving information cannot be passed to the proper authorities.

8. *Constructing line without safe anchor point.* This is extremely dangerous practice can place crews in positions of being outflanked and possibly being surrounded completely by the fire. Any individual who decides to do this should evaluate the risks against the potential gains carefully.

9. *Building fire line downhill with fire below.* A good general rule to follow is NEVER to advance a line downhill in the green toward an advancing fire.

10. *Attempting frontal assault on fire.* Attempting a frontal assault on a fire from the green area can place a crew in a very dangerous position. It should not be done unless the crews have an adequate number of lines of the proper size with plenty of water to support them. Even then, there is a risk involved as the fire might overrun the firefighters or spot behind them.

11. *Unburned fuel between you and fire.* This is a position in which firefighters should not be placed. It is dangerous regardless of the type of fuel in the unburned area. There is always the possibility of the firefighters being overrun as they attempt to reach a safe position. If they are placed into such a position, it is critical to have escape routes and safety zones readily available.

12. *Cannot see main fire, not in contact with someone who can.* There are many locations at a wildland fire where firefighters work without being able to see the main fire. This places them in a position in which they could be caught by an unseen blowup and be overrun. It is critical when placed in such a position that one or more lookouts are posted in positions where they can see the progress of the fire.

13. *On a hillside where rolling material can ignite fuel below.* Working in the green area on a slope opposite to the fire area can be dangerous. As the fire reaches the ridge, it is possible that burning material can be released that will roll downhill and start a spot fire below the working crew. Fires travel uphill very rapidly. This could catch a crew before any of the members are able to reach a place of safety.

14. *Weather becoming hotter and drier.* Hotter and drier means faster burning, easier ignition, and stronger winds. These changes increase the chance of spot fires developing, the wind fanning smoldering fires back to life, and a greater risk of hazard developing in canyons and ravines.

15. *Wind increases and/or changes direction.* A general rule is to be on the lookout for flames that stand straight up. This is an indication that the wind direction is about to change. A change in wind direction is capable of blowing hot materials or flames into unburned fuel. Be aware also that an increase in the wind speed will increase the rate of the flame spread and most likely involve new fuels.

16. *Getting frequent spot fires across the line.* Spot fires that are not caught in the early stage are capable of combining into a larger fire area, or they may develop into two or more fires and trap firefighters between them.

17. *Terrain and fuels make escape to safety zones difficult.* Climbing up a slope can be difficult under the best of circumstances. The difficulty is increased when a firefighter is tired or exhausted, the terrain is slippery, or the fuel is a hindrance to travel. It is important when working on a terrain with these problems that a spotter always be positioned who can see the fire to alert the crew in the event that any changes occur that are critical to an escape plan. Keep in mind the principle to ALWAYS HAVE AN ESCAPE ROUTE.

18. *Taking nap near fire line.* Crew members will get tired. This is unavoidable. However, crew members should not be permitted to sleep on the fire line except in safety zones. However, even with permission, crew members should not be allowed to sleep unless an adequate number of spotters are in position to alert them in the event trouble develops. Keep in mind that drowsiness and being tired are not necessarily the same. Because crew members most likely will be working without the protection of breathing apparatus, they are subject to the possibility of carbon monoxide poisoning.

Additional Safety Practices.   While the 10 standard firefighting orders and the 18 "watch out" situations alert firefighters to the potential dangers of fighting wildland fires, other factors are also important to their survival. Some of these are listed here:

1. Continually check the environment. Get in the habit of looking up, looking down, and looking around.
2. Always be prepared to escape to a predetermined safety zone.
3. Remember that situations that appear to be insignificant can quickly claim the life of a firefighter.
4. Do not use land-based telephones during a thunderstorm.
5. Fire shelters should be used only as a last resort to survive a fire entrapment. All other means of escaping to a safety zone should be tried prior to deploying a fire shelter.
6. Do not ride on the outside of a moving fire apparatus. This is dangerous and could result in serious injuries. It is extremely important never to ride on the fenders or the front bumpers of fire apparatus.

7. Backfiring should be done only by trained, experienced personnel.
8. Except in an emergency, firefighters should not be wet down. The moisture from this procedure increases the likelihood of a firefighter being burned. One emergency involves heatstroke victims. Any firefighter suffering from heatstroke should be cooled down immediately using a copious amount of water and then transported to a medical facility as soon as possible.
9. All downed electrical wires should be considered as energized.

USE OF THE INCIDENT COMMAND SYSTEM

No other type of fire might require the expansion of the organization chart of the Incident Command System to the extent required at a major wildland fire. The organization chart will expand even beyond that illustrated in the Model Incident Command System in Chapter Three. However, the expanded system does not appear suddenly at a fire but usually develops over a period of time. The time frame expansion can best be illustrated by example. In most cases, the duties of the positions referred to were covered in Chapter Three and will not be repeated in this chapter.

Initial Attack. About 95 percent of all wildland fires are controlled and extinguished by the initial attack dispatch. The initial assignment varies according to various elements such as the location of the fire, the weather conditions, the travel time for the first company to reach the incident, the type of fuel involved, and the amount and type of available equipment. The assignment may be a single engine company or it may consist of a number of companies, supporting mechanized equipment, and aircraft. Figure 8.45 is an example of the organization of an initial alarm response.

The Incident Commander in this organization also functions as the safety officer. He or she directs and controls all firefighting resources until it is apparent that the situation is beyond his or her control or span of control. A rapid intervention team is not normally used at a wildland fire.

Extended Attack. When it is apparent that the fire situation is beyond the control of the initial response assignment, the Incident Commander will determine what is needed and call for the needed assistance. The extended response may be filled within the organization or it may involve the assignment of mutual-aid units. In most cases, this is developed on a pre-incident plan and all that the Incident Commander has to do is request what is needed.

As far as the command function is concerned, it is important that the principle of span of control be maintained. It probably will be required that an operations chief be appointed and that the fire be broken down into geographic divisions/sectors and/or functional groups/sectors. If broken down into geographic areas, the

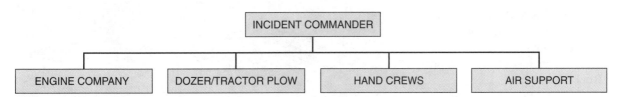

FIGURE 8.45 ◆ Organizational chart.

FIGURE 8.46 ◆ Dividing the fire into sectors.

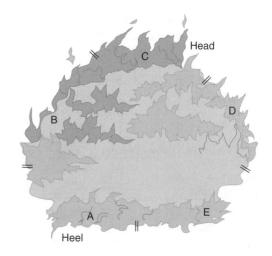

breakdown generally starts at the heel and progresses clockwise around the entire fire. (See Figure 8.46.)

Another feature that is added to the organization is staging. Staging is assigned to Operations and the operations chief will decide on how many staging areas are needed. Some of the additional responding units will report to locations designated under the pre-incident plan and others will report to staging. In staging they will be organized into strike teams or task force units depending on the needs of the situation.

A Major Fire Attack.   When the fire situation extends beyond the capabilities of the extended attack assignment, it is considered as having developed into a major fire attack. This usually occurs after the fire has been burning for at least 24 hours.

Not only are additional resources needed to cope with an expanding and intensified fire, but additional personnel also are required to provide relief for those firefighters on the line. Most of them will have been on the fire line for a considerable length of time and need food, showers, and sleep. Their equipment may need to be resupplied with hose, foam, and so on. (See Figure 8.47.)

FIGURE 8.47 ◆ Castle Rock, Colorado, June 13, 2002—Wildland firefighters working on the Hayman fire try to get some rest at the incident command post in Castle Rock, Colorado.
(Photo by Michael Reiger/FEMA News, Photo Courtesy of FEMA)

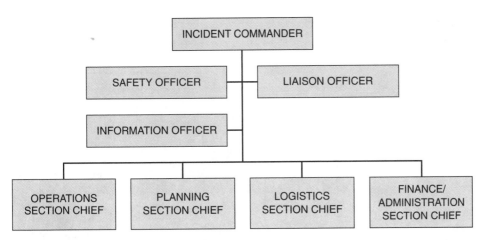

FIGURE 8.48 ◆ Organizational chart—ICS.

The need has also arisen to expand the organization into a full-blown organization. The Command portion of the organization would probably appear as shown in Figure 8.48. If a unified command has been organized, and it most likely will have been, the term unified command can be placed on the chart in lieu of Incident Commander.

It will be required that various individuals be assigned to perform the duties shown on the Incident Command System organization chart in Chapter Three. Additional personnel will be required to staff the branches. Branches are not common to all structural ICS organizations. The branches in a wildfire organization are the ground operation branch and the air operations branch. The organization of the ground operation branch would appear as shown in Figure 8.49. The organization of the air operations branch would appear as shown in Figure 8.50.

The Operations section is responsible for all personnel and equipment assigned to the fire that is directly involved in the extinguishment and control of the fire. In accepted organizational language, this section would be known as the line portion of the organization. The officers assigned to this section would be referred to as line officers. The other portions of the organization are designed to support the line. Officers in these sections are referred to as staff officers.

The Operations section of a wildland fire is the largest portion of the organization. The organization chart for this section would appear as shown in Figure 8.51.

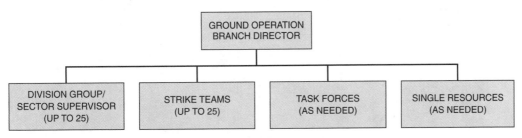

FIGURE 8.49 ◆ Organizational chart—ground operation.

FIGURE 8.50 ◆ Organizational chart—air operations.

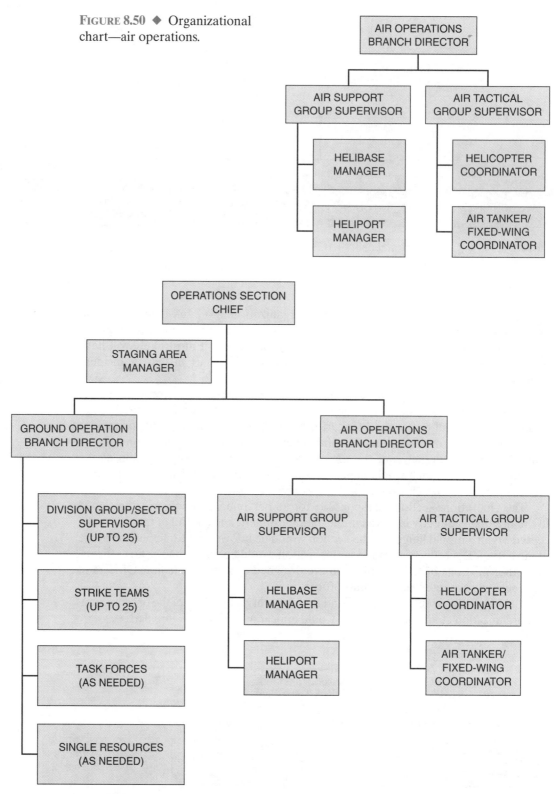

FIGURE 8.51 ◆ Organizational chart—Operations.

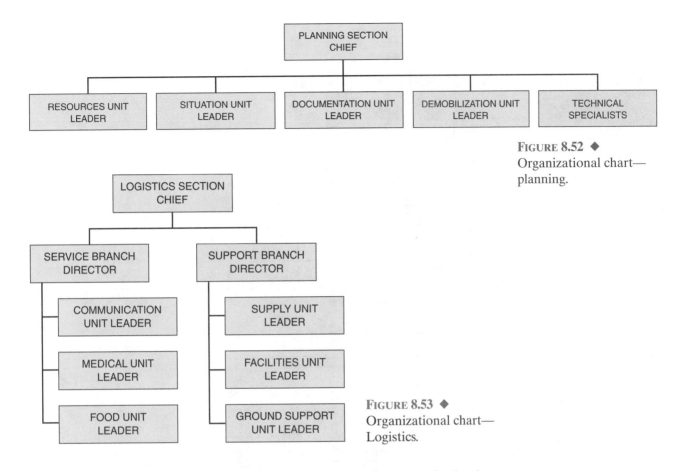

FIGURE 8.52 ◆ Organizational chart—planning.

FIGURE 8.53 ◆ Organizational chart—Logistics.

The Planning section contains the resources unit, the situation unit, the documentation unit, the demobilization unit, and the technical specialists. The organization of this section would appear as shown in Figure 8.52.

The Logistics section is divided into two branches—the service branch and the support branch. The service branch includes the communication unit, the medical unit, and the Food Unit. The responder rehab unit is a subunit of the medical unit. The support branch includes the supply unit, the facilities unit, and the ground support unit. The entire organization of the Logistic section appears as shown in Figure 8.53.

The Finance/Administration section includes the time unit, the procurement unit, the compensation/claims unit, and the cost unit. The organization appears as shown in Figure 8.54.

When the organization charts for the four sections are grouped together, the completed chart would look like Figure 8.55. It is very unlikely that any other type of fire will ever result in a more extensive organization.

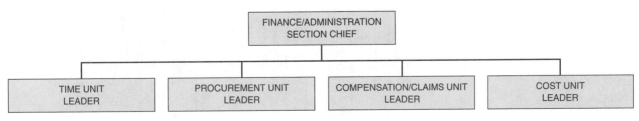

FIGURE 8.54 ◆ Organizational chart—Finance/Administration.

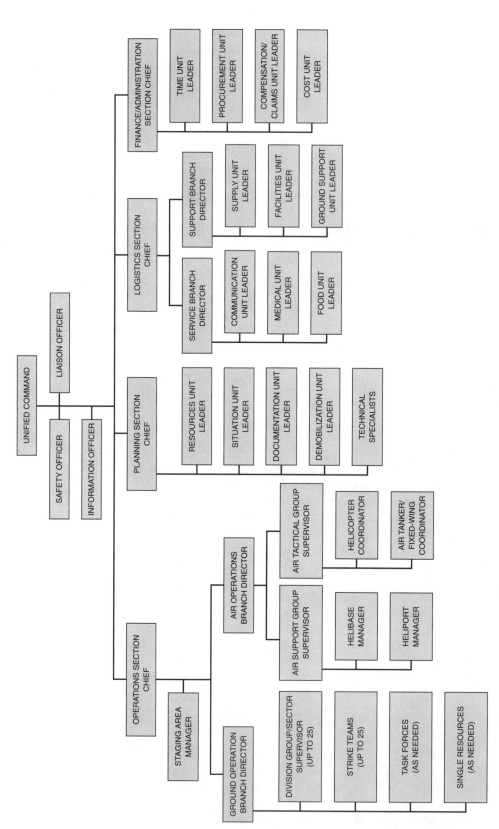

FIGURE 8.55 ◆ Organizational chart—Unified Command.

On Scene Scenario

The following is a short article from the *Joplin Globe*, Joplin, Missouri, dated August 14, 2003.

Firefighters Try to Contain Wildfire

WEST GLACIER—Firefighters set backfires along the western side of Lake McDonald on Wednesday to try to keep a massive wildfire in Glacier National Park from burning a historic lodge and private homes.

Fire officials said they were confident the intentional fires—which together could create a barrier five miles long—will burn up fuel and help keep the 39,750-acre blaze at a safe distance from Lake McDonald Lodge and about a dozen private homes.

It is interesting to read the follow-up article that appeared the next day.

Officials: Wildfire Under Control

WEST GLACIER, Mont.—Glacier National Park's popular Apgar Village was expected to reopen to visitors Friday, with officials confident that a fire set in the path of a massive wildfire had succeeded in keeping the larger blaze away from the tourist attraction.

The backfire was set near the southern end of Lake McDonald. Meanwhile, the historic Lake McDonald Lodge and about a dozen homes on the northern end of the lake remained evacuated Thursday, and officials cautioned that the blaze was still dangerous.

◆ BRUSH FIRES

Brush fires are a particular type of wildland fire. There are many similarities between a wildland fire as presented in the previous section and the brush fires presented in this section. (See Figure 8.56.) Consequently, this section contains some subject matter that was previously presented. However, certain facets peculiar to

brush fire   A particular type of wildland fire.

FIGURE 8.56 ◆ Firefighting units waiting for an assignment at a brush fire. *(Courtesy of Rick McClure, LAFD)*

brush fires will be of assistance to those interested primarily in wildland fires as previously introduced. More knowledge about wildfires can be gained by reading both sections.

Brush fires, as used in this text, are those common to the Southern California area. Southern California brush fires are different from wildland fires found elsewhere in the United States. The primary difference is due to the type of ground cover and the weather. During the hot summer months, the chaparral covering the hills and canyons is dried out and becomes explosive. **Chaparral** is a ground cover that is known as one of the fastest burning ground covers in the world. Then come the Santa Ana winds. The combination of explosive ground cover, gale strength winds, and homes being built in canyons and brush areas forms a design for disaster. The problem can be summed up in the excerpts from an article that appeared in the *Joplin Globe*, Joplin, Missouri, on October 26, 2003.

chaparral A ground cover that is known as one of the fastest burning ground covers in the world.

On Scene Scenario

A wildfire roaring through the foothills of the San Bernardino Mountains burned as many as 20 homes and forced thousands to flee Saturday as it hopscotched through dense housing tracts.

The fire, which erupted around 9 A.M. about 30 miles east of downtown Los Angeles, was fed by fierce Santa Ana winds and in a matter of hours had devoured 6,000 acres of chaparral. The cause of the blaze was unknown.

The fire was the largest of a half-dozen burning throughout Southern California, where much of the region was under a "red flag" alert because of low humidity and blustery Santa Ana winds that forecasters expected to reach 60 mph at times in canyons.

Statistically, these fires have played a prominent role when evaluating those fires that have caused losses of $1 million or more. An example is the fire that occurred in Malibu, California, in 1987. The summary of this fire appeared in Table 2.1.8 of the nineteenth edition of the *NFPA Fire Protection Handbook*.

Urban/wildfire interface Forest fires or brush fires that spread to nearby buildings.

Urban/Wildfire Interface, Malibu, California, 1987

A fire of incendiary origin with multiple points of origin was driven by Santa Ana winds gusting over 60 MPH. The fire moved 12 mi in 10 hours, entering Malibu where it damaged or destroyed 74 dwellings and mobile homes and 54,000 acres of property. Losses were estimated at nearly $9 million.

In addition to the tremendous property loss, brush fires have also been responsible for the death of a number of firefighters. Although this chapter discusses brush fires that are common to southern California, the tactics involved and the safety precautions required are applicable to similar types of ground cover fires that occur throughout the country.

Brush fires can occur during any month of the year; however, the most critical period is one following an extensive number of days of low humidity. Relative humidity is the amount of moisture in the air compared to the amount of moisture the air can hold at that temperature. The lower the humidity, the drier the air. Low humidity allows the brush to become timber-dry, making it extremely susceptible to any source of ignition. Some of the common sources of ignition are carelessness, an act of nature, or a deliberate application of flame. The overall result is the same. If the fire is reported in its infancy and properly handled, the loss will be minor. If the report is received late or the fire is improperly handled, it can develop quickly into a major conflagration.

FACTORS TO CONSIDER

A number of factors should be considered when planning an attack on a brush fire. Each of these should be carefully evaluated, as they all will have a direct bearing on where the fire should be attacked and what method of attack will prove to be most successful.

Wind and Draft Conditions. This is the first and most important factor to be considered in fighting brush fires. The fire will act completely differently when the air is still than it will when a stiff breeze is blowing. The wind has a dual adverse effect on the fire. It causes the fire to move rapidly and it also brings in fresh oxygen to feed the fire. Not only should the wind be evaluated when deciding how to make the initial attack, but it should be watched continuously during the entire fire operation. Winds have a habit of rapidly changing direction. This action has trapped firefighters by suddenly blocking their avenue of retreat. Everyone on the fire line should be on the alert continuously for flames that suddenly stand straight up. This is a strong indication that the wind direction is about to shift. (See Figure 8.57.)

Canyons have a definite and varying effect on the winds. Each canyon should be considered as a flue. Canyons act similarly to vertical openings in a building. The wind can be pulled in one direction in the main canyon and can be flowing in different directions in the lateral canyons. This can result in the fire burning in various directions at the same time. These changing conditions can make it extremely hazardous for firefighters. Personnel should *never* be sent into a canyon ahead of the fire when the fire is close to a steep slope, as the fire will rush rapidly toward a ridge and likely trap the firefighters.

The Ridges. A **ridge** is the top of the hill, the place where two opposing hill slopes come together. The ridge generally is the safest place to work on the fire and the most likely place of stopping it. When going in on a ridge, it is safer to take the one on the windward side. It should be remembered, however, that the fire will travel rapidly uphill, which makes it almost impossible to stop until it reaches the ridge. The steeper

ridge The top of the hill, the place where two opposing hill slopes come together.

FIGURE 8.57 ◆ Be careful of a wind shift when the flames stand straight up.

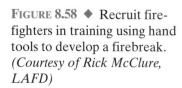

FIGURE 8.58 ◆ Recruit fire-fighters in training using hand tools to develop a firebreak. *(Courtesy of Rick McClure, LAFD)*

the slope, the faster the fire will travel. There is usually a burst of flame as the fire reaches a ridge due to fresh oxygen being pulled into the fire from the opposite side of the ridge. Because of this action, it is best that hose lines and firefighters be positioned just over the ridge from an advancing fire.

Firebreaks.   Firebreaks serve a dual purpose at a brush fire. Most importantly, they may stop the fire. Second, they provide a means of getting into areas that would be inaccessible otherwise.

Firebreaks are of two types. One is referred to as a **permanent break.** In many parts of the country, permanent breaks are constructed by governmental workers in strategic locations along the ridges well in advance of the fire occurrence. A well-designed firebreak system will create a patchwork of hazardous areas and provide protection against fire spread, much as firewalls do in buildings.

The second type is referred to as an **emergency break.** These breaks are made at the time of the fire. (See Figure 8.58.) They vary in width, depending on the sharpness of the ridge, the brush conditions, and other factors contributing to the seriousness of the fire. Under average conditions, a break from 10 to 12 feet in width is sufficient. If possible, the break should be started on the point of the ridge where it is anticipated that the fire will first reach. The break should cover the top of the ridge and extend down partway on the opposite side from the fire approach. All cut brush should also be thrown down this side. (See Figure 8.59.)

permanent break   A permanent break in an area subject to a wildfire is one made long before the fire occurs. It should be of sufficient width to stop the fire, if possible.

emergency break   A fire break that is made at a wildfire or brush fire at the time of the fire.

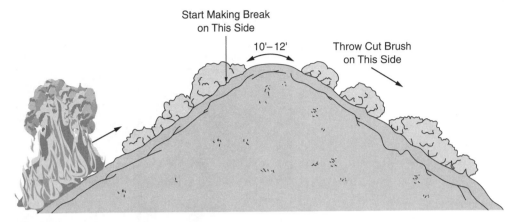

FIGURE 8.59 ◆ Emergency firebreaks should be 10 to 12 feet in width.

Available Equipment.   Both motorized and hand equipment are used on brush fires, but a good portion of the work is done with hand tools. Probably the most useful universal tool is the long-handled, round-point shovel. It is primarily used to throw dirt on the fire, but it can also be used to swat the fire, scrape the fire line, and dig a safety hole, if necessary. If things get critical, it also can be used as a face shield on a hot fire.

Some of the cutting tools used for making an emergency firebreak are the brush hook, the axe, and the Pulaski tool. The popularity of each of these tools varies in different parts of the country. Care should be taken when using these tools so as not to injure other members of the crew.

Air Support.   The amount and type of air support is a prime consideration when developing plans for stopping a brush fire. Helicopters are extremely useful for reconnaissance work, dropping smoke jumpers into strategic locations, delivering equipment, making rescues, and serving as an elevated command post. Both fixed-wing aircraft and helicopters are useful for water drops. The aerial drops may be regular or viscous water. Although it is very unlikely that water drops from aircraft or helicopters will completely extinguish a fire, they do play an important part in slowing down the fire to allow ground crews to complete the job.

Air support is readily available to some departments in the country, whereas it is a long way away from others. However, if it is apparent that the fire will last over a long period and the use of aircraft will be extremely beneficial, it should be requested if there is a chance that it is available. (See Figure 8.60.)

The most effective use of helicopters for making water drops is when a team is established, consisting of three helicopters and a land-based support group. The support group will seek out a location for a helipad that is not too far away from the fire and in close proximity to a good water source. Many of these helipads are determined long in advance of the actual fire. A line is laid from the water source to the helipad, using large diameter hose. The helicopters land at the helipad in a planned coordinated operation. The ideal is for one chopper to be dropping water at the head of the fire, the second about to land to pick up a fresh supply of water, and the third just taking off and heading in to make its drop on the fire. (See Figure 8.61.)

FIGURE 8.60 ◆ A helicopter picking up water for the next drop.
(Courtesy of Rick McClure, LAFD)

FIGURE 8.61 ◆ A helicopter making a water drop.
(Courtesy of Rick McClure, LAFD)

FIGURE 8.62 ◆ An engine company in position to protect a house from an advancing brush fire.
(Courtesy of Rick McClure, LAFD)

Exposed Structure.   Any structure in the path of a brush fire should be considered as an **exposed structure.** When evaluating the potential path of the fire travel, constant consideration should be given to the possibility of the direction of travel changing due to wind or terrain conditions. In those areas where fire apparatus can be parked safely and firefighters can work safely, companies should be dispatched to protect the exposures and lines should be laid long before the arrival of the fire. Any movable burnable objects outside of the house, such as lawn furniture, should be moved into the garage or house. All windows and drapes in the house should be closed. After all outside preparations have been made and the brush in front of the house is thoroughly wetted, the lines should be placed in a safe location and the crew members should go inside the house and wait for the approaching fire. After the fire has swept by, the firefighters can go outside and knock down any fires that may have started. (See Figure 8.62.)

exposed structure (wildfire or brush fire)   Any structure in the path of the fire.

Brush Conditions.   The density of the brush growth, or the ground cover, always should be considered when planning firefighting operations. This, together with the wind and draft conditions, will have a definite effect on how the fire is to be fought. Under favorable conditions, a direct attack may be practical in light and medium brush; however, if the brush is heavy or the winds adverse, other means of attack probably will have to be made.

FIRE EXTINGUISHMENT

The initial response to a reported brush fire will vary according to the jurisdiction in which it occurs. Some departments include the movement of tractors and helicopters, whereas others dispatch only engine companies. The first arriving officer should be familiar with the number and types of companies dispatched on the first alarm. This officer is the Incident Commander and is responsible for determining whether additional help may be required and for requesting what may be needed. (See Figure 8.63.)

Many departments use engine companies that have tank wagons as part of the routine dispatch to brush fires. A tank wagon normally carries 700 or more gallons of

FIGURE 8.63 ◆ Engine companies are always dispatched to brush fires as part of the initial response team.
(Courtesy of Rick McClure, LAFD)

water and a good supply of small lines. Small brush fires generally can be handled with two or three of these units. If the fire is of any size whatsoever, additional help probably will be required. The first-in officer should evaluate all conditions and then call for the help that is necessary, including a sufficient number of units to allow the officer to **"fail safe."** The officer should then place his or her developed plan into operation.

Fire extinguishment of larger brush fires is normally made using a combination of the direct and indirect attacks. The direct attack on a brush fire is similar to the offensive mode in a structure fire. The direct attack is one in which action is taken directly against the flames of the fire. It may be made using hose lines, hand tools, or a combination of the two.

The indirect attack is similar to the defensive mode in a structure fire. The indirect attack means applying control techniques some distance ahead of the fire to stop its progress. It generally consists of creating an area ahead of the fire through which the fire cannot burn. This can be done by using a natural barrier such as an old burn, a permanent firebreak, creating an emergency firebreak, or wetting down the brush. Usually a combination of these is employed.

After calling for the required equipment and personnel, the Incident Commander should request that any available air support make a quick survey of the area ahead of the fire's progress to determine whether any structures may have to be protected. If an aircraft is available, it should then be directed to make a water drop

fail safe The first-in officer at an emergency should evaluate all conditions and then call for the help that is necessary, including a sufficient number of units to allow the officer to "fail safe." This means that if a mistake is made, it should be made with the officer ending up with more equipment and personnel than is needed rather than not having a sufficient amount.

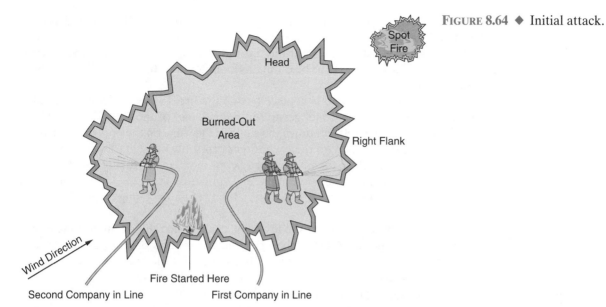

FIGURE 8.64 ◆ Initial attack.

on the head of the fire. Instructions should be given to incoming tractor units as to where to start making an emergency firebreak. Sometimes their most useful work can be done on a flank, working on the fire's edge.

If the brush conditions are such that it is possible to do so, the first-in company officer should have his or her crew lay lines to the hot side of the fire. The second-in company officer should lay lines to the opposite flank. The lines need to be advanced into the burned area so as not to endanger the crews. Both officers must carry portable radios to provide directions for other units and to relay information regarding the fire's progress to the chief officer who should arrive within a short period of time. (See Figure 8.64.)

Both companies should continue to advance their lines along the flanks of the fire in an effort to contain the main body of the fire while it moves toward the projected areas where it can be stopped. Additional arriving units can lay lines to support them in their progress.

It may be possible that water is not available or that the fire is beyond the reach of hose lines. In these cases it will be necessary to make the direct attack by using hand tools. The objective will be to separate the fire from the fuel by cutting a fire line along the edge of the fire. Dirt can be used to extinguish the fire during this process.

Shoveling dirt on a fire is hard, backbreaking work. It is important that every shovel full of dirt extinguish as much fire as possible. The most effective method is to get a good shovelful of dirt and throw it with a sweeping motion. This will generally kill several feet of fire. If the shovel is used for clearing brush, it should be done in such a manner that the brush will not have to be handled more than once.

Water drops from aircraft can be extremely effective for extinguishing or controlling spot fires. If air support is available, there should be no hesitation in calling for it.

It is possible for sparks to be blown in advance of the main fire to start spot fires. A constant lookout should be maintained for these spot fires, for they can rapidly develop into larger and more dangerous fires. Consequently, they should not be neglected. Action should be taken as quickly as possible to extinguish them. However, crews should not be put into a precarious position in order to do so. It is good practice to have roving units available to be on the lookout for spot fires and to handle them where possible.

The ridge is the safest and best place to stop the fire. To stop the fire at the ridge requires that sufficient personnel get to the ridge ahead of the fire. Crew members should be stationed 40 to 50 feet down the slope of the ridge on the opposite side from which the fire is traveling. With the exception of spot fires that start on their side of the ridge, which could place them in jeopardy, crew members normally should be safe at this location. It is important to handle any spot fires that develop as quickly as possible. Failure to do so is likely to result in the firefighters being trapped between two fires.

Two conditions must exist for the fire to be considered under control. One is that the main body of the fire reaches the prepared breaks and diminishes in size. The other is that the spread at the flanks has been cut off. Cold trailing and patrolling should be started once the fire is considered to be under control.

cold trailing A term used for brush fires that refers to the process of making a clear break between the burned-out area and unburned brush while extinguishing any burning embers.

Cold trailing is the process of making a clear break between the burned-out area and unburned brush while extinguishing any burning embers. The width of the break should be at least as wide as the brush is high and preferably wider. It is the only positive means of ensuring that the fire is out. It is done through the combined use of hand tools and mechanized equipment. The area should be raked clean of leaf mold and other combustible material. All material should be thrown back into the burned area. If water is available, it can be used effectively as an aid in cold trailing. All stumps or other burning material must be completely extinguished. Care must be taken when water is used to ensure that all sparks or embers are not driven into the unburned brush. (See Figure 8.65.)

patrolling (wildfire or brush fire) The process of maintaining a vigilant watch over the entire area to ensure that the fire does not rekindle.

Patrolling consists of maintaining a vigilant watch over the entire area to ensure that the fire does not rekindle. Extra precaution should be taken at daybreak, as this is the period when the danger of rekindle is the greatest. The danger is caused by a change in the wind conditions. The wind normally dies down during the night and starts picking up at daybreak.

Occasionally it may be necessary to backfire as a portion of the indirect method of attack. Backfiring is the intentional setting of a fire to establish a defense perimeter for the fire or as an emergency safeguard for personnel. Backfiring may become necessary due to extra heavy brush, adverse wind conditions, or on those occasions when it is not possible or practical to get ahead of the fire and establish a firebreak. Backfiring is a risky business and should be done only as a last resort. Except in extreme emergencies, backfiring should not be done without

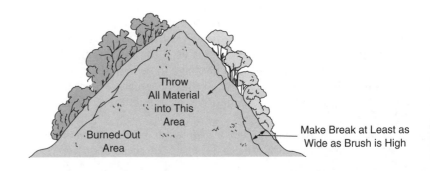

FIGURE 8.65 ◆ A cold trail break.

the direct approval of the Incident Commander or the person having the authority to issue this order.

Except in an emergency in which personnel are in danger, it is good practice not to start a backfire unless control lines are in existence around the entire area to be consumed by the backfire and all people and animals have been removed from the area. The backfire should be commenced at the nearest safe natural or personnel-made firebreak of sufficient width that will permit the backfire to burn against the wind. The backfire should be started in a wide line with the ends secured at a safe break whenever possible. It is good practice always to have a contingency plan in the event the backfire procedure fails. However, if the winds hold steady and the backfiring is done properly, it should prove to be a successful operation.

PERSONNEL CONSIDERATIONS

1. Occasionally firefighters are trapped at a brush fire. It may be possible that the only way to save themselves is by starting a backfire. If this is the case, there should be no hesitation to do so.
2. Every person going into a brush fire should be equipped with a canteen. This point cannot be overstressed. It is good practice, however, not to drink too much water when overheated or exhausted. This practice could cause severe nausea.
3. Operating in the burned area is safer than operating in an unburned area.
4. Personnel should remain calm if caught in a heavy smoke situation. It should be remembered that the freshest air is close to the ground. Good air is also normally trapped underneath a firefighter's clothing. It is important not to overlook this fact in a critical situation.
5. If necessary to retreat from a fire, it is best to work toward the flank and to the rear of the fire front.
6. If personnel are trapped on an apparatus ahead of an advancing fire, it is best to climb under the apparatus and remain there until the fire has swept by. *Do not* take a line off the apparatus and try to protect the apparatus from the advancing fire. Firefighters have been killed trying to do so.
7. Firefighters should be brought off the fire line, fed, and allowed to rest at appropriate intervals. Those replacing them should be thoroughly briefed on what has been done and what remains to be done. It is best that reliefs be made prior to nightfall. Those working on the line at night should, at a minimum, be equipped with warm clothing, a good flashlight, a canteen, and the necessary tools to complete the assigned tasks.
8. If available, fire shelters should be carried by all personnel trained in their use.

APPARATUS CONSIDERATIONS

1. If it becomes necessary to park an apparatus and leave it, the apparatus should be parked so as not to block other apparatus and be pointed in a direction of safe retreat.
2. An apparatus should not be driven through heavy smoke unless there is positive knowledge of where the pathway leads.
3. It is well to remember that it is possible for the engine of an apparatus to stop if engulfed in heavy smoke. The engine of the apparatus is operated by combining air with fuel. No air, no ignition. This point is particularly important to remember if lines are being supplied from a pump on the apparatus.
4. The headlights should be on and the windows closed when traveling through smoke. It is also good practice to sound the horn intermittently.
5. A line should be laid to protect the apparatus whenever the apparatus is being operated at a brush fire.

◆ PIER AND WHARF FIRES

dock An open body of water in which a vessel floats when tied up to a wharf or pier.

pier A structure that projects out into navigable waters so that vessels may be moored alongside of it.

wharf Similar to a pier but is built along and parallel to navigable waters.

It is not unusual to find an experienced firefighter who has recently been transferred to a harbor battalion to start talking about dock fires. One of the first things this individual learns comes as a shock—there is no such thing as a dock fire. Much to the surprise of many people, a **dock** is an open body of water in which a vessel floats when tied up to a wharf or pier. A **pier** is a structure that projects out into navigable waters so that vessels may be moored alongside of it. It is usually longer than it is wide. A **wharf,** on the other hand, is a similar type of structure that is built along and parallel to navigable waters. For the purpose of firefighting, piers and wharfs may be considered the same. In this section of the chapter, both will be referred to as wharfs. (See Figure 8.66.)

There is no universal standard for the construction of wharfs. Some are built of fire-resistive construction; others are constructed of wood. Those constructed of wood present a problem to firefighters. Although the construction of wharfs will vary in different parts of the country, an explanation of the construction features of those found in some locations of Los Angeles harbor will provide a general idea of the problems involved.

For the most part, the wharfs are built on pilings that have been driven into the sand by pile drivers. The diameter of a piling is approximately that of a telephone pole. The pilings are capped by 10" by 14" stringers. The deck is constructed of 4" by

FIGURE 8.66 ◆ A wharf fire presents many challenges to firefighters. *(Courtesy of LAFD)*

14" planking laid over 4" by 14" joists. Sometimes the deck is double thickness; at other times it is paved with a 3-inch coating of asphalt. On the land side of the wharf is a concrete bulkhead, which prevents access to the underside of the wharf from the land side. Large pipes are often attached to the joists, which provide a barrier to the effective use of streams from the waterside.

Access to the underside of the wharf can be made by climbing down a permanent wood ladder that is found on the waterside of the wharf. These ladders are strategically placed approximately every 100 feet. The ladders extend from the top of the wharf to the fender logs that are chained to the pilings at water level. Partway down the ladder are stringers and bracing, which can be walked on to reach various parts of the underside of the wharf. The pilings are heavily coated with creosote before being driven into the sand, and most of them are covered with heavy oil that has accumulated over the years. The oil on the pilings can be ignited very easily and will give off a heavy, acrid smoke when burning.

The first arriving company at a reported wharf fire is generally safe in driving directly out onto the wharf. If a small amount of smoke is rising from under the wharf, it is generally possible to take a small line to the underside and extinguish the fire. If a permanently mounted ladder is not in the vicinity of the smoke, a roof ladder may be used effectively to reach the underside. If complete extinguishment is not possible, the line can be used to control the fire while planking is pried up. Small lines from fireboats can also be used effectively for this purpose. If **scuba** (self-contained underwater breathing apparatus) **team** members are available, it is good practice to have them in the water any time a member of a land company is working under the wharf. This measure is a safety factor in the event a member of a land company falls into the water.

scuba team A self-contained underwater breathing apparatus team.

A large amount of smoke rising from under the wharf indicates the potential for a rapid spread of the fire. In this case it is still probably safe to drive the apparatus out onto the wharf, as collapse of the wharf is unlikely for a long period of time. A sufficient amount of hose should be removed from the apparatus and a layout made back to a hydrant on the land side of the wharf. It is important that the apparatus be removed from the wharf because the wharf construction and the prevailing winds could rapidly spread the fire in a horizontal direction with the wind and trap the apparatus on the wharf. If it appears that the fire may become severe, it is important that boats, ships, or other mobile equipment be moved as quickly as possible. This is important not only for the purpose of protecting these units but also to provide access to the fire for the responding fireboats.

The overall strategy when there is a strong potential for the fire spreading is to get ahead of the fire on both sides and stop its advance by setting up water curtains while attempting to extinguish the fire. Although it is important to try to set up water curtains on both sides of the fire at the same time, priority should be given to the leeward side if a decision between the two must be made.

The tactics used to set up the water curtain will depend on whether the department has a well-trained scuba team. The scuba team should be given priority if one is available, as they are able to set up the protective curtain quicker and more effectively than can be done by the alternative. Most scuba teams are now equipped with floats on which 1½-inch spray nozzles have been permanently attached. These floats can be interconnected and floated under the wharf and tied into place. The nozzles can be adjusted by team members to ensure that a solid protective screen is being provided. Once this equipment is in place, the scuba

FIGURE 8.67 ◆ Floating 2½-inch monitor developed by LAFD in early the 1960s. These were attached with short bypasses in groups of three and placed as a water curtain at each end of the fire area. They could also be used individually to attack the fire directly. *(Courtesy of LAFD)*

team can be removed and used to make a direct attack on the fire. (See Figures 8.67 and 8.68.)

If a scuba team is not available, it will be necessary to take a position well in advance of the fire spread and pry up planking or cut holes in the decking to set up distributor nozzles to create the water curtain. A number of distributors will be needed to provide an effective stop. The distributors should be set up perpendicular to the wharf's edge. The first should be placed near the land side of the wharf and others

FIGURE 8.68 ◆ SCUBA members leading in 2½-inch floating monitors to provide a water curtain using nozzle reaction to propel the monitors. *(Courtesy of LAFD)*

located a maximum of 10 feet apart until the wharf's edge is reached. The last one should be placed over the edge of the wharf to prevent the fire from getting around the outside of the curtain.

This method is time-consuming and is usually not successful if the fire moves fast. In fact, if the wharf is paved, it is almost impossible to get a water curtain in place in the time available.

Some wharfs have openings permanently installed that may be used for the placement of distributors. These are not normally located exactly where needed, but they might serve some useful purpose while the attempt is being made to cut through the wharf's decking. Regardless of whether holes are cut or the permanent installed holes are used, it is important that small lines be laid to protect those working prior to opening the deck. It is good practice that breathing apparatus also be worn.

A direct attack on the fire should be made while the water curtains are being put in place and continued until the fire is extinguished. The first attack is generally made by a fireboat approaching from the leeward side and running parallel to the wharf while sweeping the underside with a battery of heavy streams. Firefighters on the attacking boat should wear breathing apparatus, as an approach from the leeward side places them directly in the path of the heavy smoke being given off by the fire. Unfortunately, due to the mass of pilings, timbers, piping, and so on, it is very unlikely that the fire will be extinguished on the original sweep. It can be expected that some of the fire will remain inaccessible to the fireboat's streams regardless of how many sweeps are made. This is not only due to the hidden areas created by the construction but also because the fireboat's streams cannot be projected from a low enough location to reach all parts of the underside of the wharf.

Similar to a structure fire in which it is necessary to go inside to extinguish the fire completely, it is necessary to go under the wharf to get at all of the fire. The best method of doing this is to put members of the scuba team in the water and send them under the wharf. Most scuba teams are equipped with floats that can be attached to the hose lines to keep them from sinking while the advancement is being made. Scuba members working under a burning wharf are probably in a safer working environment than a hose handler working a line inside a burning building. If the wharf begins to collapse or if fire spreads over the water due to a flammable liquid spill, a scuba member merely has to roll over, dive down 10 or 15 feet, and swim to a place of safety.

If a scuba team is not available, it may be possible to use a small vessel such as a rowboat to take lines under the wharf to complete the extinguishment.

The primary task of land-based companies, other than trying to establish a water curtain, is to protect the exposures. The exposures include any cargo on the wharf, the structure located above the wharf, and possibly adjacent wharfs. If possible to do so, the best method of protecting any cargo on the wharf is to have it moved to a safe location.

Overhauling a wharf fire may become a difficult task. In addition to getting lines under the wharf to wet it down thoroughly, it may be necessary to pry up planking to get at hidden spots, generally using heavy equipment. Jackhammers can be employed if it becomes necessary to remove paving.

◆ SUMMARY

This last chapter describes the types of fires that firefighters may not necessarily encounter each day, but must be prepared to handle. Over the past few years we have seen on television and read in the newspapers about the destructive nature of wildland fires and wildfires. Firefighters must be trained to deal with these types of fires and understand their destructive nature. Firefighters also must know how to wear wildland firefighting personal protective clothing properly and understand the tactics and strategy of these sometimes very large fires.

Regardless of the type of emergency that a dispatch center receives, firefighters must always be ready to respond. When people need help, even in nonfire emergencies, firefighters are all called for assistance. Fire departments are now "all-risk" departments, and there is not an event yet that occurs that the fire service does not respond to in order to save lives and property.

Review Questions

Flammable Liquid Spill Fires

1. From which side of a flammable liquid spill should firefighters approach the spill?
2. Which is more important in determining the size of a fire, the amount of the liquid spilled or the area covered by the spill?
3. What size spill fires should be attacked with foam?
4. What is the most effective method of knocking down a large flammable liquid fire?
5. How can water be used effectively on a flammable liquid spill fire?
6. How should the advance be made to stop the flow of liquid feeding a flammable liquid spill fire?

Vertical Storage Tank Fires

1. What factors will determine the best method of extinguishing a fire in a vertical storage tank?
2. When can a fire in a vertical storage tank be extinguished with the use of water?
3. What are some of the problems encountered when extinguishing fires in large vertical storage tanks?

Horizontal Storage Tank Fires

1. How should a fire in the vent of a horizontal storage tank be extinguished?

2. What is the primary firefighting problem associated with horizontal storage tanks?
3. What is the basic principle that should not be ignored when playing water on an exposed horizontal tank?

Fires Resulting from Breaks in Transportation Lines

1. What is the basic procedure to use for extinguishing fires in transportation lines?
2. What method should be used for stopping the liquid flow in a transportation line once the fire is extinguished?

Oil Tank Fires

1. What are the two primary problems of oil tank fires?
2. What principle should be followed once water has been played on an adjacent tank to keep it cool?
3. What is the most effective extinguishing agent for oil tank fires?
4. What type of oil fires can be extinguished by the use of water alone?
5. What are perhaps the greatest dangers to firefighters at an oil tank fire?

Flammable Gas Emergencies

1. What are the primary control tactics used on LPG emergencies?

2. What do the letters in the term BLEVE stand for?
3. What generally causes BLEVEs?
4. What are the two positive warning signs that should not be ignored regarding the potential for a BLEVE?

Electrical Fires

1. What is the primary problem for fire-fighters when fighting electrical fires?
2. What should be used to extinguish a fire in a small energized electrical appliance?
3. What are some of the precautions that should be used in fires in underground equipment?

Silo Fires

1. What are the three types of silos?
2. With which type of silo is it recommended that no attempt be made to extinguish the fire?
3. What is the primary cause of silo fires?
4. What should be the first action of a fire-fighter who is placed inside a silo for the purpose of extinguishing the fire?
5. What action should be taken if the fire in an oxygen-limiting silo is still burning after three weeks?

Wildfires

1. What are the various classifications of wildland fires?
2. Define each of the following as related to wildland fires: origin, head, fingers, rear, flanks, perimeter, island, spot fires, slopover, flare-up, blowup, firestorm, green area, and black area.
3. What are some of the factors affecting fire behavior at wildland fires?
4. What are some of the common fuel types common to wildland fires?

5. Define each of the following as related to wildland fires: relative humidity, local winds, sea breezes, offshore breezes, surface winds, gradient winds, and gravity winds.
6. Define each of the following as related to wildland fires: slope, aspect, chute, saddle, and canyons.
7. What tactics are used to make a direct attack on a wildland fire?
8. What tactics are used to make an indirect attack on a wildland fire?
9. What is the difference between backfiring and burning out?
10. How are mop-ups completed?
11. What are the 10 standard firefighting orders for wildland fires?
12. What are the 18 "watch out" situations for wildland firefighting?

Brush Fires

1. What are some of the factors that should be considered in fighting brush fires?
2. What tactics are used to protect a structure at a brush fire?
3. What are the tactics used for extinguishing brush fires?
4. What are some of the personnel considerations for brush fires?
5. What are some of the apparatus considerations for brush fires?

Pier and Wharf Fires

1. What is the difference between a dock fire, a pier fire, and a wharf fire?
2. What are the best tactics to use for extinguishing the underside of a wharf fire?

Glossary

abnormal smoke colors Colors such as yellow and red that usually indicate that chemicals are involved.

adiabatic compression When gases that are compressed develop heat.

aerial fuels (wildfire) Those fuels that are physically separated from the ground.

air agitation method A subsurface injection of air used as a means of extinguishing an oil tank fire.

aircraft rescues The control of a life-threatening fire in the critical area during the time that it takes all physically able occupants to leave the aircraft on their own.

aspect (wildfire or brush fire) Refers to compass direction.

automatic nozzles Automatic nozzles are designed to maintain a constant nozzle pressure regardless of the flow.

backdraft A condition referred to as a fire or smoke or smoke explosion.

backfiring (wildfire or brush fire) The intentional setting of a fire to establish a defense perimeter for the fire or as an emergency safeguard for personnel.

Bangor ladders Bangor ladders are extension ladders that use stay poles for added leverage and stability.

base The base is the reporting point for incoming companies and serves as the collection point for personnel and equipment pending transfer to the staging area.

basement The lowest story of a building or the one just below the main floor, usually wholly or partially lower than the surface of the ground.

bilge The lower part of a ship's hull or hold.

biological weapons Weapons that contain toxins that can enter the body through inhalation or skin absorption.

black area The burn area of a wildfire or brush fire.

BLEVE A boiling-liquid, expanding vapor explosion.

blowup A situation that is similar to a wildfire or brush fire flare-up but is of significant magnitude to force a change in existing control plans.

boat deck The second deck above the main deck in the superstructure of a cargo ship.

boilover The expulsion of the contents of an oil tank by the expansion of water vapor that has been trapped under the oil and heated by the burning oil.

bow The front portion of a ship.

branches Branches are portions of an incident used for management purposes.

bridge deck The third deck above the main deck in the superstructure of a cargo ship.

brush fire A particular type of wildland fire.

buddy breathing A means for two firefighters to share the air supply from a single air chamber.

buddy system A situation in which two firefighters work together and monitor each other to enforce safety procedures at a fire.

bulkhead Similar to a wall except it is made of metal.

burnout (wildfire or brush fire) A method of widening a control line by eliminating fuel between the control line and the fire.

CAD Letters refer to a computer-aided dispatch system.

CAFS Letters stand for compressed air foam system.

canyons Canyons can be thought of as topography chimneys. The heat, smoke, and fire gases can be channeled through canyons at a very rapid rate.

carbon dioxide The product of complete combustion.

carbon monoxide The product of incomplete combustion.

cellar A room or group of rooms below ground level and usually under a building, often used for storing fuel, provisions, and so on.

center-core design The center-core design concept features office areas or living areas surrounding a "core" containing stairwells, elevators, and utilities.

chain reaction A process that occurs during the growth of a fire that can be compared with the process that occurs with the transmission of heat during the process of conduction.

chaparral A ground cover that is known as one of the fastest burning ground covers in the world.

charcoal alley A term used at civil disturbances to identify an area that is completely or nearly completely destroyed by fire.

chemical weapons Weapons designed to attack an individual's nervous system, eyes, skin, intestinal tract and mucous membranes.

chute (wildfire or brush fire) A steep V-shaped drainage.

civil disorder See "civil disturbance."

civil disturbance An intentional act of disobedient behavior by a group of people in violation of public policy or established laws or regulations that results in acts of violence directed at persons or property.

Class A fire A fire involving ordinary combustible material such as wood, paper, cloth, and so on.

Class B fire A fire involving flammable liquids, gases, and greases.

Class C fire A fire involving energized electrical equipment.

Class D fire A fire involving a combustible metal such as magnesium, titanium, sodium, and potassium.

coamings The raised sides of the hatchways of a ship at deck level.

cold trailing A term used for brush fires that refers to the process of making a clear break between the burned-out area and unburned brush while extinguishing any burning embers.

collapse zone A safety zone in which no operations should be performed or anyone allowed within it.

collapsible ladders Collapsible ladders are versatile ladders that are easy to carry and position by one firefighter.

combination ladders Combination ladders are designed so that they can be used as extension ladders or converted to step ladders.

combustible liquid A liquid that has a flash point of 100°F or more.

command function The command function refers to the Incident Commander or to the unified command, depending on the incident and the number of agencies involved.

command mode An operational mode whereby there is a need for an immediate strong, direct attack on the fire due to its size, complexity, or the potential for a rapid expansion.

compensation/claims unit The compensation/claims unit is responsible for the financial concerns over serious injuries and deaths that occur as a result of operations at the incident.

confining the fire Refers to preventing a fire from spreading beyond the point of origin.

conflagration A major building-to-building flame spread over a large area that crosses natural or man-made barriers.

constant flow nozzles Nozzles designed to provide the same flow at a specific nozzle pressure regardless of the setting of the flow pattern. Also known as fixed-gallon age nozzles.

continuous sizeup The continuous and comprehensive estimate of the situation as firefighting operations proceed.

control line (wildfire or brush fire) Any natural or constructed barrier used to control a fire.

conventional silo A conventional silo is usually equipped with a three-foot diameter chute that runs the full length of the height of the silo. The chute allows the silage to fall down into the barn, into a loading wagon or into a conveyer during unloading operations.

cost unit The cost unit is responsible for tracking cost data, analyzing the data, making cost estimates, and recommending cost-saving measures.

critical temperature for structural steel The temperature at which steel can no longer be considered as a supporting element.

dead smoke Nonbuoyant smoke that remains relatively still and may collect in pockets.

deckhead Similar to a ceiling except it is made of metal.

decomposition The breaking of a compound into its basic elements.

deep tanks Tanks that are located under the cargo hold of a cargo ship that are used to trim the ship.

defensive mode A fire situation in which the fire is fought from outside of the building.

defensive/offensive mode A transition mode whereby operations are changed from a defensive mode to an offensive mode. It is generally used where the initial size-up of the fire indicates that the fire is beyond the capability of hand lines.

demobilization unit The demobilization unit is responsible for the orderly, safe, and efficient stand-down of an incident.

diked area An area around a storage tank that has been constructed to contain any spill from the tank.

direct attack (wildfire or brush fire) Refers to fighting the fire itself with the use of water, throwing dirt on it, using beaters, or building a fire line at the burning edge and throwing the material into the burned-out area.

dirty bomb An explosive device that spreads radioactive materials over a specific area.

division A portion of the command function that can be allocated by geographical area or by functional sectoring depending on the type and extent of the emergency.

dock An open body of water in which a vessel floats when tied up to a wharf or pier.

drafting Taking water from a nonpressure source such as the ocean or other open body of water.

dwelling fires The term dwelling fires refers to all types of structures in which people live.

electrical fire A fire in energized electrical equipment.

emergency break A firebreak that is made at a wildfire or brush fire at the time of the fire.

endothermic reaction When heat is absorbed during a reaction.

engine company The company responsible for locating, confining, and extinguishing the fire.

ensiling A method of preserving green fodder.

enunciator A piece of equipment that indicates on what floor or floors a problem exists and whether it was tripped by a manual pull station, a smoke detector, a heat detector, or a water flow.

escape hatch An escape hatch consists of an iron ladder inside a metal tube. A small door from the tube opens into the ship's hold at each of the various deck levels.

exothermic reaction When heat is given off during a reaction.

explosive limits The limits within which a vapor will burn.

exposed structure (wildfire or brush fire) Any structure in the path of the fire.

exposure Anything in close proximity to the fire that is not burning but that might start burning if some type of corrective action is not taken quickly.

extension ladders Extension ladders are adjustable ladders that can be raised to various heights.

external exposures Those exposures that are located outside the fire building.

facilities unit The facilities unit is responsible for providing the layout and activation of those fixed facilities required at the incident.

fail safe The first-in officer at an emergency should evaluate all conditions and then call for the help that is necessary, including a sufficient number of units to allow the officer to "fail safe." This means that if a mistake is made, it should be made with the officer ending up with more equipment and personnel than is needed rather than not having a sufficient amount.

fantail A deck house on the aft of a cargo ship that contains extra crew quarters.

fast attack mode A term used for marginal offensive fire attacks or when the safety of responding firefighters is a major concern or when there are rescues to be made immediately.

fender log A log that is attached to the wharf at the water level to prevent approaching ships from damaging the wharf.

fingers Long, narrow strips of fire that extend out from the main body of a wildfire or brush fire.

finger piers At a storage anchorage for small boats, there are floating walks that extend out for varying distances into the body of water. Extending out from these main floating walks are finger piers to which the boats are secured.

fire Rapid oxidation accompanied by heat and flame.

fire containment (wildfire or brush fire) A fire is contained when it is surrounded on all sides by some kind of boundary but is still burning and has the potential to jump a boundary line.

fire controlled (wildfire or brush fire) A fire is controlled when there is no further threat of it jumping a containment line.

fire curtain A fire curtain in a theater is designed to prevent fire, heat, and smoke from entering the auditorium area.

fire edge The boundary of a wildfire or brush fire.

fire line (wildfire or brush fire) A portion of a control line that is constructed near the fire's edge.

fire out (wildfire or brush fire) The fire is out when mop-up is finished and all crews can be released.

fire point The temperature at which a flash will occur above a liquid and the resultant fire will continue to burn.

fire stream A stream of water from the time it leaves a nozzle until it reaches the point of intended use, or until it reaches the limit of its projection, whichever occurs first.

firestorm The intense burning of the fuel at a wildfire or brush fire over a large area, which results in a huge convection column over the fire area.

first-in district A district into which a company should be the first company to arrive at an emergency.

fixed-gallonage nozzles See "constant flow nozzles."

flame The visible and colorful portion of a fire.

flammable limits The limits within which a vapor will burn.

flammable liquid A liquid that has a flash point below 100°F.

flanks The sides of a wildfire or brush fire.

flare-up The sudden increase in the intensity or speed of the fire spread of a wildfire or brush fire.

flash point The temperature at which a flash will occur across the face of a liquid and go out.

flashover When all of the combustibles in a room break into flames.

flying bridge The fourth deck above the main deck in the superstructure of a cargo ship.

food unit The food unit is responsible for feeding those at the emergency.

footing Stationing a firefighter at the bottom of a ladder with his or her toes placed against the heel of the ladder or one foot placed on the bottom rung for the purpose of securing the ladder.

forage (crops) Food of any kind used for feeding horses and cattle.

forcible entry The process of using force to gain entrance into a building or secured area.

freelancing The problem of company commanders committing the company to a certain task without orders from the Incident Commander.

fuel (wildfire or brush fire) The combustible or flammable material that is available for feeding a fire.

fuel gas A flammable gas customarily used for burning with air to produce heat.

GIS Refers to a Geographical Information System.

gravity wind A downslope wind.

green area The unburned area adjacent to the involved area at a wildfire or brush fire.

group fire A building-to-building flame spread over a small area.

groups Groups are units of personnel generally assigned to a function such as ventilation, search and rescue, salvage, and so on.

hatch covers The hold of a ship is secured by placing metal support beams across the hatchways and attaching them to the coamings. Hatch boards (referred to as hatch covers) are then laid on the beams to cover the opening.

hatchways The openings in the deck of a ship through which the cargo is loaded and unloaded.

head That part of a wildfire or brush fire that is spreading or traveling most rapidly.

heat of compression When gases that are compressed develop heat.

heat A form of energy in motion.

heat capacity The ability of an extinguishing agent to absorb heat.

heel The side opposite the head at a wildfire or brush fire.

"here and now" era The "here and now" era refers to the fact that a terrorist attack can occur at any time, at any place, and possibly involve weapons of mass destruction (WMD).

high-rise Any multiple-story building that requires the use of high-rise firefighting tactics for effective extinguishment.

high-rise fire attack team The first company to arrive at a high-rise fire that has the primary responsibility to locate and identify the emergency and determine its scope.

high-rise staging area The collection point for the equipment and personnel that will be used on the fire.

high-temperature thermometer A thermometer used for locating hot spots in silage.

hit-and-run tactics Where the fire is knocked down by the use of heavy streams and the task force moves on to another assignment without doing salvage work or overhaul.

holds Where the cargo is carried on cargo ships.

horizontal storage tank A term used to include tank trucks and railroad tank cars as well as permanently installed tanks.

hot zone An area designated as containing dangerous chemicals.

humidity The amount of moisture in the air.

hydrocarbons Those flammable liquids that are not miscible.

IDLH Refers to firefighting atmospheres that are considered dangerous to life and health.

ignition temperature The minimum temperature that will cause self-sustained combustion of a material.

incident action plan (IAP) The IAP is the basis for determining when and where resources will be assigned at the incident.

Incident Commander The individual in charge of an emergency.

indirect attack With an indirect attack, the water is not directed at the seat of the fire as with the direct attack, but rather above the fire and into the heat that has built up at the ceiling level.

indirect attack (on a wildfire) The indirect attack is similar to the defensive mode on a structure fire. The indirect attack means applying control techniques some distance ahead of the fire to stop its progress. It generally consists of creating an area ahead of the fire that contains no fuel.

Information Officer The individual responsible for the production and release of all information about the incident to the media and other appropriate agencies.

interior (wet) standpipe system A built-in fire protection system that contains water under pressure at all times and has racks attached containing linen hose.

internal exposures Those exposures located inside a building.

investigation mode A situation in which nothing is showing on the arrival of the first unit.

island An unburned area located within the fire perimeter of a wildfire or brush fire.

Jacob's ladder A rope ladder with wooden steps used aboard ships. It is kept coiled up and lowered over the side to permit someone to climb aboard.

jet engines Jet engines draw air in through the front, compress it, mix it with fuel, and ignite it.

laddering operations Raising or using ladders for the purpose of making physical rescue, gaining entrance into a building, providing a path for hose lines, gaining access to various portions of the building, and various other phases of firefighting and fire or water control.

lapse rate   The rate of decrease of the temperature as the elevation increases.

latent heat   The heat absorbed as a material changes from a solid to a liquid or from a liquid to a gas.

latent heat of fusion   The amount of heat absorbed as a material changes from a solid to a liquid.

latent heat of vaporization   The amount of heat absorbed as a substance passes between the liquid and gaseous phases.

leeward side   The side toward which the wind is blowing.

liaison   In the Incident Command System, the functional responsibility for identifying, contacting, assisting, and cooperating with outside agencies.

live smoke   Heat will cause the smoke to be buoyant. It will rise and swirl about, the degree of which depends on the amount of heat providing the stimulus. Smoke in this condition is referred to as live smoke.

lobby control officer   At a high-rise fire, this officer is responsible for controlling all vertical access routes (including the elevators), controlling the HVAC, and coordinating the movement of personnel and equipment between the base and the staging area.

local winds   Winds caused by heating and cooling patterns.

logistics   The Logistics section is responsible for ordering the personnel, equipment, and resources required to control the incident and support the responding personnel.

LOUVER   This word stands for: ladder operations, overhaul, controlling the utilities, ventilation, forcible entry, and rescue.

louvering   An effective method of ventilating a roof.

lower deck   The bottom deck of a ship's hold.

manifest   The manifest indicates what materials are being carried on a ship and in what holds the material is stored.

manually adjustable nozzles   See "selectable-gallonage nozzles."

medical unit   The medical unit is responsible for the emergency treatment and transportation of any individual injured or taken sick at the emergency.

miscible   Capable of being mixed with water.

modified silo   A modified silo is one that has been modified from its original design.

molotov cocktail   A thin-skinned bottle filled with gasoline and equipped with a rag in the neck of the bottle that is used as a wick.

mop-up   The process of completing the extinguishment of a wildfire or brush fire. It entails finding and extinguishing all hot spots, particularly those close to the control line.

multipurpose nozzle   A nozzle that is capable of producing a straight stream and a spray stream at the same time.

mushroom effect   The vertical rise of the heat and products of combustion, their banking along the ceiling, and their travel down the walls.

nacelle   The housing around an aircraft engine.

negative ventilation   Whenever a blower is used to pull smoke and noxious gases out of a structure.

offensive mode   A fire situation in which a direct attack is made on the fire. It requires firefighters to go inside and put out the fire.

offensive/defensive mode   A transition mode whereby operations are changed from an offensive mode to a defensive mode.

offshore breeze   The movement of air from the land to the water.

operations function   The objective of the operations function is to provide a system for dividing up the incident into more manageable sectors or areas.

origin   The area where ignition first occurs at a wildfire or brush fire.

overhaul   Overhaul is the final task performed by firefighters at the scene of a fire. The primary objective of overhaul is to ensure that the fire is out; however, it generally includes doing whatever is necessary to leave the premises in as safe and secure a condition as possible.

oxidation A chemical process whereby an atom from one material combines with an atom of oxygen from another material to form a new material.

oxidizers Chemicals that contain oxygen in their makeup and release it under the process of decomposition.

oxygen-enriched atmosphere When the oxygen concentration is above 21 percent.

oxygen-limiting silos Oxygen-limiting silos are constructed of steel or poured concrete. The most common of these silos is the blue-colored Harvestore brand.

panic An emotional reaction to fear.

partition fire A concealed fire in a wall.

PASS A personnel alert safety system that can assist rescuers in locating a missing or trapped firefighter.

patrolling (wildfire or brush fire) The process of maintaining a vigilant watch over the entire area to ensure that the fire does not rekindle.

percent of a grade The rise in feet for every 100 feet of horizontal distance.

perimeter The boundary of a wildfire or brush fire.

permanent break A permanent break in an area subject to a wildfire is one made long before the fire occurs. It should be of sufficient width to stop the fire, if possible.

personnel accountability system A system designed to ensure that the Incident Commander has consistent and up-to-date information on the location of every member and unit at an emergency.

pier A structure that projects out into navigable waters so that vessels may be moored alongside of it.

piston engines Internal-combustion reciprocating engines.

polar solvents Those flammable liquids that are miscible.

port Left, looking from the stern toward the bow of a ship.

positive pressure ventilation A method of clearing an area of smoke and gases for the purpose of gaining entry to extinguish the fire.

pre-alarm size-up A size-up made prior to the alarm sounding. It normally first manifests itself when any information on the fire building or its exposures that could affect operations on the fire ground is obtained.

pre-incident planning Planning to ensure that personnel responding to a fire incident know as much as possible about the building or group of buildings to which they are responding.

pre-incident planning inspection An inspection of a building to gather as much information as possible about the building for the purpose of fighting a fire before it occurs.

preliminary size-up The immediate estimate of the situation made by the fire officer who arrives on the scene first.

procurement unit The procurement unit is responsible for all financial matters pertaining to vendor contracts.

pyrophoric reaction When the mixing of two materials causes instant ignition.

rapid intervention team A standby team that is available for the rescuing of firefighters who become lost, incapacitated, or trapped in a fire due to a backdraft, a flashover, a collapse, an SCBA malfunction, an injury, or some other similar event.

rear The side opposite the head at a wildfire or brush fire.

rekindle A situation in which the fire department leaves the scene, thinking it has extinguished the fire, and is called a second time to the location because the fire had not been completely extinguished the first time.

relative humidity The amount of moisture in the air compared with the amount the air can hold at a given temperature.

rescue The process of removing people from burning buildings or buildings likely to become involved to a place of safety.

resource unit This unit is responsible for tracking all of the resources that have been requested, dispatched, or are on the scene at an incident.

responder rehab section This section is responsible for the rehabilitation of firefighters.

ridge The top of the hill, the place where two opposing hill slopes come together.

riots Riots can generally be defined as a spontaneous outburst of group violence characterized by extreme excitement mixed with rage.

rollover When the resultant fire manifests itself in a rolling motion, normally at ceiling level and ahead of the main fire.

roof ladders Roof ladders are single ladders that are equipped at the tip with folding hooks.

saddle (wildfire or brush fire) A depression between two adjacent hilltops.

safe location This is in reference to a ship that enters a harbor when the cargo is explosive or considered too dangerous. The ship should be anchored in a safe location away from congested areas.

safety officer An officer at the scene of an emergency who has the authority to take corrective action at the scene, if possible, and to monitor the actions of those at the scene to ensure that safety procedures are followed.

scuba team A self-contained underwater breathing apparatus team.

scuttle hatch An opening in the roof that extends directly into the attic.

sea breeze A movement of air from the water to the land.

second alarm The number and type of companies that will be received when the Incident Commander requests a second alarm dispatch. The assignment will vary from one city to another.

selectable-gallonage nozzles Nozzles that provide a constant flow regardless of the stream pattern. Also known as manually adjustable nozzles.

shaft alley A fairly large tunnel that houses the propeller shaft of a ship. It extends from the engine room to the propeller.

shaft tunnel hatch A vertical, tunnel-like opening that extends from the main deck of a ship to the shaft alley.

shelter deck The first deck above the main deck in the superstructure of a cargo ship.

silos Large farm structures made of poured concrete, concrete staves, or steel.

single ladders Single ladders are nonadjustable in length and are constructed as a one-section ladder.

situation unit This unit is responsible for the collection, tracking, and displacing

of all information relative to the status of the incident.

skylights Used to provide light to areas below the roof.

slash The material that is left on the ground after logging operations are complete.

slope Slope is the same as grade. The percent of a grade (or the slope) can be defined as the rise in feet for every 100 feet of horizontal distance.

slopover (tank) A small spill of oil out of an oil storage tank.

slopover (wildfire) A fire that has extended over the control line or natural barrier into an unburned area at a wildfire or brush fire.

smoke One of the primary products of combustion.

solid stream Another term for a straight stream.

span of control The number of personnel that one supervisor can manage effectively.

specific heat The ability of an extinguishing agent to absorb heat.

spot fires New fires that have been started by sparks or flying embers carried aloft by convection or winds of a wildfire or brush fire.

spray stream One of the two basic types of streams produced by a nozzle.

stable air condition A condition in which the smoke will rise nearly straight up and then spread out nearly equally in all directions.

staging The staging area is a location to which incoming units report and can be rapidly deployed.

staging area At a high-rise fire, the staging area is the collection point for the equipment and personnel that will be used on the fire. It normally is located two floors below the fire floor.

starboard Right, looking from the stern toward the bow of a ship.

staves Curved concrete blocks held in place by steel rings on silos.

stern The rear portion of a ship.

strike team A combination of five identical units under the command of an individual designated as the strike team leader.

strip mall A row of small businesses housed under a common roof.

subsurface injection of foam Refers to injecting foam at the bottom of an oil tank fire. The foam and the cool oil both rise to the surface. A foam blanket is eventually built over the surface to obtain complete extinguishment.

superstructure The decks of a cargo ship above the main deck.

supply unit The supply unit is responsible for the ordering of all personnel, equipment, and supplies required at the incident.

support branch The support branch is a staff function responsible for the development and implementation of logistics plans designed to support the incident's action plans.

target hazard An occupancy that constitutes a large collection of burnable valuables or an occupancy in which the life hazard is severe.

task force A combination of resources assembled to carry out a specific assignment.

taxpayer building A cheaply constructed building that was constructed in the early 1900s for the purpose of saving money.

temperature The amount of heat present or given off.

temperature inversion A condition in which the temperature increases instead of decreasing when a given elevation is reached.

terrorism A type of act that is designed to kill innocent people or disrupt the lives of the survivors of those killed and create panic in a community or nation.

thermal heat The ability of an extinguishing agent to absorb heat.

thermal imaging unit A unit designed to detect variations in temperature.

time unit The time unit is responsible for tracking and recording the on- and off-duty time for all personnel working at the incident.

trench A trench is a cut that is placed in the roof of a structure that divides a portion of the attic that is burning from a portion of the attic that is not burning.

turbine engines Jet engines.

'tween deck The middle of three decks in a ship's hold.

unified command A situation in which agencies from several jurisdictions or in which several departments from the same jurisdiction work together for a common cause.

unity of command The principle that an individual can work effectively for only one boss and that everyone within an organization has a designated supervisor to whom he or she reports.

unstable air condition A condition in which the rising smoke will drift downwind.

upper deck The upper of three decks in a ship's hold.

urban/wildfire interface Forest fires or brush fires that spread to nearby buildings.

vapor density The weight of a volume of vapor as compared with an equal volume of air.

ventilation The process of replacing a bad atmosphere with a good atmosphere.

vertical ventilation The creation of an opening above a fire that allows the heat and smoke to escape.

viscosity Viscosity is the ability to flow. A liquid with a low viscosity will flow easily. A liquid with a high viscosity flows relatively slowly.

weather (wildfire or brush fire) The state of the atmosphere over the fire area.

wharf Similar to a pier but is built along and parallel to navigable waters.

winds Air in motion.

windward side The side from which the wind is blowing.

worker A fire in progress.

Bibliography

Brunacini, Alan V., *Fire Command,* 2nd edition, Quincy MA: NFPA, 2002.

Ertel, Mike & Berk, Gregory C., *Firefighting Basic Skills and Techniques,* Atlanta GA: Goodheart-Wilcox Company, Inc., 1998.

FEMA. Retrieved on June 20, 2006 from www.fema.gov/emergency/nims/index.shtm.

Firefighter's Handbook: Basic Essentials of Firefighting, Clifton Park, NY: Thomson Delmar Learning, 2000.

Goodson, Barbara & Adams, Carl (eds.), *Fundamentals of Wildland Fire Fighting,* 3rd edition, Oklahoma, OK: IFSTA, 1998.

Klinoff, Robert, *Introduction to Fire Protection,* 2nd edition, Clifton Park, NY: Thomson Delmar Learning, 2003.

International Association of Fire Chiefs and National Fire Protection Association, *Fundamentals of Fire Fighter Skills,* Sudbury, MA: Jones and Bartlett, 2004.

Mahoney, Eugene F., *Fire Department Hydraulics,* Upper Saddle River, NJ: Brady, 2004.

National Resource, Agriculture, and Engineering Service (NRAES). P. O. Box 4557, Ithaca, NY 14852-4557.

Smith, James P., *Strategic and Tactical Considerations on the Fireground,* Upper Saddle River, NJ: Prentice Hall, Brady Fire, 2002.

Abbreviations

AFFF   Aqueous film forming foam
BLEVE   Boiling-liquid, expanding vapor explosion
BTU   British thermal unit
CAD   Computer-aided dispatch
CAFS   Compressed air foam system
CO   Carbon monoxide
CO_2 Carbon dioxide
DEMOB   Demobilization unit
DOC   Documentation unit
EMS   Emergency medical services
FAST   Firefighter assist and search team
FEMA   Federal Emergency Management Association
FPF   Fluoroprotein foam
GIS   Geographical Information System
GPS   Geographical Positioning System
HAVC   Heating, ventilation, and air conditioning
IAFC   International Association of Fire Chiefs
IAFF   International Association of Fire Fighters
IAP   Incident action plan
IC   Incident Commander
ICS   Incident Command System
IDLH   Immediately dangerous to life or health
IMS   Incident Management System
ISO   Incident safety officer

LOUVER   Ladder operations overhaul, controlling the utilities, ventilation, forcible entry, and rescue
NFA   National Fire Academy
NFPA   National Fire Protection Association
NRAES   Natural Resource, Agriculture, and Engineering Service
OSHA   Occupational Safety and Health Administration
PAR   Personnel accountability report
PASS   Personal alert safety system
PF   Protein foam
PIO   Public information officer
PPE   Personal protective equipment
PRV   Pressure reducing valve
QAP   Quick action plan
RECEO VS   Rescue, exposures, confinement, extinguishments, overhaul, ventilation, salvage
RESTAT   Resources status unit
RIC   Rapid intervention crew
RIT   Rapid intervention team
SAR   Supplied-air respirator
SCBA   Self-contained breathing apparatus
SCUBA   Self-contained underwater breathing apparatus
SITSTAT   Situation status
SO   Safety officer
SOP   Standard operating procedure
WMD   Weapons of mass destruction

Index